To all professionals dedicated to improving the health and well-being of their clientele and expanding the knowledge and understanding of their students.

Applied Body Composition Assessment

Second Edition

Vivian H. Heyward, PhD
Regents Professor Emeritus, University of New Mexico

Dale R. Wagner, PhD

Human Kinetics

Library of Congress Cataloging-in-Publication Data

Heyward, Vivian H.
  Applied body composition assessment / Vivian H. Heyward, Dale R.
Wagner.-- 2nd ed.
      p. ; cm.
Includes bibliographical references and index.
  ISBN 0-7360-4630-5 (hardcover)
  1. Body composition--Measurement. 2. Anthropometry.
  [DNLM: 1. Body Composition. 2. Anthropometry. QU 100 H622a 2004]
I. Wagner, Dale R., 1966- II. Title.
  QP33.5.H49 2004
  612--dc22

                                                                2003019596

ISBN-10: 0-7360-4630-5
ISBN-13: 978-0-7360-4630-5

The Web addresses cited in this text were current as of 8/22/2003, unless otherwise noted.

Acquisitions Editor: Loarn D. Robertson, PhD; **Developmental Editor:** Elaine H. Mustain; **Assistant Editor:** Maggie Schwarzentraub; **Copyeditor:** Joyce Sexton; **Proofreader:** Red Inc.; **Indexer:** Sharon Duffy; **Permission Manager:** Dalene Reeder; **Graphic Designer:** Andrew Tietz; **Graphic Artist:** Yvonne Griffith; **Photo Manager:** Kareema McLendon; **Cover Designer:** Keith Blomberg; **Photographer (interior):** Linda K. Gilkey; **Art Manager:** Kelly Hendren; **Illustrators:** Argosy and Robert Reuther; **Printer:** Edwards Brothers

Printed in the United States of America   10 9 8 7 6 5 4 3 2

Human Kinetics
Web site: www.HumanKinetics.com

United States: Human Kinetics, P.O. Box 5076, Champaign, IL 61825-5076
800-747-4457
e-mail: humank@hkusa.com

Canada: Human Kinetics, 475 Devonshire Road Unit 100, Windsor, ON N8Y 2L5
800-465-7301 (in Canada only)
e-mail: info@hkcanada.com

Europe: Human Kinetics, 107 Bradford Road, Stanningley, Leeds LS28 6AT, United Kingdom
+44 (0) 113 255 5665
e-mail: hk@hkeurope.com

Australia: Human Kinetics, 57A Price Avenue, Lower Mitcham, South Australia 5062
08 8372 0999
e-mail: info@hkaustralia.com

New Zealand: Human Kinetics, Division of Sports Distributors NZ Ltd., P.O. Box 300 226 Albany
North Shore City, Auckland
0064 9 448 1207
e-mail: info@humankinetics.co.nz

Contents

Preface

The first edition of *Applied Body Composition Assessment* was written primarily for use by practitioners working with healthy clients in health and fitness settings. That edition focused on identifying appropriate field methods and accurate prediction equations for estimating the body composition of clients and refining techniques to improve the technical skills of body composition practitioners. In addition to updating this practical information, the second edition of this book includes more comprehensive coverage of the theoretical and scientific aspects of body composition assessment as well as recommendations for measuring body composition of clinical populations. This book serves as a resource guide for practitioners working in either health/fitness or clinical settings and as a textbook for advanced undergraduate and graduate students enrolled in exercise science, nutrition, or allied health courses dealing with applied aspects of body composition assessment. Although some of the material may be easily understood and useful to practitioners with limited scientific backgrounds, this book is written primarily for those who have a basic knowledge and understanding of anatomy, physiology, physics, statistics, and measurement.

The second edition of *Applied Body Composition Assessment* is divided into three major sections. Part I, "Body Composition Methods," includes three new chapters that address in more detail theoretical body composition models (chapter 1); the use of regression analysis in body composition research (chapter 2); and body composition reference methods including hydrodensitometry, air displacement plethysmography, dual-energy X-ray absorptiometry, and hydrometry (chapter 3). The remaining chapters in this part address field methods ("Skinfold Method," chapter 4; "Additional Anthropometric Methods," chapter 5; "Bioelectrical Impedance Analysis Method," chapter 6; and "Near-Infrared Interactance Method," chapter 7) and provide an up-to-date synthesis of research pertaining to each method.

In part II, "Body Composition Methods and Equations for Healthy Populations," you will find updated information about measuring the body composition of children (chapter 8), older adults (chapter 9), individuals from diverse ethnic groups (chapter 10), and athletes and physically active individuals (chapter 11). The tables in each chapter summarize suitable methods and prediction equations based on your client's age, gender, ethnicity, and physical activity level. Recommended methods and equations are also summarized in appendixes B.1 through B.3.

Part III, "Body Composition Methods and Equations for Clinical Populations," is completely new and is designed for practitioners working in clinical settings. These chapters identify appropriate methods and prediction equations to measure the body composition of individuals with:

- cardiopulmonary diseases such as coronary artery disease, congestive heart failure, chronic obstructive pulmonary disease, and cystic fibrosis (chapter 12);
- metabolic diseases including obesity, type 1 and 2 diabetes, and thyroid diseases (chapter 13); and
- wasting diseases such as anorexia nervosa, HIV-AIDS, cancer, kidney failure, cirrhosis, spinal cord injury, and neuromuscular diseases (chapter 14).

Tables within each chapter and appendix B.4 summarize suitable methods and prediction equations for assessing the body composition of persons with these diseases. The final chapter updates recommendations for assessing changes in body composition (chapter 15).

Appendix A has step-by-step explanations for the derivation of constants for two-component model conversion formulas. Appendixes B.1 through B.4 help you easily identify field methods and prediction equations for healthy adults, children and older adults, athletes, and clinical populations. Using these summary tables, you will be able to quickly identify suitable methods and equations for assessing your client's body composition. Appendix C presents updated contact information for various equipment manufacturers and suppliers, including Web site addresses.

This edition of *Applied Body Composition Assessment* also incorporates pedagogical features designed to enhance its use as a textbook for university and college courses. At the beginning of each chapter, key questions are identified to help students focus on the most important concepts and principles. At the end of each chapter, key points, key terms, and review questions are provided. Also, a web-based instructor guide is made available to instructors who adopt this book for their courses. The guide contains the following elements:

- Sample course syllabus
- Ideas for class projects
- Guidelines for writing and presenting research abstracts
- Evaluation form to assess students' oral presentations
- Laboratory activities (demonstrations and experiments)

- Checklists for evaluating practical laboratory skills
- 450-question test package
- 40 figures and tables that can be used to supplement class lectures (e.g., transparencies or computerized PowerPoint® presentations)

It is our hope that this book increases your understanding and appreciation of both the art and the science of body composition assessment, further develops your knowledge and skill as a body composition practitioner, and helps you provide your clients and students with the most accurate information about body composition.

Acknowledgments

In addition to all of the individuals acknowledged for their contributions to the first edition, we would like to recognize especially the important contribution that Lisa M. Stolarczyk, PhD, made by coauthoring the first edition of this book. Dr. Stolacrzyk very much wanted to coauthor the second edition, but was unable to do so while attending medical school. We also thank her for strongly encouraging us to include a clinical applications section in this latest edition of the book. We also are indebted to Loarn Robertson, Elaine Mustain, and Maggie Schwarzentraub of Human Kinetics for suggesting that we write this second edition and for their critical comments and meticulous editing of this book. Thanks to each of you; we truly appreciate your effort, cooperation, and support.

PART 1

Body Composition Methods

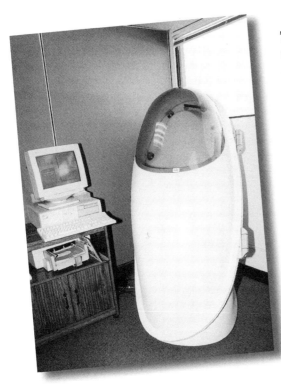

This section provides basic information about theoretical body composition models (chapter 1) used to develop and validate reference and field methods for assessing body composition in research and clinical settings. To fully understand how body composition methods are validated, chapter 2 addresses the use of regression analysis in body composition research. Chapter 3 reviews the basic principles and assumptions of reference methods, as well as testing procedures and sources of measurement error for the following reference methods:

- Hydrodensitometry
- Air displacement plethysmography
- Hydrometry
- Dual-energy X-ray absorptiometry

In chapters 4 through 7, the basic principles and assumptions, testing procedures, and sources of measurement error for field methods are addressed. The following field methods are presented:

- Skinfold method (chapter 4)
- Additional anthropometric methods (chapter 5)
- Bioelectrical impedance analysis (chapter 6)
- Near-infrared interactance method (chapter 7)

BODY COMPOSITION DEFINITIONS, CLASSIFICATION, AND MODELS

Key Questions

- What is healthy body weight and what are the health risks associated with being over- or underweight?
- How can health and fitness professionals use body composition measures?
- What are the standards for classifying relative body fatness?
- What is the difference between two-component and multicomponent models of body composition?
- What models should be used to develop and validate body composition methods and prediction equations?

Maintaining a healthy body weight and level of body fatness is key to a healthier and longer life. Overweight and underweight individuals with body fat levels falling at or near the extremes of the body fat continuum are likely to have serious health problems that reduce life expectancy and threaten the quality of life. Individuals who are overweight or obese have a higher risk of developing cardiovascular, pulmonary, and metabolic diseases, as well as osteoarthritis and certain types of cancer (U.S. Dept. of Health and Human Services 2000a). Underweight individuals with low body fat levels tend to be malnourished and have a relatively high risk of fluid-electrolyte imbalances, renal and reproductive disorders, osteoporosis and osteopenia, and muscle wasting (Fohlin 1977; Mazess, Barden, and Ohlrich 1990).

At present, in large-scale epidemiological studies, the body mass index (BMI; weight in kilograms divided by the square of height in meters) is used to identify persons who are overweight, obese, or underweight. There is, however, considerable variability in body composition for any given BMI. Older people, for example, have more relative body fat at

any given BMI than younger people (Baumgartner et al. 1995). Also, some individuals with low BMIs may have as much relative body fat as those with high BMIs. Because the BMI does not take into account the composition of the individual's body weight, misclassifications of underweight, overweight, and obesity may result when this index is used. Thus, a major goal of researchers working in the field of applied body composition assessment is to develop valid field methods that practitioners and clinicians can use to accurately estimate the body fat of their clients and identify those at risk for diseases.

In addition, there are other important ways in which body composition measures may be used by professionals working in health/fitness, clinical, or school settings (see "Body Composition Applications," p. 4). In weight management clinics, estimates of body fat and lean body mass may be used to determine a healthy body weight, to formulate dietary recommendations and exercise prescriptions, and to monitor changes in body composition for clients participating in weight loss or weight gain programs.

```
┌─────────────────────────────────────────────────┐
│ Body Composition Applications                     │
```

• To identify the client's health risk associated with excessively low or high levels of total body fat

• To promote the client's understanding of health risks associated with too little or too much body fat

• To monitor changes in body composition that are associated with certain diseases

• To assess the effectiveness of nutrition and exercise interventions in altering body composition

• To estimate a healthy body weight for clients and athletes

• To formulate dietary recommendations and exercise prescriptions

• To monitor growth, development, maturation, and age-related changes in body composition

Also, body composition of patients with cardiovascular, pulmonary, metabolic, or other diseases (see chapters 12-14) is often assessed in clinical settings. Monitoring changes in body composition can further our understanding of how various diseases affect energy metabolism and alter body composition of clinical populations. This could lead to the development of more effective nutrition and exercise intervention strategies to counteract the loss of lean body tissues caused by malnutrition, aging, injury, and disease in clinical populations.

Accurate assessment of a healthy body weight is also important for athletes. Although methods are available to estimate healthy body weight and body composition for athletes in various sports (see chapter 11), many coaches and athletic trainers do not use them, thereby jeopardizing the health and well-being of their athletes. Estimating healthy body weight and body composition for athletes is especially important in sports such as wrestling, powerlifting, and bodybuilding, which use body weight classifications for competition. Many of these athletes engage in acute weight loss practices (e.g., saunas, fluid deprivation, diuretics, and fasting) to dehydrate the body and reach a specific body weight for competition.

In school settings, physical educators and allied health professionals can use body composition measures to monitor growth and maturation of children as well as to identify those at risk because of over- or underfatness. The prevalence of overweight and obesity in children is increasing at an alarming rate worldwide. Over the last two decades, the prevalence of obesity (10.9%) has doubled and the prevalence of overweight (22%) has increased 50% for children in the United States (Styne 2001). Similar trends also have been reported for

children in Canada (Tremblay & Willms 2000), Europe (Livingstone 2000), China (Wang et al. 2000), and several developing countries (DeOnis & Blossner 2000). In the United States, childhood obesity is rapidly increasing, particularly among African Americans and Hispanics (Strauss & Pollack 2001). Excess body weight and body fat are linked to coronary heart disease risk factors (i.e., blood pressure, total cholesterol, and lipoprotein ratios) in children and adolescents (Williams, Going, Lohman, Harsha et al. 1992). In addition, many children have inaccurate perceptions of their body fatness and distorted body images, which may lead to eating disorders.

To enable readers to use the increasingly important skills of body composition assessment, this chapter presents basic concepts and definitions of terms used in the study of body composition, provides standards for classification of body fatness, and describes models and equations used to assess body composition.

Definitions and Classification of Body Fatness

As previously mentioned, the **body mass index (BMI)** (BMI = weight/height^2) is often used to classify individuals as overweight, obese, or underweight. **Overweight** is defined as a body mass between 25 and 29.9 kg/m^2; **obesity** is defined as a BMI of 30 kg/m^2 or more; and **underweight** is defined by a BMI of less than 18.5 kg/m^2 (U.S. Dept. of Health and Human Services 2000a). The BMI does not take into account the composition of the individual's body weight. For example, individuals with a high BMI value may have either excess fat or a large lean body mass. *Obesity, therefore, may be better defined as an excessive amount of body fat relative to body weight.*

The absolute amount of body fat, termed **fat mass (FM)**, includes all extractable lipids from adipose and other tissues. The **fat-free mass (FFM)** consists of all residual chemicals and tissues including water, muscle, bone, connective tissues, and internal organs. Although the terms FFM and **lean body mass (LBM)** are sometimes used interchangeably, there is a distinction. Unlike FFM, which contains no lipids, the LBM includes a small amount of essential lipids (Lohman 1992). Table 1.1 defines additional basic terms commonly used in the study of body composition.

To classify levels of body fatness, **relative body fat (%BF)**, also called **percent body fat,** is used. These two terms are used interchangeably. Relative body fat is the fat mass expressed as a percentage of total body weight (%BF = fat mass/body weight × 100). Table 1.2 presents recommended %BF standards for men, women, and children, as well as physically active adults. The minimal, average, and obesity fat values vary with age, gender, and activity status. For example, the average or median %BF values for adult men and women (18-34 yr) are 13% for men and 28% for women; the minimal fat values are 8% and 20%, respectively; and the standard for obesity is >22% BF for men and >35% BF for women.

Table 1.1 Definitions of Basic Terms

Term	Definition
Adipose tissue mass	Fat (~83%) plus its supporting structures (~2% protein and ~15% water)
Body density (Db)	Total body mass expressed relative to total body volume
Body mass (BM)	Measure of the size of the body; body weight
Body volume (BV)	Measure of body size estimated by water or air displacement
Densitometry	Measurement of body density
Dual-energy X-ray absorptiometry (DXA)	Method used to measure total body bone mineral density, bone mineral content, fat, and lean soft-tissue mass
Essential lipids	Compound lipids (phospholipids) needed for cell membrane formation, ~10% of total body lipid
Fat-free body density (FFBd)	Overall density of the fat-free body calculated from the proportions and respective densities of the water, mineral, and protein components of the body
Fat-free mass (FFM) or fat-free body (FFB)	All residual lipid-free chemicals and tissues including water, muscle, bone, connective tissue, and internal organs
Fat mass (FM)	All extractable lipids from adipose and other tissues in the body
Healthy body weight	A body weight that does not increase an individual's disease risk
Hydrometry	Measurement of body water
Lean body mass (LBM)	Fat-free mass plus essential lipids
Nonessential lipids	Triglycerides found primarily in adipose tissue, ~90% of total body lipid
Reference method	Gold standard or criterion method; typically a direct measure of a body composition component
Relative body fat or percent body fat (%BF)	Fat mass expressed as a percentage of total body weight
Total body bone mineral (TBBM)	A measure of the osseous (bone) mineral content of the body
Total body mineral (TBM)	A measure of the osseous (bone) and non-osseous (cell) mineral content of the body
Total body water (TBW)	A measure of the intracellular and extracellular fluid compartments of the body

Table 1.2   Percent Body Fat Standards for Adults, Children, and Physically Active Adults

| | NR* | RECOMMENDED %BF LEVELS FOR ADULTS AND CHILDREN | | | |
		Low	Mid	Upper	Obesity
MALES					
6-17 yr	<5	5-10	11-25	26-31	>31
18-34 yr	<8	8	13	22	>22
35-55 yr	<10	10	18	25	>25
55+ yr	<10	10	16	23	>23
FEMALES					
6-17 yr	<12	12-15	16-30	31-36	>36
18-35 yr	<20	20	28	35	>35
34-55 yr	<25	25	32	38	>38
55+ yr	<25	25	30	35	>35

| RECOMMENDED %BF LEVELS FOR PHYSICALLY ACTIVE ADULTS | | | | | |
	Low	Mid	Upper		
MALES					
18-34 yr	5	10	15		
35-55 yr	7	11	18		
55+ yr	9	12	18		
FEMALES					
18-34 yr	16	23	28		
35-55 yr	20	27	33		
55+ yr	20	27	33		

*NR = not recommended.

Data from Lohman, Houtkooper, & Going (1997).

Body Composition Models

To fully appreciate the science of body composition assessment, you must first understand the theoretical models underlying the measurement of human body composition. Information about the composition of the human body has been based primarily on the chemical analysis of organs and a limited number of human **cadaver analyses** (Forbes et al. 1953, 1956; Widdowson et al. 1951) that quantified the fat, total body water, mineral (bone and soft tissue), and protein content of the body. These studies provided reference data for the development of body composition models that divide the body weight into two or more components.

Two-component (2-C) models describe the human body as the sum of the fat and **fat-free body (FFB)** compartments, whereas **multicomponent models** divide the body into three or more components (see figure 1.1). According to a schema devised by Wang, Pierson et al. (1992) to organize and define multicomponent models, the body may be viewed in terms of five distinct levels: atomic (level I), molecular (level II), cellular (level III), tissue system (level IV), and whole body (level V). The complexity of each level progressively increases from level I to level V. The sum of the components composing each level is equal to the body weight.

The classic 2-C molecular model has been more widely used than multicomponent models to obtain reference measures of body composition for the validation of laboratory and field methods

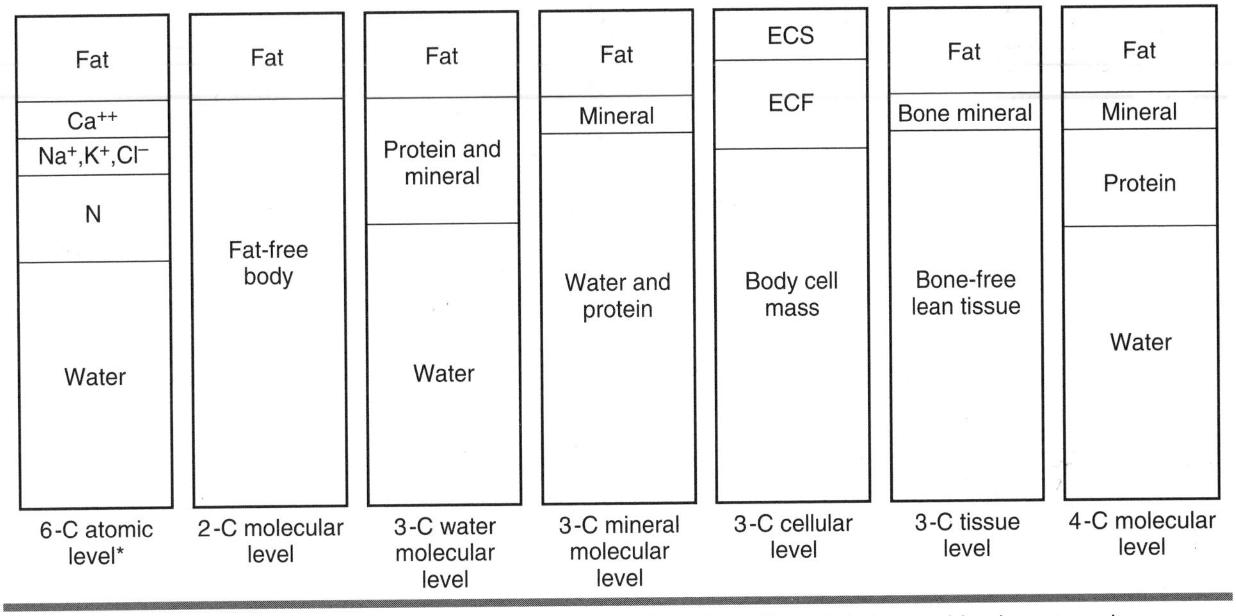

6-C atomic level*	2-C molecular level	3-C water molecular level	3-C mineral molecular level	3-C cellular level	3-C tissue level	4-C molecular level
Fat	Fat	Fat	Fat	ECS	Fat	Fat
Ca⁺⁺		Protein and mineral	Mineral	ECF	Bone mineral	Mineral
Na⁺,K⁺,Cl⁻	Fat-free body		Water and protein	Body cell mass	Bone-free lean tissue	Protein
N		Water				
Water	Water					Water

FIGURE 1.1 Examples of 2-C and multicomponent models. *In this 6-C model, the total body water, nitrogen (N), calcium (CA⁺⁺), sodium (Na⁺), potassium (K⁺), and chloride (Cl⁻) are used to predict fat-free weight. ECS = extracellular solids; ECF = extracellular fluid.

and the development of prediction equations. In the past, many researchers did not have the capability to measure the components of the FFB and therefore were restricted to using the simpler 2-C model. Due to recent technological advances for measuring water (isotope dilution), mineral (dual-energy X-ray absorptiometry), and protein (neutron activation analysis), researchers are now able to use multicomponent models to

- quantify the composition of the FFB,
- obtain more accurate reference measures of body composition,
- test the validity and applicability of methods and prediction equations derived from 2-C models, and
- derive field method prediction equations from multicomponent model reference measures.

This section focuses primarily on 2-C and multi-component models at the molecular level because these models are widely used in the study of body composition assessment.

Two-Component Models

The earliest 2-C model, pioneered by Behnke et al. in 1942, was based on the measurement of total **body density** (Db = body mass expressed relative to body volume) using hydrodensitometry. Behnke et al. established an inverse relationship between Db and adiposity and concluded that excess fat is the main factor affecting Db. Later, Behnke and colleagues (1953) developed the concept of a **reference body** that consisted of FM and LBM. This 2-C model assumed that the density of the LBM was constant for all individuals.

At the same time, Keys and Brozek (1953) developed a 2-C model equation to estimate %BF from Db. This equation was based on a reference body consisting of 14% BF and assumed that the density of fat was 0.9478 g/cc. Ten years later, the equation was revised using a reference body with an assumed Db of 1.064 g/cc and 15.3% BF and a more accurate fat density of 0.9007 g/cc (Brozek et al. 1963). In this model, any variation in measured Db from the reference body density (1.064 g/cc) is assumed to be due to a difference in obesity (adipose) tissue. The Brozek et al. (1963) equation, %BF = (4.57 / Db − 4.142) × 100, has been widely used over the years and continues to be used to obtain 2-C model estimates of %BF.

In 1956, Siri developed another 2-C model equation to convert Db to %BF. This equation, %BF = (4.95 / Db − 4.50) × 100, has constants (i.e., 4.95 and 4.50) that differ from those in the Brozek et al. (1963) equation (i.e., 4.57 and 4.142) because the Siri equation assumes that any variation in measured Db from the reference body is due to a difference in triglyceride content instead of adipose tissue. These 2-C model equations, however, yield nearly identical %BF estimates (varying by only 0.5-1.0% BF) for body densities ranging from 1.0300 to 1.0900 g/cc. For example, if a client's measured

00 g/cc, the %BF estimated by the Siri k et al. equations is 21.4% and 21.0%, ly. For individuals with more than 30% Siri equation gives relatively higher body fat estimates than the Brozek et al. equation.

Both of these 2-C model equations use assumed values for the relative (%) composition of the FFM (i.e., water, protein, and mineral) or reference body and for the respective densities of the constituents of the body (see table 1.3). Specifically, the following assumptions are made to estimate %BF from Db using these two equations:

- The densities of the fat and FFB components (water, mineral, and protein) are additive and are the same for all individuals.

- The proportions of water, mineral, and protein in the FFB or reference body are constant within and between individuals.

- The individual being measured differs from the reference body only in the amount of body fat (triglyceride) or obesity (adipose) tissue.

To understand how the proportions and respective densities of the fat and FFB components are used to derive the constants for the Siri and Brozek et al. 2-C model equations and to calculate the assumed value for **FFB density** (FFBd = 1.10 g/cc), see appendix A (p. 215).

Siri (1956) estimated the errors associated with using 2-C model equations to estimate %BF from Db. These errors are caused by biological variability in the relative (%) hydration (body water) and protein-to-mineral ratio in the FFB, as well as the composition of adipose tissue. Siri estimated that a 2% variation in body water produces a 2.7% error in estimation of %BF. Error estimates due to variability in the protein-to-mineral ratio and composition of adipose tissue were 2.1% BF and 1.9% BF, respectively. These three sources of error were combined by taking the square root of the sum of the squares of each error (i.e., law of propagation of errors, see p. 10). Siri calculated a total error of 3.9% BF for the general population

$$[\text{total error} = \sqrt{(2.7\%^2) + (2.1\%^2) + (1.9\%^2)}],$$

which is equivalent to a 0.0084 g/cc variation in the density of the FFB.

Generally, these 2-C model equations provide reasonable estimates of %BF as long as the assumptions of the model are met. However, there is no guarantee that the FFB composition of an individual or a specific population subgroup will exactly match the values assumed for the FFB or reference body. In fact, researchers have reported that these equations produce systematic prediction error when they are applied to population subgroups or individuals whose FFBd varies from the assumed value (1.10 g/cc) used to derive these equations (see table 1.3). Age, gender, ethnicity, level of body fatness, and physical activity level affect the relative proportions of water, mineral, and protein in the FFB and therefore the overall FFBd (Baumgartner et al. 1991; Deurenberg, Leenan et al. 1989; Mazariegos et al. 1994; Modlesky, Cureton et al. 1996; Wagner & Heyward 2000, 2001). For example, the average FFBd of African American women and men (1.106 g/cc) is greater than 1.10 g/cc because of their higher mineral content or relative body protein (Cote & Adams 1993; Ortiz et al. 1992). Because of this difference in FFBd, the body fat of blacks will be systematically underestimated when 2-C model equations are used to estimate %BF.

In an attempt to surmount this shortcoming of the classic 2-C molecular model equations, scientists have used a multicomponent approach to obtain the data in order to develop 2-C model equations for certain population subgroups (see table 1.4). These **population-specific conversion formulas** were

Table 1.3 Assumed Values for Components of the Fat-Free Body and Reference Body

Component	Density (g/cc)	Fat-free body (%)	Reference body (%)
Water	0.9937	73.8	
Mineral	3.038	6.8	
Protein	1.34	19.4	
Fat-free body	1.1000	100.0	84.7
Fat	0.9007		15.3
Reference body	1.064		100.0

Data from Brozek et al. 1963.

Table 1.4 Population-Specific Two-Component Model Conversion Formulas

	Population	Age	Gender	%BF	FFBd (g/cc)*
ETHNICITY	African American	9-17	Female	(5.24 / Db) − 4.82	1.088
		19-45	Male	(4.86 / Db) − 4.39	1.106
		24-79	Female	(4.86 / Db) − 4.39	1.106
	American Indian	18-62	Male	(4.97 / Db) − 4.52	1.099
		18-60	Female	(4.81 / Db) − 4.34	1.108
	Asian Japanese Native	18-48	Male	(4.97 / Db) − 4.52	1.099
			Female	(4.76 / Db) − 4.28	1.111
		61-78	Male	(4.87 / Db) − 4.41	1.105
			Female	(4.95 / Db) − 4.50	1.100
	Singaporean (Chinese, Indian, Malay)		Male	(4.94 / Db) − 4.48	1.102
			Female	(4.84 / Db) − 4.37	1.107
	Caucasian	8-12	Male	(5.27 / Db) − 4.85	1.086
			Female	(5.27 / Db) − 4.85	1.086
		13-17	Male	(5.12 / Db) − 4.69	1.092
			Female	(5.19 / Db) − 4.76	1.090
		18-59	Male	(4.95 / Db) − 4.50	1.100
			Female	(4.96 / Db) − 4.51	1.101
		60-90	Male	(4.97 / Db) − 4.52	1.099
			Female	(5.02 / Db) − 4.57	1.098
	Hispanic		Male	NA	NA
		20-40	Female	(4.87 / Db) − 4.41	1.105
ATHLETES	Resistance trained	24 ± 4	Male	(5.21 / Db) − 4.78	1.089
		35 ± 6	Female	(4.97 / Db) − 4.52	1.099
	Endurance trained	21 ± 2	Male	(5.03 / Db) − 4.59	1.097
		21 ± 4	Female	(4.95 / Db) − 4.50	1.100
	All sports	18-22	Male	(5.12 / Db) − 4.68	1.093
		18-22	Female	(4.97 / Db) − 4.52	1.099
CLINICAL POPULATIONS**	Anorexia nervosa	15-44	Female	(4.96 / Db) − 4.51	1.101
	Cirrhosis Childs A			(5.33 / Db) − 4.91	1.084
	Childs B			(5.48 / Db) − 5.08	1.078
	Childs C			(5.69 / Db) − 5.32	1.070
	Obesity	17-62	Female	(4.95 / Db) − 4.50	1.100
	Spinal cord injury (paraplegic/ quadriplegic)	18-73	Male	(4.67 / Db) − 4.18	1.116
		18-73	Female	(4.70 / Db) − 4.22	1.114

*FFBd = fat-free body density based on average values reported in selected research articles.

NA = no data available for this population subgroup.

**There are insufficient multicomponent model data to estimate the average FFBd of the following clinical populations: coronary artery disease, heart/lung transplants, chronic obstructive pulmonary disease, cystic fibrosis, diabetes mellitus, thyroid disease, HIV/AIDS, cancer, kidney failure (dialysis), multiple sclerosis, and muscular dystrophy.

derived using measured total body water or bone mineral values, or both, to estimate the average FFBd for each population subgroup. Use of these 2-C model equations may help to control some of the systematic error associated with the Siri and Brozek equations when one is estimating body fat for different age, gender, and ethnic groups. You will note in table 1.4 that population-specific conversion formulas have not yet been developed for all age groups within each ethnic group.

Three-Component Models

To account for interindividual variability in the hydration of the FFB, Siri (1961) developed a **three-component (3-C) model** equation that adjusts Db for the relative proportion of water in the body (see table 1.5). This model divides the body into three components—fat, water, and solids (i.e., protein and mineral fractions of the FFB are combined)—and assumes a constant density for the protein-to-mineral ratio. The Siri 3-C equation may yield more accurate estimates of %BF for individuals or population subgroups such as children and obese adults whose relative hydration of the FFB deviates from the assumed value (73.8% FFB) for the 2-C model (Lohman, Boileau et al. 1984; Segal et al. 1987). With this model, both the Db and **total body water (TBW)** of the individual need to be measured using one of the reference methods described in chapter 3.

Similarly, Lohman (1986) devised a 3-C model that accounts for variability in the relative mineral content of the FFB and divides the body into fat, mineral, and protein + water fractions (see table 1.5). This model assumes a constant density (1.0486 g/cc) for the protein + water fraction. For this model, densitometry and dual-energy X-ray absorptiometry (see chapter 3) are used to measure Db and estimate **total body mineral (TBM)** (TBM = osseous + non-osseous mineral) from bone ash (TBM = bone ash × 1.279). The constant (1.279) in this equation is based on the ratio of tissue mineral to bone ash established from cadaver studies. Compared to 2-C models, Lohman's 3-C model equation provides a more accurate estimate of %BF for individuals whose relative TBM deviates from the assumed value (6.8% FFB), as may occur in children, American Indians, and African American men and women (Cote & Adams 1993; Lohman, Slaughter et al. 1984; Schutte et al. 1984; Stolarczyk et al. 1994; Wagner & Heyward 2001).

In addition to these 3-C molecular-level models, the use of dual-energy X-ray absorptiometry (DXA) for whole-body body composition analysis is based on a 3-C tissue-level model (see table 1.5). This model divides the body into three compartments: **bone-free lean tissue mass (LTM)**, FM, and bone.

The DXA model is composed of two separate sets of 2-C model equations (Ellis 2000). The first set of equations is used to derive bone and the **soft-tissue mass (STM)** (STM = fat + bone-free lean tissues). The second set of equations divides the STM into lean and fat tissue.

It appears that DXA is capable of partitioning the body into bone and STM, but the accuracy of this method for differentiating the fat and lean compartments of the STM has been questioned. Lohman and colleagues, however, concluded that DXA estimates of %BF were within 1% to 3% of multicomponent molecular model estimates (Lohman, Harris et al. 2000). Compared to 6-C chemical model estimates of FM, the prediction error of the DXA model for estimating FM ranges between 1.7 and 2.0 kg (Ellis 2001; Wang et al. 1998).

Four-Component Models

Ideally, **four-component (4-C) model** equations (see table 1.5) should be used to estimate %BF from Db, especially in cases in which both the hydration and the relative mineral content of the body vary greatly. This approach divides the body into fat, water, mineral, and protein components, thereby eliminating the need to make assumptions about the relative proportions of these constituents in the body. Although the constants in these 4-C models differ slightly, most of these equations yield similar estimates of %BF (Heymsfield et al. 1996). For these models, reference methods (see chapter 3) are used to measure Db, TBW, and **total body bone mineral (TBBM)** from bone ash (TBBM = bone ash × 1.0436).

Compared to 2-C models, the 4-C models have greater accuracy in estimating fat but require measurement of more variables. Each variable has an inherent measurement error that reflects the precision of the method used to assess it. Do the cumulative errors associated with measuring multiple variables offset the improved accuracy of 4-C model estimates of %BF? To answer this question, you need to understand some basic information about measurement error. The total measurement error is a function of the error (degree of precision) associated with measuring each variable in the model (e.g., Db, TBW, and bone mineral for the 4-C molecular model). This is known as the **law of propagation of errors.** These sources of error are independent and additive. To estimate the total variance due to measurement errors, each source of error must be squared before the

Table 1.5 Body Composition Models and Equations

	Model	Equation	Reference
TWO-COMPONENT MOLECULAR LEVEL	BW = fat + fat-free body	%BF = [(4.57 / Db) − 4.142] × 100	Brozek 1963
		%BF = [(4.95 / Db) − 4.50)] × 100	Siri 1956
THREE-COMPONENT MOLECULAR LEVEL	BW = fat + water + (mineral and protein combined)	%BF = [(2.118 / Db) − 0.78W − 1.354] × 100	Siri 1961
	BW = fat + mineral + (water and protein combined)	%BF = [(6.386 / Db) + 3.961M − 6.090] × 100	Lohman 1986
THREE-COMPONENT TISSUE LEVEL (DXA MODEL)	BW = bone + bone-free lean tissue + fat	%BF = FM / BW × 100	Ellis 2000
FOUR-COMPONENT MOLECULAR LEVEL	BW = fat + water + bone mineral + protein	%BF = [(2.559 / Db) − 0.734W + 0.983B − 1.841] × 100	Friedl 1992
		%BF = [(2.747 / Db) − 0.714W + 1.146B − 2.053] × 100	Selinger 1977
		%BF = [(2.513 / Db) − 0.739W + 0.947B − 1.790] × 100	Heymsfield 1996
		%BF = [(2.747 / Db) − 0.718W + 1.148B − 2.050] × 100	Baumgartner 1991
SIX-COMPONENT ATOMIC LEVEL	BW = TBW + TBN + TBCa + TBK + TBNa + TBCl	FM (kg) = BW − (TBW + 6.525 TBN + 2.709 TBCa + 2.76 TBK + TBNa + 1.43 TBCl)	Wang et al. 1998

Key: %BF = relative body fat.

Db = total body density (g/cc).

FM = fat mass (kg).

W = TBW (kg)/BW (kg), where TBW = total body water and BW = body weight.

M = TBM (kg)/BW (kg), where TBM = total body mineral (osseous + cell mineral) and BW = body weight.

B = TBBM (kg)/BW (kg), where TBBM = total body bone mineral (osseous mineral only) and BW = body weight.

Constants: TBBM = bone ash × 1.0436; TBM = bone ash × 1.279.

TBN = total body nitrogen; TBCa = total body calcium; TBK = total body potassium; TBNa = total body sodium; TBCl = total body chloride.

errors are summed. For example, given estimates of measurement error (i.e., standard error of estimate for a single determination) of 0.43% BF for Db, 0.71% BF for TBW, and 0.05% BF for bone mineral (Heymsfield et al. 1996), the standard deviation (SD) of the **total error of measurement (TEM)** is calculated as follows:

$$TEM\ (SD) = \sqrt{[(0.43)^2 + (0.71)^2 + 0.05)^2}$$
$$TEM = \sqrt{0.1849 + 0.5041 + 0.0025}\ or\ \sqrt{0.6915}$$
$$TEM = 0.83\%\ BF$$

The value reflects the overall degree of precision (±0.8% BF) that can be expected with use of these methods to obtain a 4-C model estimate of %BF for an individual.

Using this approach, researchers compared the measurement errors for 2-C and 4-C molecular models (Friedl et al. 1992). They reported errors of ±1.0% BF and ±1.1% BF, respectively, for the 2-C and 4-C models. Thus, the cumulative measurement error associated with the multiple measures used to assess the various components of the 4-C model does not offset the improved accuracy in estimating %BF (Friedl et al. 1992; Heymsfield et al. 1996). At present, experts agree that a multicomponent approach should be used whenever possible, especially for development and validation of body composition methods and prediction equations (Going 1996; Heymsfield et al. 1996).

Six-Component Model

Atomic models (see table 1.5) require the direct analysis of the chemical composition of the body in vivo. Using neutron activation analysis (NAA), the total body content of the major elements (i.e., calcium, sodium, chloride, phosphorus, nitrogen, hydrogen, oxygen, and carbon) can be measured (see chapter 3). One such atomic, **six-component (6-C) model** (Wang et al. 1998) divides the body into the following compartments: water + nitrogen + calcium + potassium + sodium + chloride. Although atomic models provide criterion body composition measures for evaluating the accuracy of other "reference" models and methods, the lack of NAA facilities, the high expense, and client exposure to radiation limit their use.

Key Points

- Body composition is key to a healthier and longer life; %BF is related to disease risk.
- Overweight is defined as a BMI (weight/height squared) between 25 and 29.9 kg/m^2; obesity is defined as a BMI of 30 kg/m^2 or more; and underweight is defined by a BMI of less than 18.5 kg/m^2.
- Standards for percent body fat can be used to classify body composition.
- Average %BF and standards for obesity vary according to age, gender, and physical activity levels.
- The five levels for body composition models are the atomic (level I), molecular (level II), cellular (level III), tissue system (level IV), and whole body (level V).
- Two-component models divide the body into fat and FFB compartments.
- Multicomponent models divide the body into three or more compartments.
- The classic 2-C molecular model equations of Siri and Brozek are used to estimate %BF from Db; these equations assume that the densities of the FFB and fat are 1.10 g/cc and 0.9007 g/cc, respectively.
- Two-component molecular model equations may systematically over- or underestimate %BF of individuals or population subgroups whose FFB composition differs from the assumed values (i.e., 73.8% water, 6.8% mineral, and 19.4% protein).
- Two-component model equations (population-specific conversion formulas) have been developed for various age and ethnic groups based on the average FFB composition values reported in the literature for these groups.
- Multicomponent (3-C and 4-C) molecular model equations take into account interindividual variability in the hydration or mineral content of the FFB, or both; therefore, they generally yield more accurate estimates of body fat than 2-C model equations.
- The 3-C molecular model equations adjust Db for variability in either the relative TBW or relative TBM.

- The 3-C tissue-level model, used in conjunction with DXA, divides the body into bone, bone-free lean tissue, and FM.
- The 4-C molecular model equations adjust Db for variability in both relative TBW and relative TBM or TBBM.
- Researchers should use multicomponent models (4-C) to obtain the most accurate estimate of an individual's %BF and to develop and validate body composition models and prediction equations.
- The 6-C atomic-level model requires direct, in vivo measurement of the major elements of the body; this model is used to evaluate the validity of other types of body composition models and reference methods.

Key Terms

Learn the definition for each of the following key terms. Definitions of terms can be found in the Glossary, page 227.

body density (Db)	**percent body fat (%BF)**
body mass index (BMI)	**population-specific conversion formulas**
bone-free lean tissue mass (LTM)	**reference body**
cadaver analysis	**relative body fat (%BF)**
fat mass (FM)	**six-component model (6-C)**
fat-free body (FFB)	**soft-tissue mass (STM)**
fat-free body density (FFBd)	**three-component model (3-C)**
fat-free mass (FFM)	**total body bone mineral (TBBM)**
four-component model (4-C)	**total body mineral (TBM)**
law of propagation of errors	**total body water (TBW)**
lean body mass (LBM)	**total error of measurement (TEM)**
multicomponent model	**two-component model (2-C)**
obesity	**underweight**
overweight	

Review Questions

1. Why is it important to assess the body composition of your clients?
2. Using BMI, what are the cutoff values for classification of obesity, overweight, healthy body weight, and underweight?
3. What are the standards for classifying obesity and minimal levels of body fat for men and women?
4. Name the five levels used to classify multicomponent body composition models.
5. What are the three assumptions underlying the use of 2-C molecular models? Identify two commonly used 2-C model equations for converting Db into %BF.
6. Distinguish between total Db and FFB density.
7. Explain how gender, ethnicity, and age may affect FFB density and therefore 2-C model estimates of %BF.
8. Identify the three major sources of biological error contributing to total error with use of 2-C model equations to estimate %BF from Db.
9. Do the Siri and Brozek 2-C molecular models yield similar estimates of %BF? Explain.
10. What are the assumed proportions and respective densities of the three molecular components constituting the FFB?

11. What is the major difference between Siri's and Lohman's 3-C models?

12. Select one 4-C molecular model and describe the components that make up this model. What reference methods are used to measure these components?

13. Why do multicomponent models potentially give a more accurate estimate of an individual's %BF than 2-C models?

14. Ideally what type of model should be used to validate body composition methods and prediction equations? Explain.

15. What type of model provides in vivo measures of body composition? Why is it impractical to use these models in most research or clinical settings?

CHAPTER 2

USE OF REGRESSION ANALYSIS IN BODY COMPOSITION

Key Questions

- How is regression analysis used in body composition research studies?
- What statistics are used to compare body composition methods and quantify the accuracy of prediction equations?
- What are the standards for evaluating prediction errors?
- What techniques are used to cross-validate methods and prediction equations?
- What are the criteria used to assess the validity of methods and accuracy of prediction equations for a group?
- How are the validity of methods and accuracy of prediction equations evaluated for individuals?

Before we describe and compare various reference and field methods for assessing body composition, it is important to understand the statistical methods used by researchers in this field. As a body composition practitioner, you need to be able to evaluate the relative worth of body composition models, methods, and prediction equations. This is especially true given that researchers are currently using multicomponent models to obtain reference measures for the development of new field method equations.

This chapter describes basic statistical concepts, the use of regression analysis for the development of prediction equations, cross-validation techniques, and guidelines for evaluating the validity of methods and the accuracy of prediction equations in assessing the body composition of groups and individuals.

Basic Statistical Concepts

To validate methods and develop prediction equations, body composition researchers commonly use a statistical method called **regression**. The terms correlation and regression are closely related

and are often used interchangeably. **Correlation** measures the strength of association or degree of relationship between two variables, whereas the aim of **regression** is to predict one variable (Y = dependent variable) from one or more other variables (X = predictor or independent variable). The **Pearson product-moment correlation coefficient, r_{xy},** is frequently used to quantify the *linear* relationship between two variables. Values of r_{xy} range between -1.00 and 1.00 (perfect negative or perfect positive relationship). When the correlation is perfect, scores in the X distribution have the same relative positions as corresponding scores in the Y distribution, and all scores fall on straight line (see figure 2.1*a*). An r_{xy} of .00 indicates no linear relationship between the X and Y variables and complete randomness (see figure 2.1*b*). A low correlation, however, does not always mean that there is no relationship between two variables. Sometimes the relationship between two variables is best described by a curved line (i.e., *curvilinear* relationship). If a curved regression line fits the data better than a straight line, the value of r_{xy} will be low and will not reflect the true relationship between the two variables (see figure 2.1*c*).

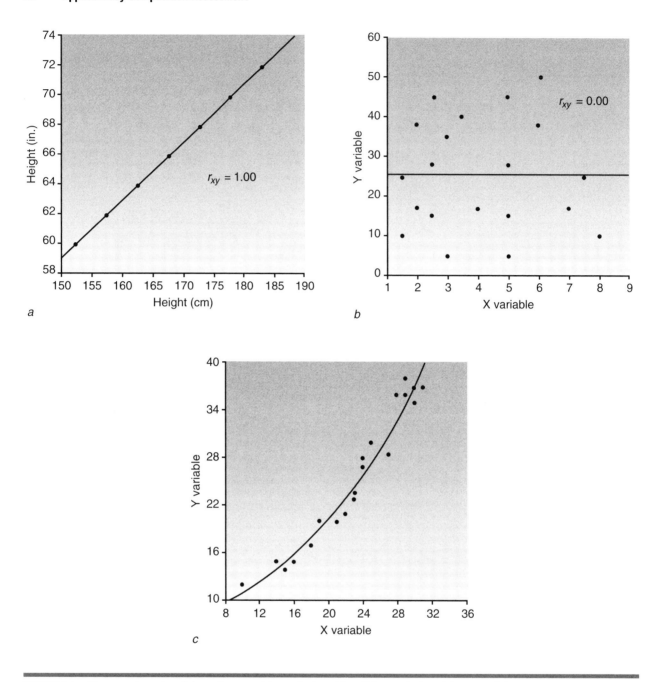

FIGURE 2.1   Scatterplots for *(a)* perfect positive relationship, *(b)* no relationship, and *(c)* curvilinear relationship.

The **coefficient of determination (r_{xy}^2)** is the square of the correlation coefficient and represents the amount of variance shared by variables X and Y. For example, if the correlation between Db estimates from hydrostatic weighing (Y) and air displacement plethysmography (X) is $r_{xy} = 0.90$, then 81% ($0.90^2 \times 100 = 81\%$) of the variability in Db is shared by these two methods. In other words, 81% of the variability in Db is explained but the remaining 19% (residual variance) is unexplained.

Bivariate Regression Analysis

To predict one variable from another variable (i.e., **bivariate regression**), one must know the relationship between the two sets of scores and find a best-fitting straight line (i.e., line of best fit) between the two variables (see figure 2.2). The **line of best fit** is the **regression line** depicting the linear relationship between the X and Y scores and is operationally defined as that line that minimizes

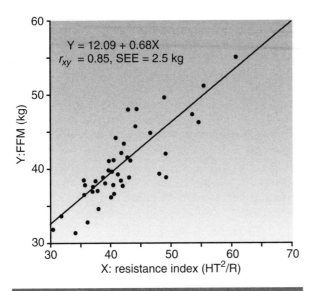

$$Y = 12.09 + 0.68X$$
$$r_{xy} = 0.85, \text{SEE} = 2.5 \text{ kg}$$

FIGURE 2.2 Line of best fit and standard error of estimate.

the squared deviations from the line (i.e., squared errors of prediction).

Using the line of best fit, values of Y can be predicted (Y') from known values of X. For example, in the scatterplot (figure 2.2), a HT^2/R of 50 yields a predicted FFM of ~46 kg. To do this mathematically, a **prediction equation** generated by the regression analysis is used.

$$Y' = a + bX$$

where Y' = the predicted score or Y value predicted from a particular X value, a = the **y-intercept** or the point at which the regression line intersects the Y axis, b = the **slope** of the regression line (change in Y divided by the change in X or the amount Y is increasing for each increase of one unit in X), and X = the value from which Y is to be predicted.

In our example, the equation is

$$\text{predicted FFM} = 12.09 + 0.68 \, (HT^2/R)$$

Plugging a HT^2/R value of 50 into this equation yields a predicted FFM of 46 kg, the same value extrapolated from the scatterplot in figure 2.2. The value of the y-intercept (12.09) is the mean of the predicted variable ($\bar{Y} = 40.0$ kg) minus the product of the regression coefficient (b = 0.68) times the mean of the predictor variable ($\bar{X} = 41.04$ kg): a = $\bar{Y} - b\bar{X}$ or a = 40.0 − 0.68 (41.04) = 12.09. The slope (b) is calculated using the correlation coefficient (r_{xy}) and the standard deviation (s) for the X and Y data sets: b = $r_{xy} (s_y/s_x)$ or b = 0.85 (4.7/5.9) = 0.68.

When the correlation is less than ±1.0, there will be a difference between the measured and predicted values of Y. The difference is called the

residual score (Y − Y'). The **standard error of estimate (SEE)** is used to quantify the accuracy of the prediction equation based on the size of the residual scores and reflects the degree of deviation of the individual data points around the line of best fit. The closer individual data points fall to the regression line, the smaller the SEE or **prediction error** (see figure 2.2). Mathematically, the SEE is calculated using one the following equivalent equations (Jackson 1989):

$$\text{SEE} = \sqrt{\Sigma(Y - Y')^2/N - p - 1} \text{ (where p =}$$
number of predictor variables) or
$$\text{SEE} = s_y \sqrt{1 - r_{xy}^2}$$

Assuming that the variances of Y scores for each value of X are equally spread (i.e., basic assumption of **homoscedasticity**), we may expect 68% of the actual Y scores to be within ±1 SEE of the regression line and 95% of the actual scores to fall within ±2 SEE of the regression line.

Multiple Regression Analysis

In contrast to bivariate regression using just one predictor, **multiple regression** uses two or more variables (i.e., **predictor variables**) simultaneously to predict the dependent variable (DV). The resulting prediction equation represents the line of best fit between the DV and all of the predictor variables:

$$Y' = a + b_1 X_1 + b_2 X_2 + b_k X_k$$

where Y' is the predicted value of the DV, X's represent the predictor variables, a is the y-intercept (the value of Y when all X values are zero), and the b's are the weights assigned to each of the predictor variables by the regression solution.

Two approaches commonly used to identify the best combination of predictor variables for a regression model are hierarchical and stepwise regression. In **hierarchical regression,** the order in which predictor variables enter the regression analysis is predetermined and specified by the researcher. The selected variables and their order of entry are usually based on theoretical models or logical considerations. In contrast, for **stepwise regression,** the order of entry of predictor variables is based on statistical rather than theoretical criteria. In the first step, the variable that shares the most variance with the DV enters the equation. Thereafter, at each subsequent step, the variable that adds the most to the prediction equation (i.e., increases the r^2 significantly) enters. The process continues until the inclusion of any of the remaining variables does not significantly improve the equation.

The multiple correlation (R_{mc}) is the correlation coefficient between the actual and predicted scores. The R_{mc}^2 (coefficient of determination) represents the proportion of variance in the DV that is predictable from the best linear combination of the predictor variables in the equation. As in bivariate regression, the accuracy of the prediction equation is estimated using the SEE:

$$SEE = \sqrt{\Sigma(Y - Y')^2 / N - p - 1}$$

where p = the number of predictor variables in the equation.

To use the prediction equation, the individual's score for each predictor variable is multiplied by its respective **regression weight** (b). All of the products are added together, along with the y-intercept value, to yield a predicted score for the individual. For example, the following skinfold (SKF) prediction equation estimates %BF using two predictor variables (triceps SKF and calf SKF) and their corresponding regression weights:

%BF = 1.0 + 0.50 (triceps SKF) + 0.31 (calf SKF)

If the individual's triceps and calf SKF values are 20 mm and 18 mm, respectively, then the estimated %BF is calculated as follows:

%BF = 1.0 + 0.50 (20) + 0.31 (18) =
1.0 + 10 + 5.6 = 16.6% BF

A good prediction equation has stable regression weights, meaning that their values do not change much from group to group. To obtain stable regression weights, the researcher must use large representative samples (100-400 subjects) from the population when developing the prediction equation. One tests the stability of regression weights for the predictor variables by applying the prediction model to independent samples from the population and comparing these regression weights to those derived on the original sample.

Polynomial Regression

As mentioned previously, sometimes the relationship between variables is curvilinear instead of rectilinear. When this is the case, a curved line (i.e., nonlinear model) will fit the data points better than a straight line (i.e., linear model). To reduce a nonlinear model to a linear model, one transforms the data by expressing variables as logarithms or by raising variables to powers.

Curvilinear regression uses a **polynomial regression** equation, meaning that the predictor variable in the regression model is raised to a certain power. For example, in a second-degree polynomial equation, the predictor variable (X) is raised to the second power:

$$Y' = a + b_1X + b_2X^2$$

A second-degree polynomial (X^2) equation describes a single bend in the regression line and is also referred to as a quadratic equation or **quadratic regression model,** whereas a third-degree polynomial (X^3) has two bends in the regression line and is called a cubic equation or **cubic regression model.** The Jackson et al. (1980) SKF equation is an example of a quadratic equation that describes the curvilinear relationship between Db and the sum of three skinfolds (Σ3SKF):

Db = 1.0994921 − 0.0009929 (Σ3SKF) +
0.0000023 (Σ3SKF)^2 − 0.0001392 (age)

Cross-Validation Techniques

Cross-validation techniques are used to determine the predictive accuracy of body composition equations. Prediction equations may be cross-validated internally, externally, or in both ways. Typically we perform **external cross-validation** by applying the prediction equation to a sample that is independent from the one used to develop the original equation. For external cross-validation, the DV and predictor variables are measured for the independent sample (cross-validation group) using the same type of instruments and procedures that were used to obtain the data for the original sample (validation group).

Typically, we perform **internal cross-validation** by randomly splitting the original sample into groups. When the original sample is large, it may be randomly divided into two equal groups. Data from each group are used to derive a prediction equation. Each equation is then applied to the other group to test its predictive accuracy. This is known as the **double cross-validation technique.** If the two equations are similar and have good predictive accuracy, the data from both groups are then combined to generate a single prediction equation for the entire sample.

Sometimes for internal cross-validation, researchers split the sample by using two-thirds of the original sample (validation sample) to develop the prediction equation, and the remaining one-third (cross-validation sample) to cross-validate it. For example, to establish the validity of a newly developed bioelectrical impedance analysis (BIA) equation for American Indian women, we tested over 150 participants (Stolarczyk et al. 1994). The sample was randomly divided into a validation group (N = 100) and a cross-validation group (N = 50). First, a prediction equation that estimated

FFM was developed for the validation group using stepwise multiple regression; then, the validity and predictive accuracy of this equation were tested through application of this equation to obtain FFM estimates for the cross-validation group.

Alternative approaches for internal cross-validation of prediction equations are the jackknife and prediction of sum of squares (PRESS) techniques (Guo & Chumlea 1996). For the **jackknife technique,** the subjects are randomly assigned to groups of equal size. The data from the first group are excluded when the equation is constructed, and residual scores (Y – Y') are calculated for the first group. This procedure is repeated for each of the groups, one at a time. The accuracy of the equation is evaluated using the sums of the squares of the residuals for each group.

The **PRESS technique** is also known as the **bootstrap technique.** For this procedure, each data point in the sample is excluded one at a time and regression analysis is performed. The value for each omitted data point is predicted, and a PRESS residual score (Y – Y') is calculated. The accuracy of the prediction equation is evaluated by measuring and averaging the sum of the squares of all PRESS residuals.

In cross-validation analysis, the **validity coefficient ($r_{y,y'}$)** represents the relationship between the actual (measured) and predicted scores. Also, the **total error (TE)** or **pure error** for the equation is evaluated using the following formula:

$$TE = \sqrt{\Sigma(Y - Y')^2 / N}$$

The TE represents the average deviation of individual scores from the line of identity (see figure 2.3). For the **line of identity,** the slope is equal to 1, and the y-intercept is equal to 0. When the equation closely predicts the actual or measured individual scores, individual values will fall close to the line of identity with a small degree of deviation from this line. Also, the line of best fit through the data points will not deviate significantly from the line of identity.

Criteria for Evaluating Methods and Prediction Equations for Groups

As a practitioner, you may have questions about which body composition methods to use. Is one method better than another method? Do the methods give similar results? How do I go about selecting the best method and prediction equation? Before making any choices, evaluate the relative worth of methods and equations by asking the following questions:

- **What body composition model and reference methods were used to develop the prediction equation?** Experts agree that multicomponent models, especially 4-C models, give fairly accurate reference measures of body composition because these models take into account interindividual variability in the water, mineral, and protein content of the body (see chapter 1). Although there are a number of highly sophisticated methods that may be used to obtain reference measures of body composition (e.g., neutron activation analysis and magnetic resonance imagery), densitometry (hydrostatic weighing or air displacement plethysmography), dual-energy X-ray absorptiometry (DXA), and hydrometry are commonly used in research settings to obtain reference measures (see chapter 3). Each of these methods, however, is subject to measurement error and has basic assumptions that do not always hold true. Therefore, none can be singled out as the **gold standard** method for body composition assessment. In fact, many researchers have combined these three methods and used them in conjunction with a 4-C molecular model to derive valid reference measures of %BF, FM, and FFM (Heyward 2001). With use of this approach, Db (measured by hydrodensitometry or air displacement plethysmography) is corrected for variations in TBW (measured by hydrometry) and total body bone mineral (measured by DXA). Whenever possible, select field method equations that used multicomponent models and valid **reference methods** (see chapter 3) to develop and cross-validate the prediction equation.

- **How large was the sample used to develop the prediction equation? What is the ratio of**

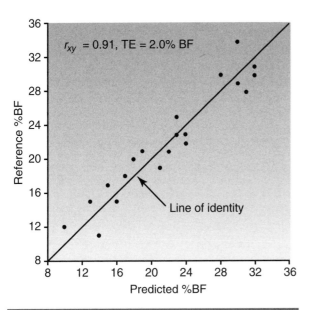

FIGURE 2.3 Line of identity and TE.

sample size to the number of predictor variables in the equation? Generally, large randomly selected samples (N = 100-400 subjects) are needed to ensure that the data are representative of the population for whom the equation was developed. Equations based on large samples tend to have more stable regression weights for each predictor variable in the equation. As previously mentioned, the **multiple correlation coefficient (R_{mc})** represents the degree of relationship between the reference body composition measure (Db, %BF, or FFM) and the predictors in the equation. The larger the R_{mc} (up to maximum value of 1.00), the stronger the relationship. The size of R_{mc}, however, will be artificially inflated if there are too many predictors in the equation compared to the total number of subjects in the sample (Jackson 1984). Statisticians recommend a minimum of 20 to 40 subjects per predictor variable (Pedhauzer 1982; Tabachnick & Fidell 1983). For example, if a SKF prediction equation has three predictors (e.g., thigh SKF, chest SKF, and age), then the minimum sample size needs to be 60 to 120 subjects. The recommended minimum subject-to-predictor variable ratio and the minimum sample size vary with the degree of conservatism that statisticians believe is necessary to ensure that the data are representative of the population. Prediction equations that are based on small samples and have a poor subject-to-predictor ratio are suspect and should not be used.

• **What were the sizes of the R_{mc} and the SEE for this equation?** In general, the R_{mc} for body composition prediction equations should exceed 0.80. This means that at least 64% [$R^2 = (0.80)^2 \times$ 100] of variance in the reference measure can be accounted for by the predictors in the equation. The larger the R_{mc}, the greater the amount of shared variance between the reference measure and predictors. When you evaluate the relative worth of a prediction equation, it is more important to focus on the size of the prediction error (SEE) than on the R_{mc} because the magnitude of R_{mc} is greatly affected by size and variability of the sample. Lohman (1992) developed standards for evaluating prediction errors of body composition methods and equations estimating %BF and FFM. These values are based on empirically derived measurement errors associated with the reference method. Standards for evaluating prediction errors are presented in table 2.1.

• **To whom is the prediction equation applicable?** To answer this question, you need to pay close attention to the physical characteristics of the sample used to derive the equation. Factors such as age, gender, ethnicity, level of body fatness, and physical activity level need to be examined carefully. Prediction equations are either population-specific or generalized. **Population-specific equations** are intended to be used only to estimate the body composition of individuals from a specific homogeneous group. For example, there are separate SKF equations for prepubescent African American males and prepubescent Caucasian males (Slaughter et al. 1988). Population-specific equations are likely to systematically over- or underestimate body composition if they are applied to individuals who do not belong to that population subgroup. There are generalized prediction equations that

Table 2.1 Standards for Evaluating Prediction Errors

%BF (SEE or TE)	Db (g/cc) (SEE or TE)	FFM (kg) (SEE or TE)		Subjective rating
Male and female	Male and female	Male	Female	
2.0	0.0045	2.0-2.5	1.5-1.8	Ideal
2.5	0.0055	2.5	1.8	Excellent
3.0	0.0070	3.0	2.3	Very good
3.5	0.0080	3.5	2.8	Good
4.0	0.0090	4.0	3.2	Fairly good
4.5	0.0100	4.5	3.6	Fair
5.0	0.0110	>4.5	>4.0	Poor

SEE = standard error of estimate; TE = total error or pure error.

Data modified from Lohman (1992, pp. 3-4).

can be applied to individuals who differ greatly in physical characteristics. **Generalized equations,** developed using diverse, heterogeneous samples, account for differences in physical characteristics by including these variables as predictors. For example, the generalized BIA equation of Van Loan and Mayclin (1987) can be applied to both men and women ranging in age from 18 to 60 years, because gender and age are predictors in this equation. Likewise, age is a predictor in some SKF equations for adults (Jackson & Pollock 1978; Jackson et al. 1980). Therefore, these equations are generalizable to men and women 18 to 61 years of age. Later chapters present population-specific and generalized body composition prediction equations for various ages (children and older adults), ethnic groups (e.g., African Americans, American Indians, Asians, Caucasians, and Hispanics), levels of fatness (in persons with anorexia or obesity), and physical activity levels (athletes).

• **How were the variables measured by the researchers who developed the prediction equation?** It is important to know not only which variables are included in a prediction equation but also how each one of these predictors was measured by the researchers developing the equation. Although it is highly recommended that standardized anthropometric procedures and measurement sites be used in all body composition studies, this does not always happen. For example, the suprailiac SKF used in the SKF equations developed by Jackson et al. (1980) is measured above the iliac crest at the anterior axillary line. In contrast, Lohman et al. (1988) recommend that the suprailiac SKF be measured above the iliac crest at the midaxillary line. For most individuals, there will be a difference between SKF thicknesses measured at these two sites. Thus, larger than expected prediction errors may result if body composition variables are not measured according to the descriptions provided by the researchers who developed the equation.

• **Was the prediction equation cross-validated?** For the predictive accuracy of a body composition equation to be determined, it needs to be cross-validated. This is done using one of the procedures described earlier. Typically, the equation is tested on other independent samples from the population, or the total sample may be split into validation and cross-validation groups. In general, one should not use prediction equations that have not been cross-validated either on the original study sample or on independent samples from other studies.

• **What was the size of the correlation ($r_{y,y'}$) between the reference measure (Y) and predicted (Y') scores (validity coefficient)? What**

was the size of the prediction errors (SEE and TE) when this equation was applied to the cross-validation sample? In general, an equation with good predictive accuracy should yield a moderately high validity coefficient ($r_{y,y'} > 0.80$) and an acceptable (i.e., good to excellent) SEE and TE (see table 2.1). Keep in mind that the SEE represents the degree of deviation of individual scores from the regression line (see figure 2.2), whereas the TE represents the degree of deviation from the line of identity (see figure 2.3). When there are small differences between individuals' reference and predicted scores, the TE will be small. Typically, the TE is larger than the SEE because the size of TE is affected by both the SEE and the difference between the average predicted and reference scores of the cross-validation sample. Also, the regression line depicting the relationship between actual and predicted scores for the cross-validation sample should not differ significantly from the line of identity. This means that the slope of the regression line should not be significantly different from 1, and the y-intercept should not differ significantly from 0.

• **Was the average predicted score similar to the average reference score for the cross-validation sample?** The prediction equation should yield similar mean values for the actual and predicted scores of the cross-validation sample. The **constant error (CE)** is the average difference between the actual and predicted means. The means are tested using a paired t-test and should not differ significantly from each other. A large significant difference between the measured and predicted means indicates a **bias** or a systematic difference (i.e., over- or underestimation) between the validation and cross-validation samples caused by technical error or biological variability (Lohman 1981).

We applied these validation and cross-validation criteria to assess the overall (i.e., group) accuracy of models, methods, and prediction equations for estimating the body composition in population subgroups. Because there is a high degree of interindividual variability within a group, it is also necessary to evaluate the predictive accuracy of these methods and equations for *individuals* within a group.

Criteria for Evaluating Methods and Prediction Equations for Individuals

Although a method or prediction equation may accurately estimate the average body composition

ecific group, it does not necessarily give
e estimates for all individuals composing
that group. Although we can expect 68% of the
predicted scores to fall within ±1 SEE and can
expect 95% of the scores to be within ±2 SEE of
the regression line, the body composition of some
individuals may be over- or underestimated by
more than these amounts.

To evaluate how well an equation works for
estimating the body composition of individuals,
researchers use the **Bland and Altman method**
(1986), which sets **limits of agreement** and **confidence intervals** around the average difference
($\bar{d}$) between the actual and predicted scores for
the sample. To use this method, the first step is
to calculate residual or difference (Y – Y') scores
for each individual in the sample. The next step is
to plot these individual difference scores against
individual average scores (see figure 2.4). We calculate the average scores by summing the actual
and predicted scores of the individual and dividing
this sum by 2. Provided that the **difference scores**
are normally distributed, we would expect most
(~95%) of these scores to lie between ±2 standard
deviation units from the overall mean difference
($\bar{d}$) for the group. In this case, the standard deviation of the difference scores (S_d) is used to set the
upper ($+2S_d$) and lower ($-2S_d$) limits of agreement.
The smaller the limits of agreement, the better the
individual predictive accuracy of the method or
equation. From the graph, we can count the number

of difference scores exceeding the upper and lower
limits of agreement. With this method, 95% confidence intervals also may be calculated to assess
the precision of the body composition estimates for
the whole population. The width of the confidence
intervals is an estimate of the degree of variability
in difference scores for the entire population, with
a small interval indicating less variability.

Another way to assess the accuracy of a prediction equation for measuring body composition of
individuals is to graph the difference scores versus
average scores of the sample as was done for the
Bland and Altman method. However, instead of
using the calculated limits of agreement, the minimal acceptable standard for prediction errors (see
table 2.1) is plotted on the graph and the percentage of individuals falling outside of these limits is
calculated. In figure 2.5 you can see that the %BF of
approximately 75% of the sample (15 of the 20 data
points) was accurately estimated within ±3.5% BF,
the minimal acceptable standard (i.e., good rating)
for estimating %BF.

To summarize, the **group predictive accuracy** for body composition models, methods, and
prediction equations presented in later chapters
was evaluated using validation and cross-validation principles and criteria. Throughout this book,
we also report the 95% limits of agreement ($\bar{d} \pm
2S_d$) to evaluate the predictive accuracy of models,
methods, and prediction equations for *individuals*
within population subgroups. Keep in mind that it

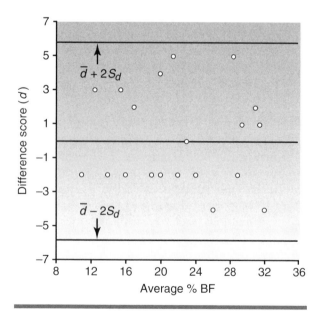

FIGURE 2.4   Bland-Altman plot with 95% limits of
agreement between two methods.

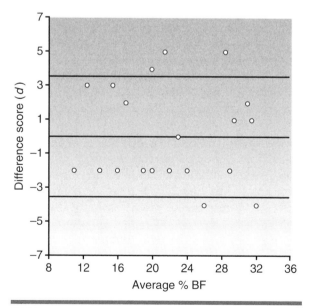

FIGURE 2.5   Comparison of two methods using minimum acceptable standard.

is possible for a method or prediction equation to accurately estimate the average body composition of a group but at the same time have wide limits of agreement for estimating body composition of individuals within that group. Whenever possible, we have selected methods and prediction equations that are accurate for both groups and individuals. We highly recommend that you apply the following criteria when evaluating newly developed equations or selecting other equations to assess body composition:

- A multicomponent model, preferably a 4-C model, is used to obtain reference measures.

- An acceptable method or combination of methods is used to derive reference measures of body composition.

- A large sample (N = 100-400) and at least 20 to 40 subjects per predictor variable are used.

- The size of the multiple correlation and validity coefficients is >0.80.

- The group prediction errors (SEE and TE) are good to excellent (see table 2.1).

- Demographic characteristics of the sample (i.e., age, gender, ethnicity, and level of body fatness) are described.

- The prediction equation was cross-validated in the original study (i.e., internal cross-validation) or on independent samples from other studies (i.e., external cross-validation).

- The line of best fit (regression line) does not deviate significantly from the line of identity; that is, the slope of the line does not differ from 1, and the y-intercept does not differ from 0.

- The CE (bias), or the average difference between the measured and predicted values for the cross-validation group, is not statistically significant.

- The 95% limits of agreement are relatively reasonable (e.g., within ±5% BF).

Key Points

- Correlation measures the strength of association or relationship between two variables; regression attempts to predict one variable from another.

- Bivariate regression is a statistical method used to predict one variable from another variable.

- Multiple regression is a statistical method used to predict one variable from two or more variables.

- All body composition methods and prediction equations need to be validated and cross-validated to determine their applicability and suitability for use in the field.

- The line of best fit is a regression line depicting a linear relationship between the DV and all of the predictor variables in the regression equation.

- The SEE is a type of prediction error that reflects the degree of deviation of individual data points around the line of best fit (regression line).

- The TE is a type of prediction error that reflects the degree of deviation of individual data points around the line of identity.

- Population-specific equations should only be used to estimate the body composition of individuals from a specific group, whereas generalized prediction equations may be used to estimate the body composition of individuals varying in age, gender, ethnicity, fatness, or physical activity level.

- To judge the relative worth of newly developed body composition methods and prediction equations, one should use standard evaluation criteria.

- The Bland and Altman method is used to compare methods and to evaluate how well an equation works for estimating body composition of individuals within a group.

Key Terms

Learn the definition for each of the following key terms. Definitions of terms can be found in the Glossary, page 227.

bias

bivariate regression analysis

Bland and Altman method

bootstrap technique

coefficient of determination (r_{xy}^2)

confidence interval

constant error (CE)

correlation

cross-validation

cubic regression model

curvilinear regression

difference score

double cross-validation technique

external cross-validation

generalized prediction equation

gold standard

group predictive accuracy

hierarchical regression

homoscedasticity

individual predictive accuracy

internal cross-validation

jackknife technique

limits of agreement

line of best fit

line of identity

multiple correlation coefficient (R_{mc})

multiple regression

Pearson product-moment correlation coefficient (r_{xy})

polynomial regression

population-specific equations

prediction error

prediction equation

predictor variables

PRESS technique

pure error

quadratic regression model

reference method

regression

regression line

regression weight

residual score

slope (b)

standard error of estimate (SEE)

stepwise regression

total error (TE)

validity coefficient ($r_{y,y'}$)

y-intercept (a)

Review Questions

1. What is the difference between bivariate and multiple regression analysis?

2. What is the coefficient of determination and what does it represent?

3. What is the difference between a Pearson product-moment correlation coefficient and a multiple correlation coefficient?

4. What is the difference between the hierarchical and stepwise regression approaches to developing a prediction model?

5. Anthropometric data were collected for a 20-yr-old woman: BW = 65 kg; neck circumference (C) = 33 cm; thigh circumference (C) = 60 cm. Calculate her estimated FFM using the following prediction equation: FFM (kg) = 0.79 + 0.757 (BW) + 0.981 (neck C) − 0.516 (thigh C).

6. What is curvilinear regression analysis and when should it be used?

7. When a prediction equation is developed on a validation sample, what statistics are used to judge the accuracy of the prediction equation for this group?

8. For equations predicting FFM, Db, and %BF of adults to be rated as "good," identify the maximum acceptable values for the prediction errors (SEE and TE).

9. Why is it important to cross-validate prediction equations?

10. What is the difference between internal and external cross-validation?

11. Identify and define three techniques for internal cross-validation of a prediction equation.

12. When a prediction equation is cross-validated, what statistics are used to judge the accuracy of the prediction equation for the cross-validation sample?

13. How can you determine if the line of best fit differs significantly from the line of identity for a cross-validation sample?

14. Identify three approaches that may be used to cross-validate a prediction equation.

15. How do population-specific and generalized prediction equations differ?

16. Briefly describe the major questions that one needs to address when evaluating the predictive accuracy of body composition methods and prediction equations for a subgroup of the population.

17. How do researchers determine the predictive accuracy of body composition methods and prediction equations for individuals within a specific group?

CHAPTER 3

BODY COMPOSITION REFERENCE METHODS

Key Questions

- In research settings, what reference methods are used to assess body composition?
- What are the basic assumptions, principles, and theory underlying each reference method?
- What are the basic testing procedures for each reference method?
- What are the major sources of error for each reference method?
- What component of body composition is being measured by each reference method?
- Which model or method is recommended to account for the greatest amount of biological variability when estimating body fat?

The primary focus of this text is on the **field methods** of body composition assessment (chapters 4-7) and their applicability to healthy (part II) and clinical (part III) populations, but it is also important to have a basic understanding of laboratory techniques in this field. **Laboratory methods** provide reference or criterion measures for the derivation and evaluation of body composition field methods and prediction equations. Laboratory methods are generally more expensive, more inconvenient, and more time-consuming than field methods, but have a greater accuracy. Nevertheless, all laboratory methods make certain assumptions and are still subject to some measurement error. Thus, a true **"gold standard,"** or perfect reference method, for in vivo body composition assessment does not exist.

This chapter addresses four commonly used reference methods: hydrodensitometry, air displacement plethysmography, hydrometry, and dual-energy X-ray absorptiometry. For each method, the principles and basic assumptions, underlying model, **validity,** testing procedures, and sources of measurement error are presented. Additionally, we discuss a multicomponent model approach that combines variables from these different reference methods. Finally, we briefly introduce some highly sophisticated, expensive, and therefore rarely used reference methods.

Hydrodensitometry

Densitometry refers to measurement of body density. To calculate Db, we divide body mass by body volume (Db = BM/BV). Body volume can be measured using either hydrodensitometry or air displacement plethysmography.

Hydrodensitometry is also known as **hydrostatic weighing (HW)** or **underwater weighing (UWW). Hydrodensitometry (HD)** provides an estimate of total body volume from the water displaced by the body when it is fully submerged. When HW is combined with a measure of residual lung volume, this method provides a good measure of body volume (BV) from which Db can be easily calculated. Over the years, the HD method has been widely used, in conjunction with 2-C molecular models (see chapter 1), by researchers and practitioners to obtain estimates of %BF from Db. These 2-C model estimates of %BF are often incorrectly regarded as "gold standard" measures by both the layperson and professionals in the fitness industry. This method, however, relies on several assumptions and therefore is not error free.

Principle and Assumptions of HD

HD is based on **Archimedes' principle.** This principle states that the volume of an object submerged in water equals the volume of water displaced by the object. Since scales, rather than volumeters, are typically used for UWW, another way to state this principle is to say that an object submerged in water is acted on by a buoyant force equal to the weight of the water it displaces. Since it is not practical to collect and weigh the displaced water, the individual's underwater weight (UWW) is used to calculate weight loss under water. This weight loss under water is directly proportional to the volume and weight of the water displaced by the BV.

To accurately estimate Db using HD, BV must be corrected for the amount of air in the lungs and gastrointestinal tract (GI) at the time of measurement. Typically, the UWW is measured at residual lung volume, but other lung volumes can also be used (see p. 31). **Residual lung volume (RV)**, or the amount of air remaining in the lungs after a maximal expiration, is fairly large (usually 1-2 L) and therefore must be measured. The volume of gas in the GI is much smaller and is assumed to be 100 ml (Buskirk 1961).

When the HD method is used, Db is estimated with few assumptions. When this method is used to estimate %BF from Db using 2-C or multi-component body composition models, however, assumptions of these model equations must also be considered (see chapter 1).

HD Model

HD was pioneered by Behnke and colleagues (1942). According to Archimedes' principle, the BV of a client is equal to the difference between the client's mass in air and mass under water. This value must be corrected for the density of water, which corresponds to the water temperature at the time of submersion. Additionally, RV and gas in the GI tract will add to the buoyancy of the client and thus must be subtracted from the apparent BV. Therefore, the equation for Db is as follows:

$$Db = M_a / \{[(M_a - M_w) / D_w] - (RV + GI \text{ volume})\}$$

where M_a = mass in air, M_w = mass in water, D_w = density of water; GI volume $\cong$ 100 ml.

Db is converted to %BF using either 2-C or multicomponent model equations (see p. 11). The derivation equation to estimate $\%BF_{2\text{-}C}$ from Db

is presented in appendix A (p. 215), and various population-specific 2-C model conversion formulas are listed in table 1.4 (p. 9).

Validity of HD

It is difficult to determine the validity of HD because it is often the standard or benchmark to which other methods are compared. Still, HD is generally accepted as a valid method for measuring BV and Db; chemically determined Db agrees within 0.6% of Db measured by HD (Heymsfield et al. 1989).

HD, however, is not entirely free of error. The combined technical error of Db measured by HD is estimated to be 0.0015 to 0.0020 g/cc (Akers & Buskirk 1969; Buskirk 1961; Lohman 1992). The largest contributor to this error is RV (0.00139 g/cc), as the combined error of body weight, UWW, and water temperature is only 0.0006 g/cc (Akers & Buskirk 1969). Although the technical error is small (<0.0020 g/cc, or <1% BF), factors such as the client's ability to comply with testing procedures, technician skill in administering the test, and use of the wrong conversion formula result in much larger %BF errors (see "Sources of Measurement Error," pp. 31-33).

Although HD is accepted as a standard method for measuring Db, the error of this method increases substantially when it is used to estimate %BF, especially when used in conjunction with 2-C model conversion formulas. Due to interindividual variability in the FFB composition, Siri (1956) calculated an error of 3.9% BF in estimating $\%BF_{2\text{-}C}$ from Db for the general population. Even among a homogeneous population, Lohman (1992) estimated this error to be 2.8% BF.

Numerous researchers have compared $\%BF_{2\text{-}C}$ estimated from HD to %BF obtained from HD estimation using 4-C models. In a meta-analysis of 54 research studies, Fogelholm and van Marken Lichtenbelt (1997) compared HD to a variety of methods. In that analysis, eight studies included estimates of %BF using 4-C models. On average, $\%BF_{4\text{-}C}$ was 0.6% BF greater than $\%BF_{2\text{-}C}$, with a 95% confidence interval (CI) of 0.1% to 1.2% BF.

In reviewing more recent studies we found that, compared to $\%BF_{4\text{-}C}$, the HD method and 2-C model conversion formulas had average errors ranging from −2.8% BF to 1.8% BF, with most having a CE of about 1% to 1.5% (Clasey et al. 1999; Evans, Arngrimsson et al. 2001; Millard-Stafford et al. 2001; Prior et al. 1997; Visser et al. 1997; Wagner & Heyward 2001; Withers et al. 1998; Yee et al. 2001). Although a CE of 1% to 1.5% BF is not large, individual differences can be much greater. Clasey et al. (1999) reported 95% limits of agreement ranging from ±8.1% to ±12.0% BF for the Siri (1961) 2-C model for different age and gender groups. In what may be the most definitive multicomponent model study to date, Wang and colleagues (1998) reported an SEE of 2.2 kg when comparing FM estimated from the Siri (1961) 2-C model to FM from a 6-C chemical model. For a 70-kg person, this corresponds to an average prediction error (SEE) of about 3.1% BF.

HD Testing Procedures

The UWW system usually consists of a chair, made of PVC tubing, that is suspended from a spring-loaded autopsy scale (figure 3.1) or a rectangular mesh platform attached to four load cells (figure 3.2). Regardless of the type of apparatus used to measure UWW, RV must be measured to accurately assess BV.

As suggested by the fact that RV is so large, significant errors in estimating BV, Db, and %BF result from using predicted RV instead of measured RV (McCrory et al. 1998; Morrow, Jackson et al. 1986). In fact, the failure to measure RV is the largest source of technical error for measuring Db via HD (Akers & Buskirk 1969). Miller et al. (1998) evaluated various equations for estimating RV and noted large prediction errors (SEE = 0.41 L). An error of this magnitude (400 ml) for RV translates into a body fat error of about 3% to 4% (Ellis 2000).

Residual volume can be measured by oxygen dilution and nitrogen washout techniques (Wilmore et al. 1980), as well as closed-circuit helium dilution (Motley 1957). All of these procedures are reliable (r ≥ .97) and highly correlated (r ≥ .91) with one another (Motley 1957; Wilmore et al. 1980). Even though there appears to be good agreement between RV measured on land and in the water (Wilmore 1969), it is preferable to measure RV under water during the UWW procedure. Simultaneous measurement of RV and UWW is less time-consuming and provides a more accurate estimate of Db. A minimum of two RV trials, within ±100 ml, should be averaged and used to calculate Db.

With use of the chair-autopsy scale system (figure 3.1), the scale needle often oscillates greatly with water movement, making it difficult for the technician to obtain an accurate reading of UWW. A load cell system that can be integrated to an analog recorder with digital readout or graphically displayed on a microcomputer makes it easier to read this measurement. With this system, originally described by Goldman and Buskirk (1961) and later refined (Akers & Buskirk 1969), the client rests in a hands-and-knees position on a rectangular mesh platform that is suspended by four load

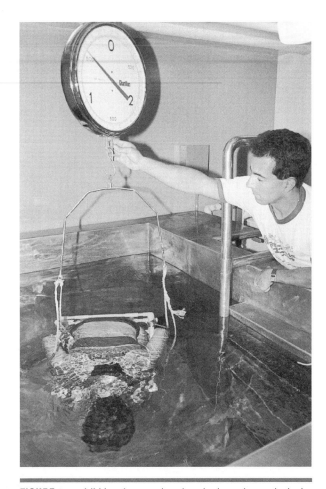

FIGURE 3.1   HW using spring-loaded scale and chair.

cells (figure 3.2). Organ, Ecklund et al. (1994) took this procedure one step further by simultaneously measuring UWW and RV for a real-time estimate of %BF.

To minimize measurement error, you should follow recommended guidelines for HW. Prior to the scheduled appointment, give your client the following pretest instructions:

HW Pretesting Client Guidelines

- Do not eat or engage in strenuous exercise for at least 4 hr prior to your scheduled appointment.
- Avoid eating any gas-producing foods (e.g., baked beans and diet sodas) for at least 12 hr prior to test.
- Bring a towel and a tight-fitting, lightweight swimsuit with you.

When the client arrives at the test center, ask the client to shower, remove all jewelry, and change into a swimsuit. Also, ask the client to urinate and eliminate as much gas and feces as possible. Prior

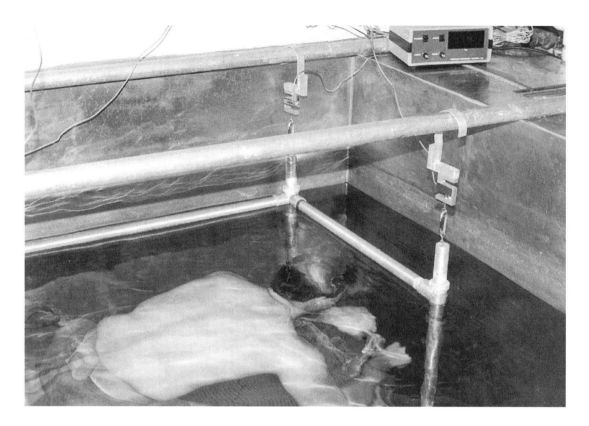

FIGURE 3.2   HW using load cell system and platform.

to the client's arrival or while your client is preparing for the test, complete the first two testing procedures.

Testing Procedures for Hydrodensitometry

1. Carefully calibrate the body weight and UWW scales. To determine the accuracy of the autopsy scale, hang calibrated weights from the scale and check the corresponding scale values. To calibrate a load cell system, place the weights on the platform and check the recorded values.

2. Measure the UWW of the chair or platform, as well as that of all of the supporting equipment and weight belt; this is the **tare weight.**

   Once your client presents him- or herself for testing, follow these procedures:

3. Measure your client's "dry weight" (weight in air) to the nearest 50 g.

4. Check and record the water temperature of the tank just prior to the test; it should range between 34° and 36° C. Use the constant values in table 3.1 to determine the density of the water at that temperature.

5. Instruct your client to enter the tank slowly, keeping the water calm. Without touching the chair or weighing platform, have the client gently submerge and rub his or her hands over the body to eliminate air bubbles from the swimsuit, skin, and hair.

6. Have the client kneel on the UWW platform or assume a sitting position in the chair. You may need to add a scuba diving weight belt around your client's waist to facilitate this position. If RV is being measured simultaneously, the mouthpiece is inserted at this time. If RV is measured outside of the tank, administer the RV test *before* the client changes clothes and showers.

7. Have the client take a few normal breaths and then exhale maximally while slowly bending forward to lower the head below the water. Check to make certain that the client's head and back are completely submerged, and that the arms and feet are not touching the sides or bottom of the tank. Instruct the client to continue exhaling until RV is reached. The client needs to remain as still as possible during this procedure. A relaxed and motionless state under water will help allow you to obtain an accurate reading of UWW.

8. Record the highest stable weight with the client fully submerged at RV; then signal the client that the trial is completed.

9. Administer as many trials as needed to obtain three readings within ±100 g. Most clients achieve a consistent and maximal UWW by the fourth or fifth trial (Bonge & Donnelly 1989; Organ, Eklund et al. 1994). Average the three highest trials and record this value as the gross UWW (Bonge & Donnelly 1989).

10. Determine the net UWW by subtracting the tare weight from the gross UWW. The net UWW is used in the calculation of BV.

Table 3.1   Density of Water

Temperature (°C)	Density (g/cc)
33	0.9947
34	0.9944
35	0.9941
36	0.9937
37	0.9934

 Sources of Measurement Error for HD

The accuracy and precision of HD are affected by client factors, technician skill and equipment, and the model and conversion formula used to estimate %BF from Db. The following questions and responses address these sources of measurement error.

Client Factors

• **What should I do when my client is unable to blow out all of the air from the lungs or to remain still while under water?** You will likely come across clients who are uncomfortable expelling all of the air from their lungs during HW. Performing a maximal expiration while completely submerged under water is a difficult and unnatural maneuver. Obviously, failure to maximally exhale will make the client more buoyant, producing a lighter UWW, lower Db, and higher %BF estimate. About 9% of the variability in RV can be attributed to learning or fatigue (Marks & Katch 1986).

When people have difficulty complying with this testing procedure, you can underwater weigh them at **functional residual capacity (FRC)** or **total lung capacity (TLC)** instead of RV. Thomas and Etheridge (1980) underwater weighed 43 males,

comparing the body densities measured at FRC (taken at the end of normal expiration while submerged) and at RV (at the end of maximal expiration). The two methods yielded similar results. Likewise, Timson and Coffman (1984) reported that Db measured by HW at TLC (vital capacity + RV) was similar (less than 0.3% BF difference) to that measured at RV if TLC was measured in the water. However, when the TLC was measured out of the water, the method significantly overestimated Db. When using these modifications of the HW method, you must still measure RV to calculate the FRC or TLC of your client. Also, be certain to substitute the appropriate lung volume (FRC or TLC) for RV in the calculation of BV.

Because of their lower-body density, clients with greater amounts of body fat are more buoyant than leaner individuals; therefore, they have more difficulty remaining motionless while under water. To correct this problem, place a weighted scuba belt around the client's waist. Be certain to include the scuba belt when measuring and subtracting the tare weight of the HW system.

• **What should I do when my clients are afraid to put their faces in the water or are not flexible enough to get their backs and heads completely submerged?** Occasionally you will encounter clients who are extremely fearful of being submerged, who dislike facial contact with water, or who are unable to bend forward to assume the proper body position for HW. In such cases, a satisfactory alternative would be to weigh your clients at TLC while their heads remain above water level. Donnelly et al. (1988) compared this method (i.e., TLCNS or total lung capacity with head not submerged) to the criterion Db obtained from HW at RV for 75 men and 67 women. Vital capacity was measured with the subject submerged in the water to shoulder level. Regression analysis yielded the following equations for predicting Db at RV, using the Db determined at TLCNS as the predictor:

Males
Db at RV = 0.5829 (Db at TLCNS) + 0.4059
r = .88 SEE = 0.0067 g/cc

Females
Db at RV = 0.4745 (Db at TLCNS) + 0.5173
r = .85 SEE = 0.0061 g/cc

The correlations (r) between the actual Db at RV and the predicted Db at RV were high, and the SEEs were within acceptable limits. These equations were cross-validated for an independent sample of 20 men and 20 women. The differences between the Db from HW at RV and the predicted Db from weighing at TLCNS were quite small (less than 0.0014 g/cc or 0.7% BF). This method may be especially useful for HW of older adults, individuals who are obese and have limited flexibility, and people with physical disabilities.

• **Will the accuracy of the HW test be affected if I estimate RV instead of measuring it?** As mentioned earlier, several prediction equations have been developed to estimate RV based on the individual's age, height, gender, and smoking status. However, these RV prediction equations have large prediction errors (SEE = 400-500 ml). When RV is measured, the precision of the HW is excellent (≤1% BF). However, the measurement error increases substantially (±2.8-3.7% BF) when RV is estimated (Morrow, Jackson et al. 1986). Therefore, always measure RV when you are using the HW method.

• **When is the best time during the menstrual cycle to hydrostatically weigh my female clients?** Some women, particularly those whose body weight fluctuates widely during their menstrual cycles, may have significantly different estimates of Db and %BF when weighed hydrostatically at different times in their cycles. Bunt et al. (1989) reported that changes in TBW values due to water retention during the menstrual cycle can partly explain the differences in body weight and Db during a cycle. On average, the %BF of the women was 24.8% BF at their lowest body weights, compared to an average of 27.6% BF at their peak body weights during their menstrual cycles. Because their low and peak body weights occurred at different times during the cycle (varied from 0 to 14 days prior to the onset of the next menses), the effect of TBW fluctuations cannot be routinely controlled by using the same day of the menstrual cycle for all women. However, when you are monitoring changes in body composition over time or establishing healthy body weight for a female client, it is recommended that you hydrostatically weigh her at the same time within her menstrual cycle and outside of the period of her perceived peak body weight.

Technician Skill and Equipment

As mentioned earlier, UWW is measured using either a chair suspended from the spring-loaded scale or a platform attached to load cells (see figures 3.1 and 3.2). Spring-loaded HW scales are very sensitive to movement. If the client submerges too quickly, the needle will oscillate greatly, making it difficult or sometimes impossible to get an accurate reading of UWW. To minimize this problem, coach

the client to move slowly and to remain motionless while submerged. Also, you can stabilize (dampen) the movement of the UWW apparatus by holding it in place; however, much practice is necessary in order to learn how to do this without affecting the scale value.

The better, but more expensive, solution to this problem is to use a load cell system (see Goldman & Buskirk 1961; Akers & Buskirk 1969) instead of the spring-loaded HW scale. With this system, the electrical output from the load cells is integrated to an analog recorder with a digital readout or is graphically displayed on a microcomputer.

Model and Conversion Formula

Earlier we mentioned that the use of 2-C model conversion formulas potentially increases the measurement error of the HD method. When the assumptions of these models are not met due to interindividual variability in FFB composition, both group and individual estimates of %BF will exceed the technical precision (<1% BF) of the HD method. For research and some clinical settings, we therefore recommend using a multicomponent model that corrects Db for body water and bone mineral to estimate %BF (see chapter 1).

In settings where a 2-C model equation must be used to estimate %BF from Db because it is impractical or impossible to measure TBW or bone mineral, we recommend using population-specific conversion formulas (table 1.4, p. 9). These 2-C model equations were derived using estimates of the average FFB density of specific population subgroups reported in the literature. Using the wrong conversion formula could result in a large prediction error. For example, if you select the appropriate age-gender conversion formula to estimate %BF of a prepubescent 10-yr-old girl with a Db of 1.0500 g/cc, her estimated body fat is 14.5% BF. If instead you use the Siri (1961) 2-C model conversion formula, her body fat estimate is 21.4% BF. This is a difference of almost 7.0% BF simply due to selecting the wrong conversion formula!

To minimize measurement error for the HD method, adhere to the recommended testing guidelines. The following tips may also increase the accuracy of your UWW measurements:

Minimizing HD Measurement Error

• Make sure that your clients adhere to all pretesting guidelines.

• Before each test session, check the calibration of BW and UWW scales or load cells and carefully calibrate gas analyzers used to measure RV.

- Precisely measure BW to ±50 g, UWW to ±100 g, and RV to ±100 ml.
- Coach the client to maximally exhale and remain motionless under the water.
- Steady the UWW apparatus as the client submerges, but remove your hand from the scale before actually reading the UWW value.
- If possible, use a load cell system and measure RV simultaneously with UWW.
- When you calculate Db, carry this value out to five decimal places. Rounding off a Db of 1.07499 g/cc to 1.07 g/cc corresponds to a difference of 2.2% BF when converted with the Siri (1961) 2-C model formula.
- If you are estimating %BF from Db with a 2-C model, use the appropriate population-specific conversion formula.

Air Displacement Plethysmography

Air displacement plethysmography (ADP) is another method used to measure BV and Db. This method uses air displacement, instead of water displacement, to estimate BV. Because ADP is quick (usually takes 5-10 min) and requires minimal compliance by the client and minimal technician skill, it may prove to be an alternative to HD.

Early attempts at using plethysmography for the assessment of body composition were not successful. In 1995, however, Dempster and Aitkens introduced a new system (i.e., the Bod Pod®) that overcame some of the limitations of earlier ADP technology. The **Bod Pod** is a large, egg-shaped fiberglass chamber that uses air displacement and pressure-volume relationships to derive BV (see figure 3.3).

In a recent study in which many of the reference methods described in this chapter were subjectively rated, the Bod Pod ranked at or near the top in categories such as cost, time, maintenance, ability to accommodate people with limitations, and ease of use (Fields et al. 2002). Consequently, this body composition tool has received much attention from researchers recently; however, its use as a reference method or as a replacement for HD is still in question.

Principles and Assumptions of ADP

The principles and assumptions of the Bod Pod center on the relationship between pressure and volume, as well as attempts to control for changes in temperature and gas pressure that occur when

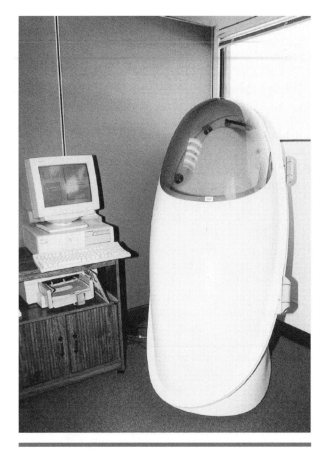

FIGURE 3.3   Air displacement plethysmograph.

a human is placed in an enclosed chamber. The Bod Pod determines BV by measuring changes in pressure within an enclosed chamber. Volume (V) and pressure (P) are inversely related according to **Boyle's law:**

$$P_1/P_2 = (V_2/V_1)$$

where P_1 and V_1 represent one paired condition of pressure and volume, and P_2 and V_2 another paired condition. P_1 and V_1 correspond to the pressure and volume of the Bod Pod chamber when it is empty; P_2 and V_2 represent the pressure and volume of the Bod Pod with the client sitting in the chamber.

Boyle's law assumes that **isothermal conditions** (air temperature remains constant as its volume changes) exist; but the majority of the air in the enclosed Bod Pod is under **adiabatic conditions**, meaning that it will compress or expand as the temperature changes within the chamber (e.g., a human being who will give off heat). Thus, the Bod Pod software uses a variation of Boyle's law that is more appropriate for the adiabatic air found within the chamber **(Poisson's law):**

$$P_1/P_2 = (V_2/V_1)^\lambda$$

where λ is the ratio of the specific heat of the gas at constant pressure to that of constant volume. For air it is equal to 1.4, representing the 40% difference between isothermic and adiabatic conditions.

This difference between isothermal and adiabatic air is significant because air under isothermal conditions is more compressible than air under adiabatic conditions, creating an apparent negative volume (Dempster & Aitkens 1995). For example, clothing is isothermal, so wearing a bulky jacket in the Bod Pod will result in an artificially reduced BV. Thus, a major assumption of the Bod Pod is that isothermal effects that could alter the BV estimate have been identified (clothing, hair, thoracic gas volume, and body surface area) and are controlled. To minimize the isothermal effects of clothing and hair, Bod Pod clients are tested wearing minimal clothing (spandex swimsuit) and a swim cap to compress the hair. An estimate of **body surface area,** calculated from the height and weight of the client, is used to correct for the isothermal effects at the body's surface. **Thoracic gas volume (TGV)**, or the volume of air in the lungs and thorax, is either directly measured or estimated by the Bod Pod to account for the isothermal conditions in the lungs.

ADP Model

The Bod Pod system actually consists of two chambers: a front chamber in which the client sits during the measurement and a rear (reference) chamber. A molded fiberglass seat forms the wall between the two chambers, and a moving diaphragm is mounted here that oscillates during testing (figure 3.4). The oscillating diaphragm creates small volume changes, equal in magnitude but opposite in sign between the two chambers, that produce small pressure fluctuations. The pressure-volume relationship (i.e., Poisson's law) is used to solve for the volume in the front chamber. This process is done twice, once with an empty chamber and once with a client in the chamber. BV is simply calculated as the difference between the chamber volume when it is empty and when the client is seated inside of the chamber.

This raw BV must be corrected for two variables having an isothermal effect that causes an erroneously low value of BV: body surface area and TGV. Body surface area is estimated using the formula

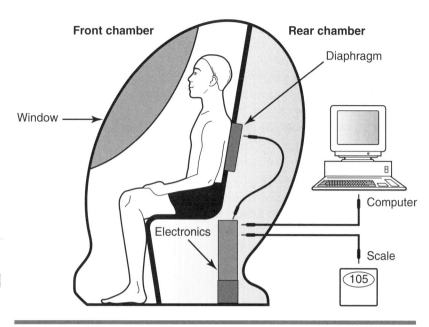

FIGURE 3.4   Two-chamber Bod Pod system.

of DuBois and DuBois (1916), and the surface area artifact is automatically accounted for by the Bod Pod software. Unlike RV, TGV may be accurately predicted; however, for research and clinical use, direct measurement of TGV is highly recommended (McCrory et al. 1998). Measuring TGV involves connecting the client to a breathing hose following the initial BV measurement. While the client is undergoing normal tidal breathing, the airway is occluded at mid-exhalation. At this time, the client gently puffs against the closed airway, alternately contracting and relaxing the diaphragm muscle. The small pressure changes in the lungs and external volume that this puffing maneuver creates are used to measure TGV. The Bod Pod software uses the following equation to derive BV (Dempster & Aitkens 1995):

$$BV (L) = BV_{raw} - \text{surface area artifact} + 40\% \text{ TGV}$$

As described for the HD method, Db is calculated from BV (i.e., Db = BM/BV). This Db value can be used in any of the 2-C or multicomponent model equations to estimate %BF from Db. The Bod Pod software defaults to the Siri 2-C model equation; however, the Brozek et al. and Schutte et al. 2-C equations, as well as the Siri 3-C and Selinger 4-C equations, can also be selected.

Validity of ADP

The accuracy of the Bod Pod was first evaluated against inanimate objects (cubes and cylinders) ranging in volume from 25 to 150 L (Dempster & Aitkens 1995). The error was estimated at <0.1% volume, and the SEE was 0.004 L. The research team of McCrory et al. (1995) was the first to test the Bod Pod on a human sample. They reported excellent validity compared to HD; the difference between Bod Pod and HD estimates of $\%BF_{2-C}$ was only –0.3% BF (SEE = 1.81% BF with 95% limits of agreement = –4.0% to 3.4% BF).

General Validation Studies

Since this initial study, numerous others have been done to assess the accuracy of the Bod Pod for measuring Db. Several researchers reported only small differences in average Db (≤0.002 g/cc) measured by the Bod Pod and HD (Fields et al. 2001; Vescovi et al. 2001; Yee et al. 2001), while others have reported slightly higher and statistically significant differences (0.003-0.007 g/cc) in adults (Collins et al. 1999; Demerath et al. 2002; Dewit et al. 2000; Millard-Stafford et al. 2001; Wagner et al. 2000). Also, several studies showed "good" group prediction errors (SEE ≤ 0.008 g/cc) in adults (Fields et al. 2000; Nunez et al. 1999; Wagner et al. 2000). However, as with other methods, the individual accuracy is not as precise. Dewit et al. (2000) reported 95% limits of agreement for Db of –0.021 to 0.024 g/cc for children and –0.018 to 0.003 g/cc for adults.

The majority of ADP validation studies compared %BF estimates from the Bod Pod ($\%BF_{ADP}$) to those obtained using HD ($\%BF_{HW}$), dual-energy X-ray absorptiometry (DXA) ($\%BF_{DXA}$), or both. In a review of 15 studies of adult participants, Fields and colleagues (2002) reported that the average difference (i.e., CE) between $\%BF_{ADP}$, $\%BF_{HW}$, or $\%BF_{DXA}$ ranged from –4.0% to 1.9%, with SEEs ranging from 2.2% to 3.7% BF. Overall, the average CE for the ADP method was estimated at <1% BF. However, the 95% limits of agreement typically were 12% BF (–7% to 5% BF) or greater, indicating large differences for some individuals. Additionally, some researchers noted that gender might systematically bias Bod Pod results, with the %BF of men being underestimated and that of women being overestimated (Biaggi et al. 1999; Levenhagen et al. 1999). In contrast, others reported that gender had little or no effect on the magnitude of the differences observed between ADP and HW estimates of %BF (McCrory et al. 1995; Nunez et al. 1999).

Several researchers assessed the predictive accuracy of ADP using multicomponent models to obtain reference measures of %BF (Collins et al. 1999; Fields et al. 2001; Millard-Stafford et al. 2001). Compared to 4-C model estimates of %BF ($\%BF_{4-C}$), the Bod Pod 2-C default calculations significantly underestimated average body fat by 1.8% to 2.8% BF. In another study, Db estimates from the Bod Pod and HW were entered into the same

4-C model equation (Millard-Stafford et al. 2001); these researchers reported a significant difference between $\%BF_{4-C}$ estimated by the Bod Pod (17.8% BF) and by HW (19.3% BF). The range for the 95% limits of agreement in these studies was about 9% BF, and the SEEs were very good to excellent (2.4-2.7% BF). Fields and colleagues (2001) concluded that the predictive accuracy of the Bod Pod and HW is similar when each method is evaluated against 4-C models.

Validation Studies on Special Populations

Because the Bod Pod is more accommodating than the HD method, there is much interest in establishing the validity of the ADP method for estimating %BF in special populations such as children and older adults as well as in clinical populations. While several researchers reported no significant differences (<1.2% BF) between %BF estimates from the Bod Pod and HD in children (Demerath et al. 2002; Dewit et al. 2000; Nunez et al. 1999; Wells et al. 2000), others reported that the average Db of children measured with the Bod Pod was either significantly overestimated (Lockner et al. 2000) or underestimated (Fields & Goran 2000) by 0.0052 to 0.0063 g/cc.

Compared to DXA, the Bod Pod underestimated average %BF by 1.9% to 2.2% BF in children (Fields & Goran 2000; Lockner et al. 2000; Nicholson et al. 2001). However, Nicholson et al. (2001) commented that the predictive accuracy of the ADP method was better than that of the SKF and bioimpedance field methods. Also, Lockner et al. (2000) noted that there was closer overall agreement between ADP and DXA methods than between HW and DXA. Furthermore, Fields and Goran (2000) concluded that the predictive accuracy of the Bod Pod in estimating FM derived from a 4-C model is better than that of HD, DXA, or TBW methods.

While several research teams have included older participants (>60 yr) in Bod Pod validation studies (Demerath et al. 2002; Koda et al. 2000; Nunez et al. 1999), we are aware of only one study that used only older adults (Yee et al. 2001). In this study of 58 men and women, aged 70 to 79 yr, the Bod Pod and HW methods were validated against 4-C model estimates of %BF. The authors reported no significant difference in Db between ADP and HD methods, and no significant difference in the estimation of %BF among 2-C, 3-C, and 4-C models using either ADP or HD for the Db measurement.

Validation Conclusions

Overall, even though results are equivocal and the CE may be slightly larger for the Bod Pod, the predictive accuracy and validity of the ADP and HD methods appear to be similar. Fields et al. (2002) suggested that the limitation of using 2-C models to estimate %BF, rather than the Bod Pod itself, is more likely the reason for slight differences in Bod Pod and 4-C model estimates of average %BF. Interlaboratory studies are needed to quantify variability between Bod Pod instruments. Also, future research should compare %BF estimates from the Bod Pod against more sophisticated methods and models (e.g., neutron activation analysis and 6-C chemical model) to further clarify the validity of ADP as a reference method.

Testing Procedures for ADP

The Bod Pod is user-friendly, with computer prompts for each step of the procedure. ADP is a faster and easier procedure than the HD method; researchers reported better compliance and a preference for ADP over HD among participants, including children (Demerath et al. 2002; Dewit et al. 2000; Lockner et al. 2000). Prior to the scheduled appointment, give your client pretesting instructions. These are the same as the instructions for HW (see p. 29), except for the addition of bringing a swim cap. The procedure for obtaining Db from the Bod Pod is as follows:

Testing Procedures for the Bod Pod

1. Instruct the client to change into a swimsuit and to completely void the bladder and bowels.

2. Measure the client's height to the nearest centimeter and body weight to the nearest 5 g using the Bod Pod scale. These measures are used to calculate body surface area.

3. Perform the two-point calibration: (a) baseline calibration with the chamber empty and (b) phantom calibration using a 50-L calibration cylinder. Be careful handling the calibration cylinder; a dent in the cylinder will alter its volume.

4. Instruct your client to sit in the chamber and close the door tightly. During this 20-sec test, ask your client to breath normally.

5. Open the door and then close it tightly; repeat the 20-sec test. If the two tests disagree by more than 150 ml, then perform additional tests until two tests are within 150 ml. These two tests are averaged and used in the calculation of raw BV.

6. Open the door and connect the client to the system's breathing circuit to begin the TGV measurement.

7. Close the door. After a few tidal volume (i.e., normal) breathing cycles, the airway is occluded

by the Bod Pod. Instruct your client to perform the puffing maneuver. If the computer-calculated "figure of merit" (indicating similar pressure signals in the airway and chamber) is not met, repeat this step.

Sources of Measurement Error for ADP

The reliability of the Bod Pod is good. For 20 repeated trials of a 50.039-L cylinder, the coefficient of variation (CV = s/M) was 0.025%, and the mean difference between two days was only 3 ml (Dempster & Aitkens 1995). In humans, the trial-to-trial CV ranges from 1.7% to 3.4% (Biaggi et al. 1999; McCrory et al. 1995; Miyatake et al. 1999; Sardinha et al. 1998; Vescovi et al. 2001). The range of between-day CVs is 2.0% to 2.3% (Levenhagen et al. 1999; Miyatake et al. 1999; Nunez et al. 1999).

Several researchers reported that the technical error of the Bod Pod is 0.4% BF (Collins et al. 1999; McCrory et al. 1995) and ≤0.0020 g/cc (Sardinha et al. 1998; Vescovi et al. 2001; Wells & Fuller 2001). These values match the standard of precision established for HD (Lohman 1992). Although the precision for ADP is equal to or better than that of HD (Dewit et al. 2000; McCrory et al. 1995; Wells & Fuller 2001), several factors related to the client, testing conditions, and %BF conversion formulas may increase the measurement error of the Bod Pod. The following questions and responses address these issues.

Client Factors

• **How are the test results affected if my client has excess body hair?** As mentioned earlier, isothermal air, trapped in body hair, may affect test results. For clients with beards, %BF may be underestimated by 1% BF; when scalp hair is exposed (i.e., no swim cap), %BF is underestimated by about 2.3% BF (Higgins et al. 2001). Wearing a tight-fitting swim cap, as recommended in the original validation study of McCrory et al. (1995), and shaving excess facial or body hair will ensure the most accurate estimate of BV and Db using this method.

• **Can I use the Bod Pod to measure the body composition of children?** During the 20-sec testing procedure, the client must remain very still. The BV estimate from the ADP method can vary if the client is moving inside the chamber during testing (Taylor et al. 1985). Fields and Goran (2000) commented that it took twice as long to measure children compared to adults, primarily because children tend to move during the test. As a result, the test-retest reliability of the Bod Pod is lower in children (r = .90) than in adults (r = .96) (Demerath et al. 2002).

- **Can I use the Bod Pod to measure the body composition of a client of any size?** Although the precision of the ADP method for estimating BV and Db is good across a wide range of body sizes (Wells & Fuller 2001), several researchers commented that body size may affect Bod Pod estimates, with the largest differences seen in the smallest clients (Demerath et al. 2002; Lockner et al. 2000; Nunez et al. 1999). The ideal chamber-to-client volume ratio may be exceeded for clients, especially children, who have a small BV (Fields & Goran 2000). This is an area that requires further investigation.

Testing Conditions

- **Is it absolutely necessary that my client wear a swimsuit and swim cap during the Bod Pod test?** The original investigators of the Bod Pod recognized that the isothermal effect of clothing leads to an underestimation of BV; they recommended that clients be tested wearing only a swimsuit and swim cap to minimize this effect (Dempster & Aitkens 1995; McCrory et al. 1995). The more clothing that is worn, the larger the layer of isothermal air and the greater the underestimation of BV. For example, wearing a hospital gown instead of a swimsuit lowers %BF by about 5% (Fields et al. 2000). Thus, the clothing recommendation of a tight-fitting swimsuit and swim cap needs to be followed.

- **Do I need to measure my client's TGV or can I use a predicted TGV value?** Although McCrory et al. (1998) reported an insignificant difference (54 ml) between measured and predicted TGV, the SEE was large (442 ml), and some researchers have reported larger mean differences (344-400 ml) and SEE (650 ml) (Collins et al. 1999; Lockner et al. 2000). Given that only 40% of the TGV value is used in the calculation of BV, using a predicted TGV has a relatively smaller effect on Db and %BF estimates compared to using a predicted RV for the HD method. Nevertheless, for maximum accuracy, it is recommended that you use a measured, rather than a predicted, TGV.

- **If I am using both HW and the Bod Pod to measure my client's body composition, which test should I give first?** The Bod Pod manufacturer recommends that clients be tested under resting conditions and when the body is dry. Although there are no published studies indicating the amount of error that may occur if these guidelines are violated, experts suggest adhering to these recommendations (Fields et al. 2002). Thus, if a test battery includes both HW and ADP, the Bod Pod test should be administered first. If this is not possible, make certain that your client is completely dry and fully recovered from the HW test before you administer the Bod Pod test.

Conversion Formula

As previously discussed for the HD method, in research and clinical settings, using a multicomponent model and conversion formula will increase the group and individual accuracy of your %BF estimates. The Siri 3-C and Selinger 4-C formulas are included in the Bod Pod software and can be used when the water and mineral components of the FFB are known. The default equation in the Bod Pod software is the Siri 2-C model formula for non-black adults, but the Schutte formula for blacks is also available. In field settings, these 2-C conversion formulas may be appropriate for some of your clients, depending on their demographic characteristics. For other clients, however, you may need to select an appropriate population-specific 2-C model formula (see table 1.4, p. 9). As mentioned for the HD method, it is critical to select the correct conversion formula if you are using the Db value obtained from the Bod Pod to estimate $\%BF_{2-C}$. Because HD and ADP are both densitometry methods and the same assumptions are made about the FFB, the magnitude of error for the Bod Pod will be the same as for HD when Db is converted to %BF.

Hydrometry

Hydrometry is the measurement of body water. Water is the most abundant constituent of the body, typically composing over 60% of the body weight (Brozek et al. 1963) and approximately 73% of the FFB (Pace & Rathbun 1945; Sheng & Huggins 1979; Wang et al. 1999). However, as you will see throughout this text, these percentages can vary considerably with age, level of body fatness, and health status.

Because water is the most abundant component of the body and is predominantly associated with the FFB, the measurement of TBW, as well as the distribution of water extracellularly (ECW) and intracellularly (ICW), is important to the assessment of body composition. TBW is typically estimated via measurement of the dilution of isotopic tracers. With this method, the concentration of either hydrogen or oxygen isotopes in biological fluids (e.g., saliva, plasma, and urine) after equilibration is measured and used to estimate TBW. As with the densitometry methods, %BF can be estimated from hydrometry using a 2-C model, but the isotope dilution technique and the conversion from TBW to $\%BF_{2-C}$ are not without assumptions.

Principles and Assumptions of Hydrometry

Hydrometry is based on the **dilution principle.** This principle states that the volume of a solvent (body water) is equal to the amount of a compound (isotopic tracer) added to the solvent divided by the concentration of the compound in that solvent (Edelman et al. 1952). Thus, if the dose of the isotopic tracer is known and the concentration of the tracer in a sample of water can be determined, body water can be calculated.

Assumptions

Four basic assumptions are applied to the tracers and the isotope dilution technique for estimating TBW, and there is an additional assumption for estimating %BF from the hydrometry method.

• **The tracer is distributed only in body water.** This assumption is violated slightly as we know that tracers exchange to a small degree with nonaqueous molecules. Because of the exchange of hydrogen in the tracers with proteins and carbohydrates in the body, the dilution space of the tracers will be slightly greater than TBW. It is estimated that the tracer oxygen-18 ($H_2^{18}O$) overestimates TBW by 1%, and that the tracers tritium oxide (^3H_2O) and deuterium oxide (^2H_2O, which is also abbreviated as D_2O) overestimate TBW by 4% (Schoeller 1996).

• **The tracer is distributed evenly throughout all water compartments.** There is very little fractionation, or change in the abundance of isotopes, between various anatomical water compartments; however, some fractionation occurs and must be accounted for especially when respiratory water vapor samples are used to estimate TBW (Wong et al. 1988).

• **Tracer equilibration is achieved relatively rapidly.** Wong et al. (1988) demonstrated that tracer equilibration occurs within 3 hr regardless of the body fluid sampled (plasma, saliva, urine, or breath), but a small amount of equilibration (0.3%) may continue over the next hour (Schoeller et al. 1985). A 4-hr equilibration time is recommended for patients with expanded extracellular water compartments (e.g., because of edema, pregnancy) (Schoeller 1996).

• **Neither the tracer nor body water is metabolized during the equilibration time.** Nonisotopic tracers, like ethanol and antipyrine, rapidly metabolize and should not be used for hydrometry. Instead, isotopic tracers like $H_2^{18}O$, ^3H_2O, and D_2O, which are not rapidly metabolized, should be used. However, even isotopically labeled water is subject to complications because body water is in a constant state of flux and is turning over daily due to ingestion of beverages, hydration of foods consumed, oxidation of fuels, exchange with atmospheric moisture, urination, defecation, breathing, and sweating.

• **With use of the hydrometry method to estimate %BF, it is assumed that the relative hydration of the FFB is 73.2%.** There is large interindividual variation in TBW. Siri (1961) estimated the biological variability in the water component of the FFB to be 2.0%. Siri (1961) speculated that biological variability in the hydration of the FFB would produce a substantial error in the estimation of body fat (2.7% BF) for the general population. Additionally, factors such as growth (chapter 8), aging (chapter 9), and disease (chapters 12-14) may alter the relative hydration of the FFB.

Hydrometry Model

Two fluid samples (typically blood, saliva, or urine) are collected: (a) The first sample provides a baseline measure and is taken just before the administration of a tracer, and (b) the second sample provides a measure of the concentration of the tracer taken after a sufficient amount of time for the tracer to equilibrate with all water spaces. The radioactive tracer tritium oxide (^3H_2O) and the stable isotopic tracers deuterium oxide (D_2O) and oxygen-18 ($H_2^{18}O$) best meet the assumptions of tracers; thus, one of these is typically used to estimate TBW. Several methods are available for assaying labeled water, but isotope ratio mass spectrometry seems to be the method preferred by most researchers. In addition to the client's fluid samples, a sample of the diluted dose must also be measured. To calculate the isotope dilution space, the following equation is used (Schoeller 1996):

$$N = [(WA / a) (Sa - St) f] / (Ss - Sp)$$

where N is the isotope dilution space, W is the mass of water used to dilute the dose, A is the dose given to the client, a is the amount of dose diluted for analysis, Sa is the measured value for the diluted dose, St is the value for tap water used in the dilution, f is the fractionation factor for the body fluid sample, Ss is the value for the client's postdose sample, and Sp is the value for the client's predose sample.

After the isotope dilution space is determined, this space needs to be corrected for the exchange of hydrogen or oxygen with the nonaqueous compartment as mentioned earlier in connection with the assumptions. If $H_2^{18}O$ is the tracer, the isotope dilution space is divided by a factor of 1.01 (i.e., 1% overestimation of this space) to estimate TBW. For ^3H_2O or D_2O tracers, the isotope dilution space is divided by 1.04 (i.e., 4% overestimation of this space).

In the past, TBW, measured by hydrometry, has been used in 2-C and multicomponent molecular models to obtain reference measures of FFM and %BF. When hydrometry is used singularly in conjunction with a 2-C model, the estimated $\%BF_{2-C}$ is based on the assumption that the relative hydration of the FFM is a constant 73.2%. In this case, FFM and %BF are calculated using the following 2-C model formulas (Pace & Rathbun 1945):

$$FFM \ (kg) = TBW \ / \ 0.732$$
$$\%BF = \{[BM - (TBW \ / \ 0.732)] \ / \ BM\} \times 100$$

The volume of **extracellular water (ECW)**, or the amount of water outside of the cells, can also be measured using the isotope dilution principle. Bromide is used as the tracer for this procedure because it is assumed to be distributed only in the ECW compartment.

The same four assumptions that are made for the measurement of TBW (see p. 38) apply to the estimation of ECW from the bromide dilution space. The bromide dilution space overestimates ECW by about 5%; therefore, the bromide dilution space must be divided by a factor of 1.05 to estimate ECW. If both TBW and ECW are known, then **intracellular water (ICW)**, or the volume of water in the body cell mass, may be calculated as the difference between TBW and ECW (i.e., ICW = TBW − ECW).

Validity of Hydrometry

It is generally agreed that the isotope dilution method is accurate for measuring TBW and ECW (Schoeller 1996; Wang et al. 1999). The precision and accuracy of this method are both estimated to be 1% to 2% (Schoeller 1996; Schoeller et al. 1985). The biggest concern in using this method is estimating the amount of tracer that exchanges with the nonaqueous compartments.

As with the densitometry methods, the validity of hydrometry as a method to estimate body composition is questionable when this method is used alone (i.e., 2-C model) to obtain reference measures of FFM. The conversion of TBW to FFM and %BF is based on the assumption that the hydration fraction of the FFB is about 73%. According to Wang et al. (1999), on average this is a fairly stable value among adults. However, even with no technical error for the hydrometry method, biological variability in the water content of the FFB ($\cong$2%) corresponds to a 3.6% BF error (Lohman,

Harris et al. 2000; Siri 1961). For a 70-kg person with 15% BF, hydration fractions of 71% and 75% FFM correspond to %BF estimates of 17.3% and 12.7%, respectively. For clients such as children and individuals with diseases that produce edema, this method should not be used singularly to derive reference measures of body composition.

Several researchers have reported only small average differences (<1% BF) for the TBW method compared to 4-C model estimates of %BF (Bergsma-Kadijk et al. 1996; Friedl et al. 1992; Fuller et al. 1994; Withers et al. 1998). Wang et al. (1998) compared FM, as estimated from TBW, to 6-C chemical model estimates and noted that the mean difference (0.9 kg) and prediction error (SEE = 0.95 kg FM) were relatively small. However, caution should be used in interpretation of studies comparing hydrometry to multicomponent models because most multicomponent models include the measurement of TBW. Ideally, to more accurately estimate the validity of hydrometry for assessing body composition with a 2-C model, models that do not rely on the measurement of TBW should be used to derive reference measures of FFM (Lohman, Harris et al. 2000; Wang et al. 1999).

The procedures for using isotope dilution to measure TBW vary depending on the type of fluid sample (i.e., saliva, urine, or plasma). Here is the procedure for saliva:

Hydrometry Using Saliva

1. Test the client in a euhydrated state. Instruct the client to fast overnight, to avoid any exercise after the last meal, and to drink no fluids after midnight.

2. Collect a baseline sample of saliva (about 1 ml). Immediately store the sample in an airtight container and refrigerate it temporarily.

3. Measure the client's body weight with the client wearing minimal clothing and no shoes.

4. Prepare a weighed dose of the isotope (e.g., 10 g of D_2O) and instruct the client to drink it. Rinse the cap and container with 50 ml of water and have the client drink this water.

5. During the equilibration period (usually 3-4 hr), do not allow your client to void or to ingest any food or water.

6. Collect a second saliva sample after the equilibration period. For clients with excess ECW, this sample should be collected at 4 to 5 hr postadministration.

7. Store all samples in airtight containers and freeze them at −10° C until they can be analyzed.

8. Assay the labeled water by isotope ratio mass spectrometry or some other acceptable method, and calculate TBW as described on page 38.

Sources of Measurement Error for Hydrometry

The precision and accuracy for estimating TBW from isotope dilution are both estimated to be between 1% and 2% (Schoeller 1996; Schoeller et al. 1985). However, Lohman (1992) noted that procedural variations such as the physiological fluid selected for sampling, equilibration time, the estimated correction factor for the isotopic dilution space, and the method selected for assaying the labeled water may all contribute to an increased technical error. The accuracy of this method for measurement of TBW depends mainly on correct estimation of the nonaqueous exchange of the isotope. As mentioned previously, if this method is used to estimate FFM_{2-C} of individuals with a relative hydration that differs from the assumed constant value (73%), large errors will result.

Dual-Energy X-Ray Absorptiometry

In the early 1980s, researchers used a method called **dual-photon absorptiometry (DPA)** to assess total body bone mineral (TBBM) and **bone mineral density (BMD)** (Gotfredsen et al. 1984; Mazess et al. 1984; Peppler & Mazess 1981). DPA uses the attenuation of photon beams from a radionuclide source to identify body tissues. **Dual-energy X-ray absorptiometry (DXA)**, which replaced DPA in the 1990s, uses an X-ray tube instead of a radioactive isotope. This improved technology provides greater precision and more accurate estimates of BMD and soft-tissue composition compared to DPA (Mazess, Barden, Bisek et al. 1990).

DXA is a versatile body composition tool. In addition to measuring TBBM and BMD, which can be important health indices for diseases such as osteoporosis, DXA can provide estimates of bone-free LTM, FM, **soft-tissue mass (STM = LTM + FM)**, FFM, and %BF. This technology is attractive because it can be used to assess regional body composition as well. Additionally, this method requires virtually no effort on the part of the participant and does not depend on technician skill. Like all body composition methods, DXA relies on certain assumptions.

Principles and Assumptions of DXA for Measuring Whole-Body Composition

The basic principle underlying DXA technology is that the attenuation of X-rays with high- and low-photon energies is measurable and is dependent on the thickness, density, and chemical composition of the underlying tissue. The **attenuation,** or weakening, of X-ray energies through fat, lean tissue, and bone varies due to differences in the densities and chemical composition of these tissues. These attenuation ratios at two different X-ray energies are thought to be constant for all individuals (Pietrobelli et al. 1996).

The major assumptions for the DXA method focus on the estimation of the soft-tissue composition using this technology.

- **The amount of fat over bone is the same as the amount of fat over bone-free tissue.** Soft-tissue composition is calculated only from pixels that do not contain bone. Approximately 40% to 45% of the pixels in a whole-body scan contain bone; thus, the relative lean-to-fat composition of the body is calculated from only about 60% of the body (Ellis 2000; Lohman 1996). In areas that have a large area of bone relative to STM (e.g., the arm), proportionately fewer pixels are used to estimate soft-tissue composition than in bone-free regions (Genton et al. 2002; Lohman 1996). Thus, the soft-tissue estimate may not be as accurate in bony regions as in bone-free regions. Also, variation in regional fat distribution among population subgroups could be an additional source of error (Deurenberg & Deurenberg-Yap 2001).

- **Measurements are not affected by the anteroposterior thickness of the body.** In theory, the attenuation of a given substance is a constant, but these values may systematically change with variations in thickness (Pietrobelli et al. 1996). Manufacturers of DXA equipment have attempted to correct this limitation using calibration phantoms. These are substances and materials of a known quantity and density that simulate fat, soft tissue, or bone used routinely to check the accuracy of the DXA scan.

- **The hydration and electrolyte content of the LTM is constant.** Changes in hydration of 1 kg are not likely to greatly affect the accuracy of DXA measurements (Going et al. 1993; Pietrobelli et al. 1996, 1998). After reviewing several studies, Lohman and colleagues theorized that a 5% change in the water content of the FFM will likely affect DXA %BF estimates by only 1% to 2.5% BF (Lohman, Harris et al. 2000).

DXA Model

There are three different manufacturers of DXA scanners: Hologic, Lunar, and Norland. Although there are slight differences among these scanners, they are all based on the same theoretical principle, and all have similar physical features such as a computer system, a scanning table, a detector, and an X-ray source (figure 3.5). The DXA method uses an X-ray tube with a filter to create low-energy (40 kV) and high-energy (70 or 100 kV) photons. When photons at these two different energies are passed through a tissue, the absorption can be expressed as a ratio (R) of the attenuation at the lower energy relative to the attenuation at the higher energy. Pietrobelli et al. (1996) summarized the attenuation coefficients and R values for various elements and calculated theoretical R values for fat and lean soft tissue.

The DXA method is sometimes referred to as a 3-C model method because it provides estimates of TBBM, LTM, and FM. However, rather than providing three independent measures, the DXA model really comprises two separate sets of 2-C model equations (Ellis 2000). The first set of equations describes the derivation of bone and STM (Sutcliffe 1996):

$$\text{soft tissue} = (\mu_{B40}\, LR_{70} - \mu_{B70}\, LR_{40})\, /$$
$$(\mu_{B70}\, \mu_{T40} - \mu_{T70}\, \mu_{B40})$$
$$\text{bone} = (\mu_{T40}\, LR_{70} - \mu_{T70}\, LR_{40})\, /$$
$$(\mu_{T70}\, \mu_{B40} - \mu_{B70}\, \mu_{T40})$$

where B = bone, T = soft tissue, μ = the attenuation coefficient, and LR = the log of the ratio of the intensity of the attenuated beam that is transmitted back to the detector to its initial photon intensity ($I_{transmitted}$ / $I_{initial}$). A second set of equations is used to separate lean tissue from fat (Sutcliffe 1996):

$$\text{lean tissue} = (\mu_{F40}\, LR_{70} - \mu_{F70}\, LR_{40})\, /$$
$$(\mu_{F70}\, \mu_{L40} - \mu_{L70}\, \mu_{F40})$$
$$\text{fat tissue} = (\mu_{L40}\, LR_{70} - \mu_{L70}\, LR_{40})\, /$$
$$(\mu_{L70}\, \mu_{F40} - \mu_{F70}\, \mu_{L40})$$

where F = fat tissue and L = lean tissue.

Validity of DXA

It is difficult to assess the validity of the DXA method because each of the three manufacturers has developed different models and software versions over the years. As many researchers and some clinicians have discovered, body composition results vary with manufacturer, model, and software version (Kistorp & Svendsen 1998; Modlesky, Lewis et al. 1996; Tataranni et al. 1996; Tothill et al. 1994; Tothill & Hannan 2000). Thus, some of the variability in results reported in DXA validation studies may be due to the different scanners and software versions. Because of this variation, experts who have reviewed DXA studies have called for more standardization among manufacturers (Genton et al. 2002; Lohman 1996).

Early DPA and DXA studies showed strong relationships between these technologies and total body

FIGURE 3.5   Dual-energy X-ray absorptiometer.

calcium measured by neutron activation analysis (Ellis et al. 1996; Heymsfield et al. 1990), as well as ash weights of lumbar vertebrae from cadavers (Ho et al. 1990). More recently, different DXA devices have been compared to in vitro chemical analysis of a femur from a cadaver (Economos et al. 1999). The bone mineral content of the femur was higher for the Lunar scanner (~3%) but lower for the Hologic (~1%) and Norland (~3%) scanners compared to values obtained through chemical analysis. Using phantoms to simulate STM, there was a trend for BMC to decrease with increasing phantom thickness, but the overall influence of thickness on the DXA measures of BMC was small. When whole-body phantoms were created to simulate average body sizes of children, BMC values were highly correlated to the known calcium content of the phantom. The BMC values, however, increased from 1% to 5% after simulated fat overlays were added (Shypailo et al. 1998).

It appears that DXA is capable of partitioning the body into bone and STM, but the accuracy of this method for differentiating this STM into FM and FFM has been questioned. Researchers reported that DXA underestimated fat in the central regions of the body when packets of lard were placed over research participants (Milliken et al. 1996; Snead et al. 1993). However, in a more recent study, Kohrt (1998) noted that improvements in software appear to have corrected this problem and reported that DXA accurately recognized exogenous fat regardless of whether it was positioned over central or peripheral regions of the body.

Compared to multicomponent model estimates of %BF, some researchers reported that the predictive accuracy of DXA was better than that of HD (Clasey et al. 1999; Fields & Goran 2000; Friedl et al. 1992; Prior et al. 1997; Wagner & Heyward 2001; Withers et al. 1998). However, the opposite finding (i.e., HD more accurate than DXA) was also reported in some studies (Bergsma-Kadijk et al. 1996; Goran et al. 1998; Millard-Stafford et al. 2001). In a review of DXA studies that used recently developed software, Lohman and colleagues concluded that DXA estimates of %BF were within 1% to 3% of multicomponent model estimates (Lohman, Harris et al. 2000). Additionally, DXA was ranked equally with HD and ADP for estimating FM, with a CE of 2 kg (Ellis 2001). Compared to a 6-C chemical model of FM, Wang et al. (1998) reported an SEE of 1.7 kg FM for the DXA method. Regarding individual accuracy, DXA accurately estimated the body fat of 26 of 30 African American men within ±3.5% BF$_{4-C}$ (Wagner & Heyward 2001). Fuller et al. (1992) reported the 95% limits of agreement of DXA to be 5% BF, slightly greater than for HD (3.8% BF) and TBW (3.9% BF).

Testing Procedures for DXA

The DXA method requires minimal cooperation from the client and technical skill. However, for precise and accurate DXA scans, proper training by the manufacturer in use of the scanner is essential. Also, many states require that a licensed X-ray technician perform the scan. The general procedures are as follows:

Basic Procedures for the DXA Method

- Prior to testing, calibrate the DXA scanner using a known calibration marker provided by the manufacturer.
- Measure your client's height and weight, with the client wearing minimal clothing and no shoes.
- Carefully position the client in a supine position on the scanner bed for a head-to-toe, anteroposterior scan.
- Use a skeletal anthropometer to accurately determine body thickness (see sagittal abdominal diameter [SAD] measurement described on p. 79).
- Set the scanner for a medium-speed whole-body scan, which usually takes about 20 min. For clients with SAD measures exceeding 27 cm, use a slow-speed scan, which typically takes 40 min.

Sources of Measurement Error for DXA

The short-term precision for DXA is excellent (Mazess, Barden, Bisek et al. 1990). The precision error for TBBM and BMD was 50 g (1.8%) and <0.01 g/cm^2 (0.8%), respectively; the errors for %BF, FM, and LTM were 1.4%, 1.0 kg, and 0.8 kg, respectively. In a research review of DXA studies, Lohman (1996) determined that the precision for measuring %BF$_{DXA}$ is generally about 1% BF. Other researchers reported that the long-term precision of DXA is comparable to its short-term reliability (Haarbo et al. 1991; Johnson & Dawson-Hughes 1991). The major sources of error for estimating body composition with DXA are client factors and instrumentation. The following questions and responses address these issues.

Client

• **Will the body size and hydration of my client affect the test results?** The DXA method should not be used to assess the body composition of large clients whose body dimensions exceed the length or width of the scanning bed. With regard to hydration, early studies suggested that the individual's hydration status might affect DXA estimates of %BF (Lohman 1992; Roubenoff et al. 1993). Although this still may be an issue, more recent research indicates that normal fluctuations in hydration have little effect on DXA estimates (Lohman, Harris et al. 2000) and that improvements in DXA software make it possible to accurately assess fat (Kohrt 1998).

- **In terms of client comfort and compliance, is DXA more suitable than other reference methods?** Compared to other reference methods, DXA requires little in the way of client participation and compliance. With DXA, the client does not need to perform breathing maneuvers that are required for measuring RV via HW and TGV for ADP. Also, unlike the case with hydrometry, no body fluid samples need to be collected. Manufacturers suggest that no calcium supplements be taken the day of testing, but otherwise there are no pretesting restrictions on eating, drinking, or exercise.

Instrumentation

- **How do the various types of DXA machines and software versions affect test results?** As mentioned previously, variability among DXA technologies is a major source of error. Although the same underlying physical principles are used by all DXA manufacturers, the instruments differ in how the high- and low-energy beams are generated (filter vs. switching voltage), imaging geometry (pencil beam vs. fan beam), X-ray detectors, calibration methodology, and the algorithms used (Genton et al. 2002). Recent software versions have improved the accuracy of DXA compared to that in the early 1990s (Kohrt 1998; Lohman, Harris et al. 2000; Tothill & Hannan 2000), however, the accuracy of these more recent DXA devices and software versions still needs to be determined (Genton et al. 2002). Because of these differences, we recommend using the same device and software version for longitudinal assessments or cross-sectional comparisons of body composition.

- **Is the DXA method safe for my clients, given that X-rays are used to measure body composition?** DXA is considered a safe method for estimating body composition. Radiation exposure is less for DXA than for its predecessor, DPA. The average skin dose of radiation is 1 to 3 mrad per DXA scan (Lang et al. 1991), which is comparable to a typical weekly exposure (3.5 mrad) of environmental background radiation (Lukaski 1993). The radiation dose for the newer fan-beam DXA devices is greater than for the pencil-beam scanners, but the radiation exposure is still considered low (Steel et al. 1998).

A Combined-Methods Approach for Reference Measures

Each of the reference methods described so far provides a reasonably accurate estimate for the components of human body composition: HD or ADP for estimating Db, hydrometry by isotope dilution for determining TBW, and DXA for assessing TBBM. Because each method yields indirect estimates of body composition, none can be singled out as the "gold standard" for in vivo body composition assessment. In fact, many researchers have obtained more valid reference measures of body composition using variables obtained from all three methods simultaneously. The multicomponent molecular models, presented in chapter 1, adjust Db from densitometry for variations in TBW (measured by hydrometry) and total body mineral (measured by DXA). These models take into account interindividual variability in the hydration or mineral content (or both) of the FFB; therefore, one may obtain more accurate estimates of body composition using this combined approach than with any one of these methods singularly. Thus, for research purposes, we recommend that all three methods be used in conjunction with a 4-C molecular model to derive valid reference measures of %BF, FM, and FFM. This position is endorsed by the American Society for Exercise Physiologists (Heyward 2001). Other researchers have found that the Siri (1961) 3-C model that includes a measure of TBW is acceptable and nearly equivalent to the 4-C model for estimating %BF (Bergsma-Kadijk et al. 1996; Clasey et al. 1999; Wang et al. 1998; Withers et al. 1998). Thus, if DXA is not available to measure bone mineral, this 3-C model can be used.

Keep in mind that the assumptions associated with the measurement of each individual component of this combined-methods approach still apply. In addition, the same sources of errors, previously discussed for each individual reference method, apply to the combined-methods approach.

Because several variables must be measured for the 4-C molecular model, one could reason that the potential for error is greater (see chapter 1, discussion of propagation of error) than with use of simpler models that require measuring fewer variables. It is true that some of the gain in accuracy that comes with greater control of biological variability with the multicomponent model is offset by the compounding of errors from multiple measurements. However, several researchers have demonstrated that the individual measurement errors are not substantially additive, and the TE for the combined-methods approach (4-C model) is only 1% BF (Friedl et al. 1992; Fuller et al. 1992; Goran et al. 1998; Withers et al. 1998). Experts agree that the error for the combined-methods approach is generally equal to or even lower than the sum of the errors for the individual methods.

Friedl et al. (1992) noted that the greatest source of technical error in the 4-C multicomponent model probably comes from the HW procedure. If the ADP method can be unequivocally proven as a valid way to measure Db, it could be used as viable alternative to HW for measuring Db. Use of this method, in place of HW, might even help to further reduce the TE of this combined-methods approach.

Additional Reference Methods

Other advanced and highly sophisticated laboratory methods (i.e., neutron activation analysis, computerized tomography, magnetic resonance imaging, and whole-body counting) are sometimes used to obtain reference measures of body composition. They are of interest because they allow for the assessment of body composition at levels other than the typically studied molecular level. Although these methods may provide valid reference measures, their general lack of availability and relatively high cost make them impractical for most researchers or clinicians involved in body composition assessment. Here we briefly introduce some of these methods for your information.

- **Neutron activation analysis.** Unlike most body composition assessment methods that measure body components at the molecular level, **neutron activation analysis (NAA)** allows for direct in vivo chemical analysis at the atomic level. Multielemental models, such as the 6-C model described in chapter 1, use NAA technology as a **criterion method** for evaluating other "reference methods." Total body content of the major body elements (calcium, sodium, chlorine, phosphorus, nitrogen, hydrogen, oxygen, and carbon) can be determined with NAA. A beam of neutrons is delivered to the client, and the atoms of the target elements capture these neutrons, creating isotopes and emitting gamma rays. The abundance of each element can be determined through measurement of its emissions. NAA may be the most sophisticated method for assessing body composition, but lack of NAA facilities and researchers skilled in this technique, coupled with high expense and client exposure to radiation, limits its use in research and clinical settings.

- **Computerized tomography. Computerized tomography (CT)** is a radiographic method that measures the differences in the attenuation, or the weakening, of X-ray beams as they pass through the participant. The attenuation differences are related to the differences in the densities of the underlying tissues. A computer-generated image of the scanned area, created from the attenuated beams, allows for separate recognition of bone, adipose tissue, and lean tissue. Because of radiation exposure and cost, CT scanning is usually limited to regional assessment of body composition and not used for whole-body assessment.

- **Magnetic resonance imaging.** Unlike CT scans, **magnetic resonance imaging (MRI)** does not use ionizing radiation; instead, it creates a computer-generated image from radio frequency signals emitted by hydrogen nuclei. Hydrogen nuclei behave like tiny magnets. Applying an external magnetic field and then a pulsed radio frequency across a body part causes the nuclei to line up and absorb energy. When the radio wave is turned off, the nuclei emit the radio signal that they absorbed, and this emitted signal is used to create the image. Computerized tomography and MRI measure body composition at the tissue level and are the preferred methods for separating total adipose tissue into its subcutaneous and visceral components. However, limited availability and high cost preclude their use on a regular basis.

- **Whole-body counting of potassium.** The majority of potassium is stored intracellularly; thus, **total body potassium (TBK)** is often used as a reference measure of **body cell mass (BCM)**, or the metabolically active tissues in the body. The TBK can be quantified through measurement of the gamma rays emitted by a naturally occurring potassium isotope (^{40}K) with a whole-body counter. A **whole-body counter** consists of a gamma ray detector, shielded room or area to reduce background radiation, and computer-based data acquisition system that can identify the unique gamma ray from ^{40}K. Although BCM is often an important indicator of health and survival in diseased states (see chapter 14), the high cost and lack of availability of whole-body counters make ^{40}K-counting an impractical method for routine, nonclinical use.

Summary

Either a 4-C model or a 3-C model that includes a reference measure of TBW, such as the Siri (1961) 3-C equation, is the preferred method of body com-

Table 3.2   Summary of Body Composition Methods Compared to a Six-Compartment Chemical Model[a]

Accuracy level	Method	Technical error (kg of fat)	Coefficient of reliability[b] (%)	Limits of agreement[c] (kg of fat)
Most accurate	4-C and 3-C methods that include TBW measure	<0.8	≥99.5	<1.1
Accurate	Individual reference methods (HD, ADP[d], hydrometry, DXA)	1-2	97-99	1.0-2.5
Least accurate	Field methods (SKF, anthropometry, BIA, NIR[d])	2-4	85-95	2.5-4.0

[a]Data from Wang et al. (1998); [b]A test-retest correlation measure of consistency; [c]±1 standard deviation; [d]This method was not included in the analysis of Wang et al. (1998) but is most likely in this category based on our review of the literature.

Key: 4-C = four-component model; 3-C = three-component model; TBW = total body water; HD = hydrodensitometry; ADP = air displacement plethysmography; DXA = dual-energy X-ray absorptiometry; SKF = skinfolds; BIA = bioelectrical impedance analysis; NIR = near-infrared interactance.

position assessment (see table 3.2). The next level of accuracy below that of a multicomponent model for estimating body fat results from use of one of the reference methods presented in this chapter. Finally, "field methods" (presented in chapters 4-7) offer the least amount of accuracy. Although the multicomponent methods and the reference methods presented in this chapter provide the greatest accuracy, they may not always be feasible, available, or practical, especially in a field or clinical setting. Thus, the remainder of this text focuses on the field methods (chapters 4-7) and how they can best be used in both healthy (part II) and clinical (part III) populations.

Key Points

- A perfect, error-free reference method for in vivo body composition assessment does not exist.
- Both the HD and ADP methods are used to obtain measures of BV and Db.
- The HD method is based on water displacement, and the underlying principle of the ADP method is air displacement.
- The conversion of Db to %BF using a 2-C model assumes constant values for the FFB. Selecting the wrong conversion formula could result in a large error in the estimate of %BF.
- The HD procedure requires a maximal exhalation while one is completely submerged in water, which may be difficult for some clients to perform.
- The largest source of error in the HD method is RV; this variable must be measured rather than estimated.
- The Bod Pod uses a pressure-volume relationship between two chambers to estimate BV.
- The Bod Pod procedure, which involves measuring TGV rather than RV and no water submersion, is quicker and easier than the HD method.
- Hydrometry is used to estimate TBW and is critically important to body composition assessment because water is the largest constituent of the body and FFM.

- Hydrometry is based on the dilution principle, and the procedure involves collecting samples of a physiological fluid both before and after the administration of an isotopic tracer.
- One must assume a constant value for the hydration fraction of the FFB when converting TBW to %BF with a 2-C model.
- DXA is based on the attenuation of two different X-ray energies as they pass through tissues having different densities and chemical composition.
- DXA estimates of whole-body composition are based on a 3-C "tissue-level" model. This technology is used to estimate TBBM and BMD and provides regional as well as whole-body estimates of body composition.
- Major sources of error for the DXA method are differences in instrumentation among manufacturers, generation of the scanner, and software versions.
- The combined, multicomponent method uses variables (Db, TBBM, and TBW) from three different reference methods (HD or ADP, DXA, and isotope dilution) to account for biological variability in the FFB and produce an estimate of %BF that is independent of age, gender, ethnicity, and health status.
- The additive errors from the combined-methods approach (propagation of measurement error) do not exceed the benefit gained by accounting for biological variability in the FFB.
- Body fat estimates from the combined-methods approach have better agreement with a criterion chemical model (NAA) than any of the individual reference methods used singularly. The combined-methods approach is recommended as the reference standard for body composition studies.
- Other highly sophisticated and accurate methods (NAA, CT, and MRI) for body composition analysis exist, but the high cost and lack of availability of these devices limit their use.

Key Terms

Learn the definition for each of the following key terms. Definitions of terms can be found in the Glossary, page 227.

adiabatic conditions

air displacement plethysmography (ADP)

Archimedes' principle

attenuation

Bod Pod

body cell mass (BCM)

body surface area

bone mineral density (BMD)

Boyle's law

computerized tomography (CT)

criterion method

densitometry

dilution principle

dual-energy X-ray absorptiometry (DXA)

dual-photon absorptiometry (DPA)

extracellular water (ECW)

field methods

functional residual capacity (FRC)

hydrodensitometry (HD)

hydrometry

hydrostatic weighing (HW)

intracellular water (ICW)

isothermal conditions

laboratory methods

magnetic resonance imaging (MRI)

neutron activation analysis (NAA)

Poisson's law

residual lung volume (RV)

tare weight

thoracic gas volume (TGV)

total body potassium (TBK)

total lung capacity (TLC)

underwater weighing (UWW)

validity

whole-body counter

Review Questions

1. Why is there no "gold standard" for in vivo body composition assessment?

2. Compare and contrast the HD and ADP methods for measuring Db.

3. Outline the basic testing procedures for the HD method.

4. What are the major sources of error for the HD and ADP methods and what steps can be taken to help minimize these errors?

5. Explain the underlying principle and model of hydrometry.

6. What are the assumptions of the hydrometry method, and how might these contribute to the error of this method?

7. Explain the underlying principle and the "3-C" model of DXA. What are some advantages or unique characteristics of DXA compared to the other reference methods?

8. What are the assumptions of the DXA method, and how might these contribute to the error of this method?

9. Identify the methods and variables used to obtain reference body composition measures for the combined-methods approach. How are these variables measured? Why is this approach advantageous?

10. Explain how NAA, CT, and MRI are used in the assessment of body composition. Why are these methods not used more frequently?

11. Which is the preferred reference method for body composition studies and why?

CHAPTER 4

SKINFOLD METHOD

Key Questions

- What are the basic assumptions and principles underlying the use of the skinfold method?
- What are population-specific and generalized skinfold prediction models?
- What are the major sources of measurement error for the skinfold method?
- Can different types of skinfold calipers be used interchangeably?
- What are the standardized procedures for measuring skinfolds?
- What steps can you take to increase your skill as a skinfold technician?

In the early 1900s, the thickness of subcutaneous adipose tissue was measured by taking skinfold (SKF) measurements (Brozek & Keys 1951). Early investigators pointed out that although SKF thicknesses varied at different sites, there were moderate to high relationships among SKF measurements (Brozek & Keys 1951; Franzen 1929). Over the years, the SKF method has been widely used in field and clinical settings to estimate total body fatness. Because the SKF test is easy to administer at a relatively low cost, it is suitable for large-scale epidemiological surveys such as the National Health and Nutrition Examination Survey-NHANES III (Kuczmarski et al. 1994). SKF measurements also may be used to monitor changes in **subcutaneous fat** deposition for clients participating in exercise training and weight management programs. This chapter addresses the basic assumptions and principles of the SKF method, SKF prediction models, techniques for measuring SKFs, and sources of measurement error.

Assumptions and Principles of the Skinfold Method

Skinfolds are an indirect measure of the thickness of subcutaneous adipose tissue. When you use the

SKF method to estimate total Db to derive %BF values, certain basic relationships are assumed:

- **The SKF is a good measure of subcutaneous fat.** The **skinfold (SKF)** is a measure of the thickness of two layers of skin and the underlying subcutaneous fat (figure 4.1). Research has demonstrated that subcutaneous fat assessed by SKF measurements at selected sites is similar to the value obtained from magnetic resonance imaging (MRI) and computed tomography (Hayes et al. 1988; Orphanidou et al. 1994). However, at some specific sites, SKF measurements yielded significantly smaller amounts of subcutaneous fat compared to that measured directly from MRI (Hayes et al. 1988). This difference may be attributed to the presence of subcutaneous fat below an interface (a layer of irregularity within the fat), which may limit the amount of subcutaneous fat that can be lifted in the fold. Part of this difference also may be due to the distortion of the MRI measurements on the posterior aspect of the body because the client must be supine in the MRI scanner.

- **The distribution of fat subcutaneously and internally is similar for all individuals within each gender.** The validity of this assumption is questionable. Older subjects of the same gender and Db have proportionately less subcutaneous fat

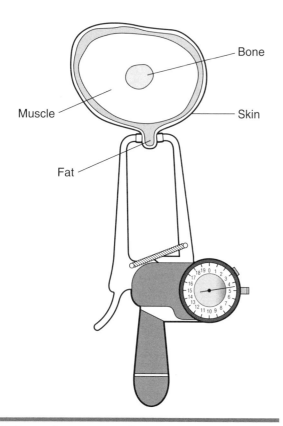

FIGURE 4.1   Anatomy of a SKF.

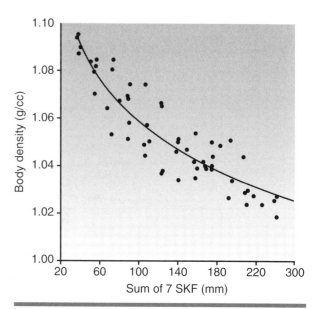

FIGURE 4.2   Relationship between sum of SKFs and Db for a heterogeneous population.

than their younger counterparts. Also, the level of body fatness affects the relative amount of fat located internally and subcutaneously. Lean individuals have a higher proportion of internal fat, and the proportion of fat located internally decreases as overall body fatness increases (Lohman 1981).

• **Because there is a relationship between subcutaneous fat and total body fat, the sum of several SKFs can be used to estimate total body fat.** Research has established that SKF thicknesses at multiple sites measure a common body fat factor (Jackson & Pollock 1976; Quatrochi et al. 1992). It is estimated that approximately 30% to 50% of the total body fat is located subcutaneously in men and women (Lohman 1981). However, there is considerable biological variation in subcutaneous, intramuscular, intermuscular, and internal organ fat deposits (Clarys et al. 1987), as well as in essential lipids in bone marrow and the central nervous system. Age, gender, and degree of fatness all affect variation in fat distribution (Lohman 1981).

• **There is a relationship between the sum of SKFs (ΣSKF) and Db.** This relationship is linear for homogeneous samples (population-specific SKF equations) but nonlinear over a wide range of Db (generalized SKF equations) for both men and women (figure 4.2). A linear regression

line depicting the relationship between the ΣSKF and Db will fit the data well only within a narrow range of body fatness values. Thus, you will get an inaccurate estimate if you use a population-specific equation to estimate the Db of a client who is not representative of the sample used to develop that equation (Jackson 1984).

• **Age is an independent predictor of Db for both men and women.** Using age and the quadratic expression of the sum of skinfolds (ΣSKF^2) accounts for more variance in Db of a heterogeneous population than using the ΣSKF^2 alone (Jackson 1984).

Skinfold Prediction Models

SKF prediction equations are developed using either linear (population-specific) or quadratic (generalized) regression models. There are well over 100 population-specific equations to predict Db from various combinations of SKFs, circumferences, and bony diameters (Jackson & Pollock 1985). *Population-specific equations* are developed for relatively homogeneous populations and are assumed to be valid only for individuals having similar characteristics such as age, gender, ethnicity, or physical activity level. For example, an equation derived specifically for 18- to 21-yr-old sedentary men would not be valid for predicting Db of 35- to 45-yr-old sedentary men. Population-specific equations are based on a linear relationship between SKF fat and Db (i.e., **linear regression model);** however, research shows that there is a curvilinear relationship (i.e., quadratic regression model) between SKFs and Db across a large range of body fatness (see figure 4.2). Thus, population-specific equations tend to underestimate %BF in fatter individuals and overestimate it in leaner individuals.

Using the quadratic model, generalized prediction equations applicable to individuals varying greatly in age (18-60 yr) and body fatness (up to 45% BF) have been developed and validated (Jackson and Pollock 1978; Jackson et al. 1980; Lohman 1981). These equations also take into account the effect of age on the distribution of subcutaneous and internal fat. An advantage of the generalized equations is that you can use one equation, instead of several, to estimate body fatness of your clients.

Most equations use two or three SKFs to predict Db. The Db is then converted to %BF using an appropriate population-specific conversion formula (see table 1.4, p. 9). Experts recommend using equations that have SKF measures from a variety of sites, including both upper- and lower-body sites (Martin et al. 1985). Boileau et al. (1985) developed SKF equations to predict %BF directly rather than Db in children (see chapter 8). They determined the reference %BF of these children using a multicomponent body composition model that included measures of Db, TBW, and bone mineral. Whenever possible, select an equation that was developed using valid reference methods and multicomponent models to obtain reference measures of body composition (see chapters 1 and 3).

Using the Skinfold Method

Commonly used population-specific and generalized SKF prediction equations are presented in chapters 8 to 14 and in appendixes B.1 through B.4. Calculating the Db or %BF is tedious and time-consuming, especially when you are assessing the body composition of many clients. In the past, nomograms and tables developed for specific SKF equations have been used to calculate %BF from SKF measures. One such nomogram (Baun & Baun 1981) was developed for the Jackson sum of three SKF equations (Jackson & Pollock 1978; Jackson et al. 1980). This nomogram and the corresponding tables (Jackson & Pollock 1985) have limited usefulness because they provide only a 2-C model (i.e., Siri's 2-C model) estimate of %BF. You should use the nomogram and tables to estimate %BF only for clients whose FFB density is likely to be close to the assumed value (1.10 g/cc) for the 2-C model (see chapter 1).

When selecting SKF equations, be certain to consider the physical demographics (e.g., age,

gender, ethnicity, and physical activity level) of your client and use an appropriate population-specific formula to estimate %BF from Db (table 1.4, p. 9). Using these equations, you can accurately estimate Db or %BF within the recommended acceptable values (i.e., ±0.0080 g/cc or ±3.5% BF) for most clients.

Skinfold Technique

It takes a great deal of time and practice to develop skill as a SKF technician. For obtaining the most accurate results, it is important to develop both your technique and your interpersonal skills.

Technical Skills

Adhering to the following standardized procedures will increase the accuracy and **reliability** (i.e., consistency) of your measurements (Harrison et al. 1988):

Skinfold Technique— Standardized Procedures

1. Take all SKF measurements on the right side of the body.
2. Carefully identify, measure, and mark the SKF site, especially if you are a novice SKF technician (see figures 4.3 through 4.11).
3. Grasp the SKF firmly between the thumb and index finger of your left hand at a distance of 1 cm above the marked site and then lift the fold. This will allow you to place the caliper jaws directly on the marked site.
4. Lift the fold by placing the thumb and index finger 8 cm (~3 in.) apart on a line that is perpendicular to the long axis of the SKF. The long axis is parallel to the natural cleavage lines of the skin. For individuals with extremely large SKFs, you will need to separate the thumb and finger more than 8 cm to lift the fold.
5. Continue grasping the fold with the left hand after the caliper jaws have been placed on it and keep the fold elevated while the measurement is taken.
6. Place the jaws of the caliper perpendicular to the fold, approximately 1 cm below the thumb and index finger and halfway between the crest and the base of the fold. Release the jaw pressure slowly.
7. Take the SKF measurement 4 sec after the pressure is released.
8. Open the jaws of the caliper to remove it from the site. Close the jaws slowly to prevent damage or loss of calibration.

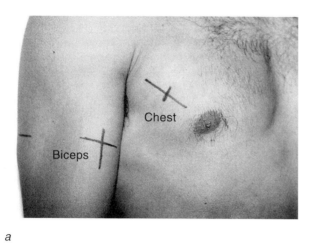
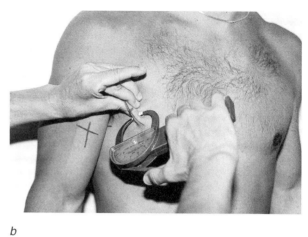

a

b

FIGURE 4.3 *(a)* Site and *(b)* measurement of the chest SKF.

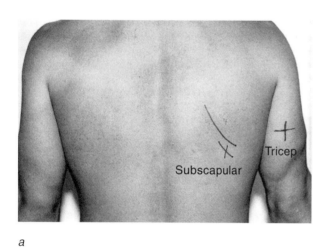
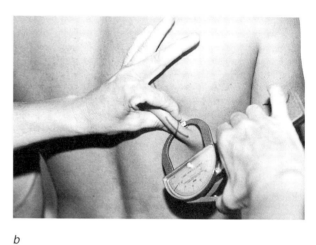

a

b

FIGURE 4.4 *(a)* Site and *(b)* measurement of the subscapular SKF.

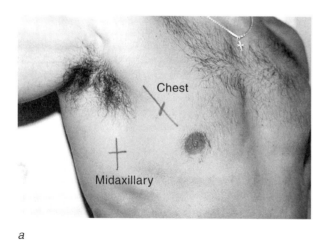
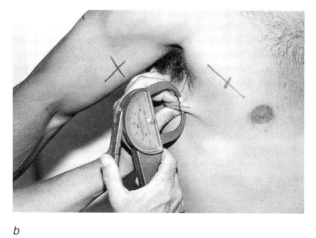

a

b

FIGURE 4.5 *(a)* Site and *(b)* measurement of the midaxillary SKF.

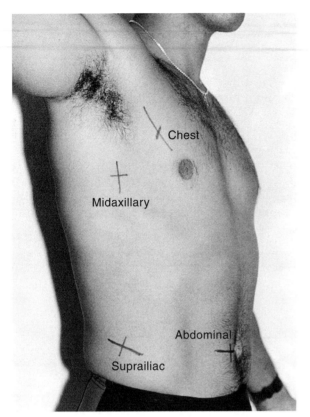

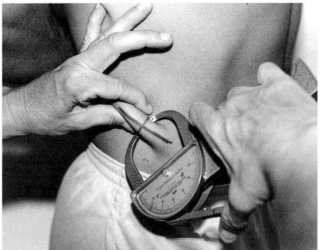

a

b

FIGURE 4.6 (a) Site and (b) measurement of the suprailiac SKF.

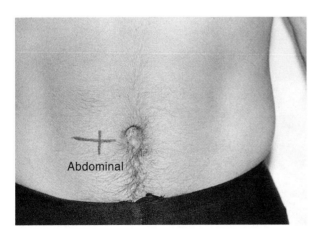

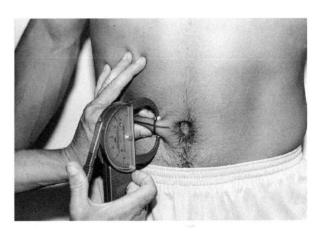

a

b

FIGURE 4.7 (a) Site and (b) measurement of the abdominal SKF.

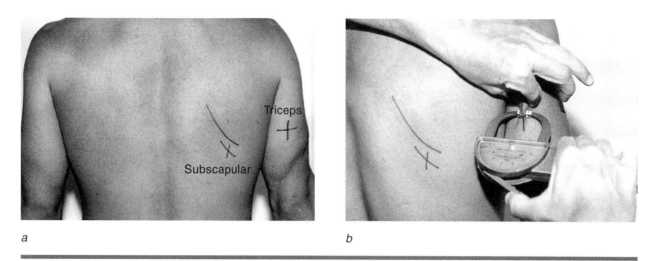

a   b

FIGURE 4.8 *(a)* Site and *(b)* measurement of the triceps SKF.

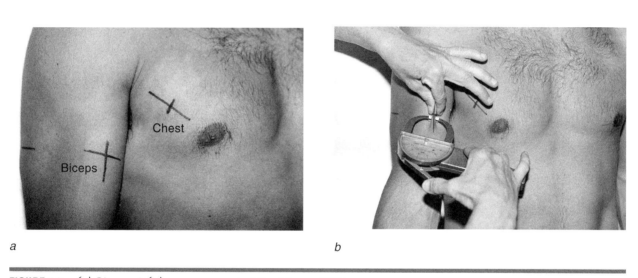

a   b

FIGURE 4.9 *(a)* Site and *(b)* measurement of the biceps SKF.

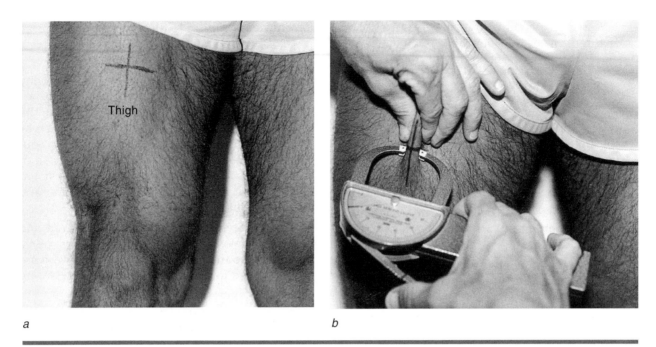

a   b

FIGURE 4.10   *(a)* Site and *(b)* measurement of the thigh SKF.

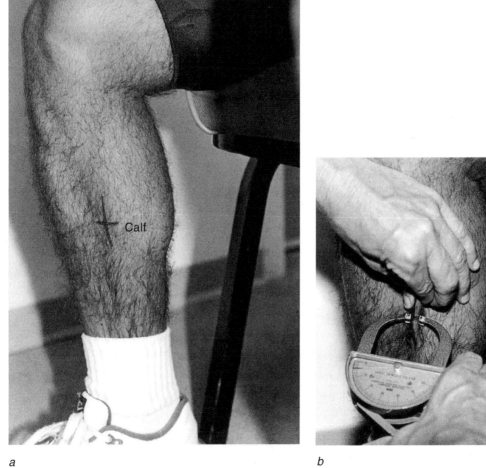

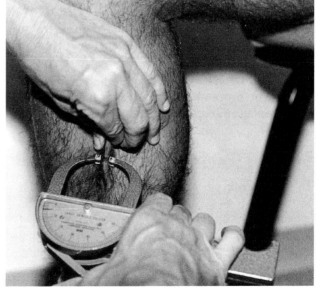

a   b

FIGURE 4.11   *(a)* Site and *(b)* measurement of the calf SKF.

You will also be able to increase your skill as a SKF technician by following recommendations made by experts in the field (Habash 2002; Jackson & Pollock 1985; Lohman, Pollock et al. 1984; Pollock & Jackson 1984):

Tips to Improve Technician Skill

- Be meticulous when locating the anatomical landmarks used to identify the SKF site, when measuring the distance, and when marking the site with a surgical marking pen.
- Read the dial of the caliper to the nearest 0.1 mm (Harpenden), 0.5 mm (Lange), or 1 mm (plastic calipers).
- Take a minimum of two measurements at each site. If values vary from each other by more than ±10%, take additional measurements.
- Take SKF measurements in a rotational order (circuits) rather than taking consecutive readings at each site.
- Take the SKF measurements when the client's skin is dry and lotion free.
- Do not measure SKFs immediately after exercise because the shift in body fluid to the skin tends to increase the size of the SKF.
- Practice taking SKFs on 50 to 100 clients who vary in size and body composition.
- Train with skilled SKF technicians and compare your results. Allow the experienced technician to demonstrate the technique on you, noting the amount of pressure used to lift the fold. Measure the experienced technician's SKF and ask for feedback about your technique.
- Use a SKF training videotape that demonstrates proper SKF technique (Lohman 1987; Human Kinetics 1995).
- Seek additional training by attending workshops held at professional meetings or by taking distance education courses (Human Kinetics 1999).

Interpersonal Skills

In addition to perfecting your technical skills, it is important to develop your interpersonal skills when administering SKF and other types of anthropometric tests (see chapter 5). To develop interpersonal skills, one expert offers the following suggestions (Habash 2002):

Tips to Improve Interpersonal Skills

- Prior to the scheduled test session, instruct clients to wear loose-fitting clothes that will allow easy access to the sites to be measured (e.g., shorts and T-shirt or two-piece exercise gear), or provide a hospital gown.
- Often clients are apprehensive about having their SKFs measured, particularly when they are meeting you for the first time. During the testing, you should put your client at ease by establishing good rapport (e.g., talk about some unrelated topic); projecting a sense of relaxed confidence; and creating a test environment that is friendly, private, safe, and comfortable.
- The private room should be uncluttered and should have a small table for calipers, pens, and clipboards and a chair for clients who are unstable standing or may need to rest during the testing.
- Some clients will feel more comfortable having their SKFs measured by a technician of the same gender. If this is not feasible, you could ask your clients if they would like another person of the same gender to observe the test.
- Educate your clients about the SKF test by talking about the purpose and use of the measurements, pointing to the SKF sites to be measured on your own body, and demonstrating on yourself how the SKF is measured.
- Limit your verbal and facial reactions to the data being collected during the test.

When interpreting results for clients, use lay language, rather than highly technical terms and jargon, to explain test scores. Whenever possible, try to phrase poor results in positive terms. For example, if a female client's body fat level is classified as obese, do not embarrass and alarm her by saying, "Your SKF test indicates that you are obese and need to lose at least 20 pounds to achieve a healthy body fat level in order to reduce your risk of diseases linked to obesity. You need to start a weight management program immediately."

Instead, use a more positive and less intimidating approach when interpreting this result. The following statement is more appropriate: "People with more than 32% body fat are at risk for disease. If you wish, I will evaluate your daily calorie intake and suggest healthy foods you like to eat that are low in fat. Also, we can discuss ways to increase your physical activity. I think we can find some activities that you will enjoy and have time for, so that you'll burn more calories each day. With these changes, you should be able to lower your body fat to a healthy level in a reasonable amount of time."

Sources of Measurement Error

The theoretical accuracy of SKF equations for predicting Db is 0.0075 g/cc or 3.3% BF due to biological variability in estimating subcutaneous fat from SKF thicknesses and interindividual differences in the relationship between subcutaneous fat and total body fat (Lohman 1981). Also, some error is attributed to the reference method (Jackson 1984). Therefore, *prediction errors* (SEE and TE) of no more than ±3.5% BF or ±0.0080 g/cc for SKF equations are considered acceptable (see table 2.1, p. 20).

The accuracy and precision of SKF measurements and the SKF method are affected by the technician's skill, the type of SKF caliper, client factors, and the prediction equation used to estimate body fatness (Lohman, Pollock et al. 1984). To increase accuracy and precision of your SKF measurements, carefully follow the standardized procedures and guidelines presented on pages 51-56. The following questions and responses address sources of **measurement error** for the SKF method.

Technician Skill

- **Is there high agreement among SKF values when measurements are taken by two different technicians?** A major source of measurement error is differences between SKF technicians. Approximately 3% to 9% of the variability in SKF measurements can be attributed to measurement error due to differences between SKF technicians (Lohman, Pollock et al. 1984; Morrow, Fridye et al. 1986). The amount of between-technician error depends on the site being measured, with larger errors reported for the abdomen (8.8%) and thigh (7.1%) SKF sites compared to the triceps (~3.0%), subscapular (~3.0% to 5.0%), and suprailiac (~4%) SKF sites (Lohman, Pollock et al. 1984; Morrow, Fridye et al. 1986).

Objectivity, or intertechnician reliability, is improved when SKF technicians follow standardized testing procedures, practice taking SKFs together, and mark the SKF site (Pollock & Jackson 1984). Although some experts believe that SKF sites do not need to be marked (Harrison et al. 1988), we highly recommend doing so, particularly if you are a novice SKF technician. A major cause of low intertechnician reliability is improper location and measurement of SKF sites (Lohman, Pollock et al. 1984).

- **Are the anatomical descriptions for specific SKF sites the same for all SKF equations?** The anatomical location and direction of the fold vary for some SKF sites. Behnke and Wilmore (1974) recommend measuring the abdominal SKF using a horizontal fold adjacent to the umbilicus; Jackson and Pollock (1978), however, recommend measuring a vertical fold taken 2 cm (0.8 in.) lateral to the umbilicus. Inconsistencies such as this have led to confusion and lack of agreement among SKF technicians. As a result, experts in the field of anthropometry have developed standardized testing procedures and detailed descriptions for identification and measurement of SKF sites (Harrison et al. 1988; Ross & Marfell-Jones 1991). Some of the most commonly used sites, described in the *Anthropometric Standardization Reference Manual* (Harrison et al. 1988), are summarized in table 4.1 and illustrated in figures 4.3 through 4.11.

Although the objective is for all SKF technicians to follow standardized procedures and recommendations for site location and SKF measurements, you may not be able to do so in some cases. For example, if you are using the generalized equations of Jackson and Pollock (1978) and Jackson et al. (1980), the chest, midaxillary, subscapular, abdominal, and suprailiac SKFs will be measured at sites that differ from those described in the *Anthropometric Standardization Reference Manual*. The descriptions for the sites used in these equations are presented in table 4.2.

- **How many measurements do I need to take at each SKF site?** A lack of **intratechnician reliability,** or consistency of measurements by the SKF technician, is another source of error for the SKF method. You need to practice your SKF technique on 50 to 100 clients to develop a high degree of skill and proficiency (Jackson & Pollock 1985). Take a minimum of two measurements at each site using a rotational order. If SKF values differ from each other by more than ±10%, take additional measurements until you meet this criterion. This ±10% criterion for duplicate measurements at each site is recommended as the standardized procedure in the *Anthropometric Standardization Reference Manual*. For example, if you initially measure a SKF thickness of 30 mm, values from 27 to 33 mm would be acceptable for the second SKF measurement (30 ± 10% or 30 ± 3 mm). Use the average value of these two measurements in the SKF prediction equation. On the other hand, some researchers suggest taking three SKF measurements at each site and using the median (middle score) instead of the mean or average (Ward & Anderson 1998).

Type of Caliper

A variety of high-quality metal and plastic calipers can be used to measure SKF thickness

Table 4.1 Standardized Sites for Skinfold Measurements

Site	Direction of fold	Anatomical reference	Measurement
Chest	Diagonal	Axilla and nipple	Fold is taken between axilla and nipple as high as possible on anterior axillary fold with measurement taken 1 cm below fingers.
Subscapular	Diagonal	Inferior angle of scapula	Fold is along natural cleavage line of skin just inferior to inferior angle of scapula, with caliper applied 1 cm below fingers.
Midaxillary	Horizontal	Xiphisternal junction (point where costal cartilage of ribs 5-6 articulates with sternum, slightly above inferior tip of xiphoid process)	Fold is taken on midaxillary line at level of xiphisternal junction.
Supraiiiac	Oblique	Iliac crest	Fold is grasped posteriorly to midaxillary line and superiorly to iliac crest along natural cleavage of skin with caliper applied 1 cm below fingers.
Abdominal	Horizontal	Umbilicus	Fold is taken 3 cm lateral and 1 cm inferior to center of the umbilicus.
Triceps	Vertical (midline)	Acromial process of scapula and olecranon process of ulna	Distance between lateral projection of acromial process and inferior margin of olecranon process is measured on lateral aspect of arm with elbow flexed 90° using a tape measure. Midpoint is marked on lateral side of arm. Fold is lifted 1 cm above marked line on posterior aspect of arm. Caliper is applied at marked level.
Biceps	Vertical (midline)	Biceps brachii	Fold is lifted over belly of the biceps brachii at the level marked for the triceps and on line with anterior border of the acromial process and the antecubital fossa. Caliper is applied 1 cm below fingers.
Thigh	Vertical (midline)	Inguinal crease and patella	Fold is lifted on anterior aspect of thigh midway between inguinal crease and proximal border of patella. Body weight is shifted to left foot and caliper is applied 1 cm below fingers.
Calf	Vertical (medial aspect)	Maximal calf circumference	Fold is lifted at level of maximal calf circumference on medial aspect of calf with knee and hip flexed to 90°.

Data from Harrison et al. (1988, pp. 55-70).

Table 4.2 Skinfold Sites for Jackson's Generalized Skinfold Equations

Site	Direction of fold	Anatomical reference	Measurement
Chest	Diagonal	Axilla and nipple	Fold is taken 1/2 the distance between the anterior axillary line and nipple for men and 1/3 of this distance for women.
Subscapular	Oblique	Vertebral border and inferior angle of scapula	Fold is taken on diagonal line coming from the vertebral border, 1-2 cm below the inferior angle.
Midaxillary	Vertical	Xiphoid process of sternum	Fold is taken at level of xiphoid process along the midaxillary line.
Suprailiac	Diagonal	Iliac crest	Fold is taken diagonally above the iliac crest along the anterior axillary line.
Abdominal	Vertical	Umbilicus	Fold is taken vertically 2 cm lateral to the umbilicus.

Data from Jackson & Pollock (1978) and Jackson et al. (1980).

(see figure 4.12). When choosing a caliper, you need to consider factors such as cost, durability, and accuracy and precision of the caliper, as well as the type of caliper that was used for developing a specific SKF equation. Accuracy refers to the extent of agreement between SKF caliper readings and standard (e.g., Vernier caliper) values. Precision refers to the extent to which a caliper is capable of measuring an exact value. Table 4.3 and figure 4.13 compare some of the basic characteristics of selected SKF calipers. This section addresses questions you may have regarding SKF calipers.

Can High-Quality Metal Calipers Be Used Interchangeably?

High-quality metal calipers yield accurate and precise measurements throughout the range of measurement. The Harpenden, Lange, Holtain, and Lafayette calipers exert constant pressure (~7 to 8 g/mm^2) throughout the range of measurement (0-60 mm). Calipers should not vary in tension by more than 2.0 g/mm^2 over the range or exceed 15 g/mm^2 (Edwards et al. 1955). Excessive tension and pressure cause client discomfort (pinching sensation) and significantly reduce the SKF measurement (Gruber et al. 1990). High-quality calipers also have excellent scale precision (e.g., 0.2 mm to 0.5 mm, respectively, for Harpenden and Lange calipers).

Although the Harpenden and Lange SKF calipers have similar pressure characteristics, a number of researchers reported that SKFs measured with the Harpenden caliper produce significantly smaller values than with Lange calipers (Gruber et al. 1990; Lohman, Pollock et al. 1984; Schmidt & Carter 1990). This difference translates into a systematic underestimation (~1.5% BF) of average %BF for women and men with use of the Harpenden caliper (Gruber et al. 1990). Even though the pressure is similar for the Lange (8.37 g/mm^2) and Harpenden (8.25 g/mm^2) calipers (Schmidt & Carter 1990), researchers noted that three times more force is required to open the jaws of the Harpenden caliper. Therefore, it is more likely that the adipose tissue will be compressed to a greater extent, resulting in smaller SKF measurements with this type of caliper.

Schmidt and Carter (1990) reported that the Skyndex and Harpenden calipers give similar values when used to measure SKF thicknesses ranging from 4 to 45 mm. However, both the Lange and Lafayette calipers gave higher values than the Skyndex and Harpenden calipers. Likewise, Zando and Robertson (1987) noted that SKFs measured with Skyndex calipers are significantly smaller than those measured with Lange calipers. Measurements made with Lange and Layfayette calipers should not be used interchangeably with those made using Harpenden or Skyndex calipers (Schmidt & Carter 1990). Also, Lohman and colleagues reported no significant difference between SKF thicknesses measured with Harpenden and Holtain calipers for the subscapula, suprailiac, abdomen, and thigh sites (Lohman, Pollock et al. 1984).

Table 4.3 Comparison of High-Quality Metal Calipers and Plastic Skinfold Calipers

	Caliper type	Average pressure (g/mm²)	Range (mm)	Scale precision (mm)	Accuracy	Durability	Approximate cost	Unique features	Supplier[c]
METAL	Harpenden (HA)	8.2	0-55	0.2	HA < LNG[a]	Excellent	$305		Creative Health Products
	Lange (LNG)	8.4	0-60	0.5	LNG > HA[a]	Excellent	$180		Creative Health Products
	Lafayette (LF)	7.5	0-100	0.5	LF > LNG[a]	Excellent	$440	Measurement range 0-100 mm	Body Trends
	Skyndex (SKN)	7.3	0-60	0.5	SKN < LNG[a] SKN ≅ HA[a]	Excellent	$450*	Skyndex I: built-in computer, Durnin & Womersley and/or Jackson & Pollock equations. Skyndex II: digital readout but no computer.	Body Trends
	Holtain (HO)	NR	0-60	0.2	HO < HA, LNG[b]	Excellent	$300		Pfister Import-Export

60

Accu-Measure (AM)	NR	0-60	1.0	NR	Fair	$20	Can be used for self-assessment of body fat	Accu-Measure, LLC
Body Caliper (BC)	NR	0-60	1.0	BC ≅ HA[b]	Good	$59	Measurement scale on both sides of caliper; suitable for right- or left-handed technicians	The Caliper Company
Fat-O-Meter (F)	5.6	0-40	2.0	F ≅ LNG[b]	Poor	$15		Creative Health Products
Fat Track (FT)	NR	0-60	0.1	NR	Good	$50	Can be used for self-assessment of body fat; digital readout; Jackson & Pollock equations	Accu-Measure, LLC
McGaw (MG)	12.0	0-40	2.0	MG ≅ HA[b] MG < LNG[b]	Fair	$5		None available
Ross adipometer (RA)	12.0	0-60	2.0	RA ≅ HA[b] RA < LNG[b]	Fair	$7		Rage Corporation
Slim Guide (SG)	7.5	0-80	1.0	SG ≅ HA ≅ SKN[a] SG < LNG[a]	Good	$25		Creative Health Products

PLASTIC

[a]Determined by comparing dynamic compression of foam rubber models of human skinfolds.

[b]Determined by comparing skinfold thicknesses of individuals measured by a technician; thus, any differences include not only instrument error but also error associated with technician skill and client factors.

[c]For supplier's address, see appendix C on page 221.

*This price is for Skyndex I. Price for the Skyndex II caliper without the computer is $285.

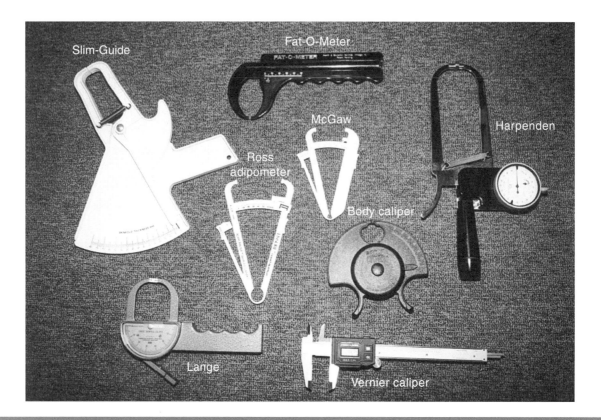

FIGURE 4.12   SKF calipers.

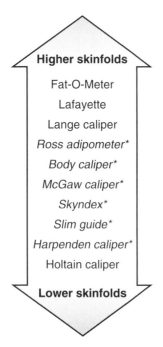

FIGURE 4.13   Comparison of SKF calipers. *Calipers in italics tend to give similar SKF readings.

How Can I Check the Accuracy of My SKF Caliper?

To calibrate SKF calipers, one checks the dial accuracy and pressure (static and dynamic) exerted by the jaws of the caliper (Schmidt & Carter 1990). You should check the dial accuracy of your caliper periodically using a high-precision **Vernier caliper** or SKF **calibration blocks** (see figure 4.14). To calibrate the SKF caliper with a Vernier caliper, set the opening of the Vernier caliper to a given width (e.g., 20 mm) and place the jaws of the SKF caliper on it (see figure 4.14). The value read from the SKF caliper (20 mm) should match the preset width of the Vernier caliper. If it does not, follow the manufacturer's calibration instructions or send the SKF caliper to the manufacturer for recalibration.

Checking static (upscale) and dynamic (downscale) caliper pressures is more difficult. For measurement of static pressure, the jaws of the caliper are opened (i.e., upscale pressure) predetermined amounts as read from the dial, and a spring scale is used to measure the force exerted

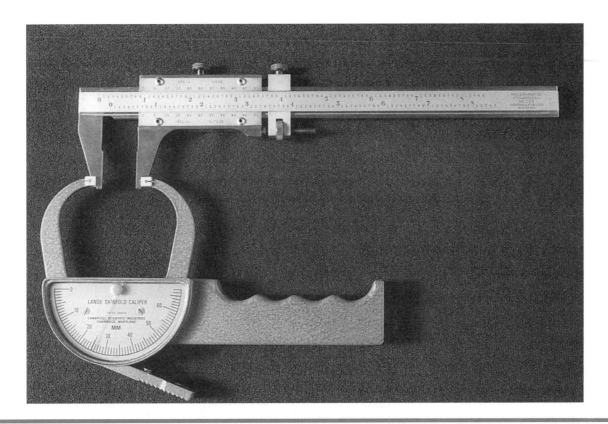

FIGURE 4.14 Checking calibration of SKF calipers using a Vernier caliper.

at several openings (see Schmidt & Carter 1990). To measure downscale pressure, compressible foam rubber blocks that simulate the density of human SKFs are used. This mode of calibration, a measure of the dynamic pressure exerted when the caliper jaws close and compress the foam rubber, closely matches the way calipers are used when measuring SKF thicknesses (see Gore et al. 1995; Schmidt & Carter 1990).

Are Plastic SKF Calipers As Accurate As Metal SKF Calipers?

Compared to high-quality calipers, some plastic SKF calipers have less scale precision (~2 mm), do not exert constant tension throughout the range of measurement (Hawkins 1983), and have a smaller range of measurement (0-40 mm). Despite these differences, some types of plastic calipers compare well with more expensive, high-quality metal calipers (see table 4.3).

• **Ross adipometer.** Several studies have compared the Ross adipometer with high-quality calipers. Researchers reported that the Ross adipometer and Harpenden caliper give similar readings at selected SKF sites (Leger et al. 1982; Lohman, Pollock et al. 1984). Also, Hawkins (1983) noted that the Ross adipometer and Lange SKF calipers produce similar average triceps SKF values for 800 elementary, secondary, and university students.

• **McGaw caliper.** In one study, the triceps SKF measured with the McGaw caliper was 8% lower on average than the value obtained with the Lange caliper, and the difference between calipers tended to increase as the thickness of the triceps SKF increased (Burgert & Anderson 1979). Although the McGaw caliper has a smaller maximum jaw aperture (40 mm) than the Ross adipometer (60 mm), there were no differences between the average triceps SKF values measured with the McGaw, Ross, and Harpenden calipers (Leger et al. 1982).

• **Fat-O-Meter.** Hawkins (1983) also compared the Fat-O-Meter and Lange SKF calipers. Although the pressure exerted by the Fat-O-Meter was lower (5.6 g/mm^2) than that of the Lange caliper (9.3 g/mm^2), the two calipers yielded similar average SKF values.

- **Slim Guide.** The pressure exerted by the Slim Guide (7.5 g/mm^2) is less than that of the Harpenden and Lange calipers (see table 4.3). Research comparing measurements made by these three types of calipers showed that the Slim Guide produces measurements significantly smaller than those obtained with the Lange caliper (Hawkins 1983; Schmidt & Carter 1990) but similar to those obtained with the Harpenden caliper (Schmidt & Carter 1990).

- **Body Caliper.** To date only one study has evaluated the accuracy of the Body Caliper. The Body Caliper is unique in that it has identical measurement scales on its front and back so that it may be held in either the right or the left hand. Cataldo and Heyward (2000) compared the Body Caliper and Harpenden caliper and reported small differences (<0.5 mm) between calipers for measuring SKF thicknesses at the medial calf, subscapula, and triceps sites in children.

- **Accu-Measure.** The Accu-Measure caliper is designed for self-assessment of SKFs, requiring individuals, instead of a technician, to measure their SKF thickness at the suprailiac site. SKF measurements made with the Accu-Measure have not been directly compared to measurements with other types of calipers. Eckerson and colleagues (1998), however, noted that this self-assessment technique produces an accurate estimate of %BF (SEE = 3.4% BF) for college-age men and women with lean to average body fatness, providing an alternative to conventional body fat testing.

- **Fat Track caliper.** The Fat Track caliper is also marketed for self-assessment of body fat. This computerized caliper digitally displays the individual's %BF as estimated from the Jackson and Pollock Σ3SKF equations (i.e., Jackson & Pollock 1978; Jackson et al. 1980). The caliper "beeps" to indicate when the correct amount of pressure has been applied to the SKF during the measurement. To date no studies have compared the Fat Track caliper to other types of calipers.

Given that caliper type is a potential source of measurement error, we recommend the following guidelines:

Tips to Reduce Caliper-Related Measurement Error

- Use the same caliper when monitoring changes in your client's SKF thicknesses.
- Calipers used to obtain body fat estimates should be the same type that was used in the development of a specific SKF prediction equation. If the same type of caliper is not available, use a caliper that gives similar readings.

- Periodically check the accuracy of your caliper and calibrate it if needed. If you use your caliper regularly to test clients, check its accuracy weekly. In research settings, we recommend checking the accuracy of your caliper prior to each test session.

Client Factors

- **Will my client's hydration level affect the size of SKF measurements?** Variability in SKF measurements among individuals may be attributed not only to differences in the amount of subcutaneous fat at the site, but also to differences in compressibility of the adipose tissue and hydration levels of clients (Ward et al. 1999). Martin et al. (1992) reported that variation in SKF compressibility may be an important limitation of the SKF method. SKF compressibility may be affected by age and gender. In addition, an accumulation of extracellular water (edema) in the subcutaneous tissue—caused by factors such as peripheral vasodilation or certain diseases—may increase SKF thicknesses (Keys & Brozek 1953). This suggests that you should not measure SKFs immediately after exercise, especially in hot environments. Also, most of the weight gain experienced by some women during their menstrual cycles is caused by water retention (Bunt et al. 1989). This theoretically could increase SKF thicknesses, particularly on the trunk and abdomen; but there are no empirical data to support or refute this hypothesis.

- **Should SKFs be measured on the right or left side of the body?** There are only small differences (1-2 mm) between SKF thicknesses measured on the right and left sides of the body for the typical individual. The standard practice in the United States, as well as in European and developing countries, is to measure SKFs on the right side of the body. This is recommended in the *Anthropometric Standardization Reference Manual* (Lohman et al. 1988) and by the International Society for the Advancement of Kinanthropometry (Norton et al. 2000).

- **Can I accurately measure SKFs of clients who are obese or heavily muscled?** It is difficult, even for highly skilled SKF technicians, to accurately measure the SKF thicknesses of persons who are extremely obese. In individuals who are obese or heavily muscled, the subcutaneous fat may not be easily separated from the underlying muscle; therefore, the fold may be more triangular, with sides that are not parallel at the base of the fold (Gray et al. 1990). In some clients who are obese, the SKF thickness exceeds the maximum aperture of the caliper, and the jaws of the caliper may slip off the fold during the measurement, resulting in a potentially embarrassing and awkward situation for you and your client. Therefore, avoid using the SKF method to measure body fat of clients with extreme obesity.

SKF Prediction Equations

As mentioned earlier in this chapter, SKF prediction equations should be selected based on the age, gender, ethnicity, and physical activity level of your client. Lohman, Pollock et al. (1984) compared %BF estimates obtained from five different SKF prediction equations applied to a sample of female collegiate basketball players. Four of these equations were developed and cross-validated on female nonathletic populations, whereas the fifth equation was developed specifically for female athletes. The average %BF estimates ranged from 16.5% (the athlete equation) to 24.2% BF, illustrating the necessity of selecting the appropriate equation in order to obtain the best estimate of a client's body fatness. Given that this is a major source of measurement error, the SKF equations that are recommended in later chapters have been carefully selected based on research substantiating their applicability for estimating levels of body fat for specific population subgroups.

Key Points

- The SKF is a measure of the thickness of two layers of skin and the underlying subcutaneous fat.
- The SKF method assumes that SKFs are a good measure of subcutaneous fat and that the relative subcutaneous and internal fat distribution is similar for all individuals within each gender.
- The sum of the SKFs is inversely related to Db and directly related to %BF.
- Population-specific SKF equations are based on linear regression models and should be used only for individuals or groups with the specific characteristics of the population used to develop them. Factors such as age, gender, ethnicity, level of body fatness, and physical activity level must be considered.
- Generalized SKF equations are based on quadratic regression models and may be applicable to individuals varying in age, gender, ethnicity, and physical activity levels because these equations usually include one or more of these factors as predictors.
- SKF prediction equations estimate either Db or %BF.
- Technician skill, type of caliper, client factors, and the prediction equation used are potential sources of measurement error for the SKF method.
- The maximum acceptable values for prediction errors of SKF equations are ±3.5% BF or ±0.0080 g/cc for Db.
- A minimum of two measurements at each SKF site should be taken in rotational order. The two measurements at each site should not vary by more than ±10%.
- Compared to high-quality metal calipers, plastic SKF calipers generally have a smaller measurement scale and less scale precision; however, some types of plastic calipers are as accurate as metal calipers.
- The Lange and Lafayette calipers give higher SKF readings than the Harpenden, Holtain, and Skyndex calipers.
- The Accu-Measure and Fat Track calipers are designed for self-assessment of body fat.
- SKF measurements are taken on the right side of the body.
- SKFs should not be measured immediately after exercise.
- The SKF method should not be used to assess the body fatness of people who are extremely obese.

Key Terms

Learn the definition for each of the following key terms. Definitions of terms can be found in the Glossary, page 227.

calibration blocks

intratechnician reliability

linear regression model

measurement error

objectivity

reliability

skinfold (SKF)

subcutaneous fat

sum of SKFs

Vernier caliper

Review Questions

1. Do SKFs provide an accurate measure of subcutaneous fat? Explain.
2. What is the difference between subcutaneous and internal fat? What factors affect the relative distribution of fat subcutaneously and internally?
3. Describe the relationship between the ΣSKFs and Db for homogeneous and heterogeneous populations.
4. What are the advantages of using generalized SKF equations instead of population-specific equations?
5. What are the major sources of measurement error for the SKF method?
6. Describe ways to increase your skill as a SKF technician.
7. What are the major differences between high-quality metal and plastic SKF calipers?
8. How can you check the accuracy of your SKF caliper?
9. Explain how the hydration level of your client may affect the accuracy of SKF measures.
10. Briefly describe the standardized procedures for measuring SKFs.
11. Explain the difficulties you may encounter when measuring SKFs of clients who are extremely obese or heavily muscled.
12. What factors should you consider when selecting a SKF prediction equation for your client?

ADDITIONAL ANTHROPOMETRIC METHODS

Anthropometry refers to the measurement of the size and proportion of the human body. Body weight and stature (standing height) are measures of body size, whereas ratios of body weight to height can be used to represent body proportion. To assess the size and proportions of body segments, one may use circumferences, SKF thicknesses, skeletal breadths, and segment lengths.

In addition to measuring body size and proportions, anthropometric measures like circumferences, SKFs, and skeletal diameters have been used to assess total body and regional body composition. Also, anthropometric indices such as BMI, **waist-to-hip ratio (WHR)**, waist circumference, and sagittal abdominal diameter are used to identify individuals at risk for disease. Compared to SKFs, other types of anthropometric measures are relatively simple and inexpensive, and they do not require a high degree of technical skill and training. Therefore, these measures are well suited for large-scale epidemiological surveys and for clinical purposes.

This chapter deals with the use of anthropometric methods for estimating body composition. It addresses basic assumptions and principles, anthropometric models and prediction equations, techniques for measuring circumferences and skeletal breadths, and sources of measurement error. Information about using anthropometric measures to classify disease risk and frame size and to establish anthropometric profiles is also presented. Although SKF measures are a part of anthropometry, the SKF method is addressed separately in chapter 4.

Assumptions and Principles of the Anthropometric Method

There are basic principles associated with the use of anthropometric measures such as circumferences, skeletal diameters, and BMI to estimate body composition:

- **Circumferences are affected by FM, muscle mass, and skeletal size; therefore, these measures are related to FM and LBM.** Jackson and Pollock (1978) reported that circumference and bony diameter measures are markers of LBM (muscle mass and skeletal size); however, some circumferences are also highly associated with the fat component. These findings confirm the fact that circumference measures reflect both the fat and FFB components of body composition.

- **Skeletal size is directly related to LBM.** Behnke (1961), proposing that LBM could be accurately estimated from skeletal diameters, developed equations for predicting LBM. Cross-validation of these equations yielded a moderately high (r = 0.80) relationship and closely estimated the average LBM obtained from hydrodensitometry (Wilmore & Behnke 1969, 1970). Behnke's hypothesis was also supported by the observation that skeletal diameters, along with circumference measures, are strong markers of LBM (Jackson & Pollock 1978).

- **To estimate total body fat from weight-to-height indices, the index should be highly related to body fat but independent of height** (Keys et al. 1972; Lee & Hinds 1981). On the basis of data from two large-scale epidemiological surveys (National Health and Nutrition Examination Surveys I and II), Micozzi et al. (1986) reported that BMI (body weight divided by height squared) is not significantly related to height of men (r = −0.06) and women (r = −0.16). However, BMI is not totally independent of height, especially in younger children (<15 yr of age). Although BMI was directly related to SKF thickness and the estimated fat area of the arm (r = 0.72-0.80) in men and women (Micozzi et al. 1986), the relationship of BMI to body fat varies with age (younger vs. older), gender, and ethnicity (Deurenberg & Deurenberg-Yap 2001; Deurenberg et al. 1998; Gallagher et al. 1996; Rush et al. 1997; Wang et al. 1994).

Using the Anthropometric Method to Estimate Body Composition

Although some anthropometric prediction models include combinations of SKFs, circumferences, and skeletal diameters to estimate body composition, only those equations using circumferences and diameters as predictors are addressed in subsequent chapters, for the following reasons:

- The predictive accuracy of anthropometric (circumference and diameter) equations is not greatly improved by adding SKF measures.

- Anthropometric equations using only circumferences as predictors estimate the body fatness of persons who are obese more accurately than SKF prediction equations (Seip & Weltman 1991).

- Compared to SKFs, circumferences and skeletal diameters can be measured with less error (Bray & Gray 1988a).

- Some practitioners may not have access to SKF calipers.

Anthropometric Models and Prediction Equations

Anthropometric prediction equations estimate total Db, %BF, or FFM from combinations of body weight, height, skeletal diameters, and circumference measures. Generally, equations with only skeletal measures as predictors have larger prediction errors than those using both circumferences and bony diameters as predictors (Boileau et al. 1981; Katch & McArdle 1973). Like SKF and bioelectrical impedance (BIA) equations, anthropometric equations are based on either population-specific or generalized models (see chapters 4 and 6).

Population-specific anthropometric equations are valid for, and can be applied only to, individuals whose physical characteristics (age, gender, ethnicity, and level of body fatness) are similar to those in a specific population subgroup. For example, anthropometric equations developed to estimate the body composition of individuals who are obese (Weltman et al. 1987, 1988) should not be applied to non-obese individuals.

On the other hand, generalized equations, applicable to individuals varying in age and body fatness, have been developed for heterogeneous populations of women (15-79 yr; 13-63% BF) and men (20-78 yr; 2-49% BF) (Tran & Weltman 1988, 1989). The predictive accuracy of these generalized equations for estimating %BF of men and women with obesity was similar to that of fatness-specific (obese) equations (Seip & Weltman 1991). Typically, generalized equations include body weight or height, along with two or three circumferences, as predictors of Db or %BF. As with generalized SKF models (see chapter 4), the relationship between some circumference measures and Db is curvilinear (Tran & Weltman 1988, 1989). Also, age was shown to be an independent predictor of Db for women (Tran & Weltman 1989).

Although BMI is related to FM and %BF (r = 0.75-0.98), the prediction errors are generally large (SEE = 3.8-5.8% BF) when BMI is used as a single predictor of body fatness (Deurenberg, Weststrate et al. 1991; Garrow & Webster 1985; Gray & Fujiko 1991; Jackson et al. 1988; Smalley et al. 1990; Strain & Zumoff 1992). BMI is limited in its ability to predict %BF and to classify levels of body fatness accurately for the following reasons:

- Individuals with a large musculoskeletal system in relation to their height can have BMI values in the obese range even though they are not overly fat. Conversely, those with relatively small musculoskeletal systems tend to have lower BMI values (Lohman 1992).

- BMI does not detect or reflect differential growth rates of muscle and bone in children or differential rates of muscle and bone loss in older individuals (Lohman 1992).
- The relationship of BMI and %BF is affected by age, gender, ethnicity, and body build (Deurenberg et al. 1998, 1999; Snijder et al. 1999).

Anthropometric prediction equations for specific healthy and clinical populations are presented in chapters 8 to 14 and are summarized in appendixes B.1 through B.4. You can use these equations to obtain body composition estimates for your clients.

Anthropometric Techniques

Anthropometric body composition techniques include measuring the circumference of body segments, skeletal diameters, body weight, and standing height.

Standard Procedures for Measuring Circumference and Skeletal Diameter

A **circumference** is a measure of the girth of body segments (e.g., arm, thigh, waist, and hip). A **skeletal diameter** is a measure of bony width or breadth (e.g., knee, ankle, and wrist). Practice is necessary to become proficient in measuring skeletal diameters and circumferences. Following these standardized procedures will increase the accuracy and reliability of your measurements (Callaway et al. 1988; Wilmore et al. 1988):

Circumference and Skeletal Diameter

- Take all circumference and bony diameter measurements of the limbs on the right side of the body.
- Carefully identify and measure the anthropometric site. Be meticulous about locating anatomical landmarks used to identify the measurement site (see tables 5.1 and 5.2 and figures 5.1 and 5.2).

Table 5.1   Standardized Sites for Circumference Measurements

Site	Anatomical reference	Position	Measurement
Neck	Laryngeal prominence ("Adam's apple")	Perpendicular to long axis of neck	Apply tape with minimal pressure just inferior to the Adam's apple.
Shoulder	Deltoid muscles and acromion processes of scapula	Horizontal	Apply tape snugly over maximum bulges of the deltoid muscles, inferior to acromion processes. Record measurement at end of normal expiration.
Chest	Fourth costosternal joints	Horizontal	Apply tape snugly around the torso at level of fourth costosternal joints. Record at end of normal expiration.
Waist	Narrowest part of torso, level of the "natural" waist between ribs and iliac crest	Horizontal	Apply tape snugly around the waist at level of narrowest part of torso. An assistant is needed to position tape behind the client. Take measurement at end of normal expiration.
Abdominal	Maximum anterior protuberance of abdomen, usually at umbilicus	Horizontal	Apply tape snugly around the abdomen at level of greatest anterior protuberance. An assistant is needed to position tape behind the client. Take measurement at end of normal expiration.
Hip (Buttocks)	Maximum posterior extension of buttocks	Horizontal	Apply tape snugly around the buttocks. An assistant is needed to position tape on opposite side of body.

(continued)

Table 5.1 *(continued)*

Site	Anatomical reference	Position	Measurement
Thigh			
Proximal	Gluteal fold	Horizontal	Apply tape snugly around thigh, just distal to the gluteal fold.
Mid	Inguinal crease and proximal border of patella	Horizontal	With client's knee flexed 90° (right foot on bench), apply tape at level midway between inguinal crease and proximal border of patella.
Distal	Femoral epicondyles	Horizontal	Apply tape just proximal to the femoral epicondyles.
Knee	Patella	Horizontal	Apply tape around the knee at midpatellar level with knee relaxed in slight flexion.
Calf	Maximum girth of calf muscle	Perpendicular to long axis of leg	With client sitting on end of table and legs hanging freely, apply tape horizontally around the maximum girth of calf.
Ankle	Malleoli of tibia and fibula	Perpendicular to long axis of leg	Apply tape snugly around minimum circumference of leg, just proximal to the malleoli.
Arm (Biceps)	Acromion process of scapula and olecranon process of ulna	Perpendicular to long axis of arm	With client's arms hanging freely at sides and palms facing thighs, apply tape snugly around the arm at level midway between the acromion process of scapula and olecranon process of ulna (as marked for triceps and biceps skinfolds).
Forearm	Maximum girth of forearm	Perpendicular to long axis of forearm	With client's arms hanging down and away from trunk and forearm supinated, apply tape snugly around the maximum girth of the proximal part of the forearm.
Wrist	Styloid processes of radius and ulna	Perpendicular to long axis of forearm	With client's elbow flexed and forearm supinated, apply tape snugly around wrist, just distal to the styloid processes of the radius and ulna.

Data from Callaway et al. (1988, pp. 41-53).

- Take a minimum of three measurements at each site in rotational order. For body segments with relatively small girths (e.g., calf, arm, and forearm), take three measurements within ±0.2 cm. For larger body segments (e.g., waist, abdomen, and buttocks), obtain three measurements within ±1.0 cm.

- To measure breadth of smaller segments, like the elbow or wrist, use a small sliding caliper (range of 30 cm) with greater scale precision instead of a larger skeletal anthropometer (range of 60 to 80 cm) (see figure 5.3). Take three measurements within ±0.1 to 1.0 cm. The acceptable limit among trials varies, depending on the site being measured and the prominence of bony landmarks at the site.

- Hold the skeletal anthropometer or caliper in both hands so the tips of the index fingers are adjacent to the tips of the caliper.

- Place the caliper on the bony landmarks and apply firm pressure to compress the underlying muscle, fat, and skin. Apply pressure to a point where the measurement no longer continues to decrease.

- Use an anthropometric tape to measure circumferences (see figure 5.3). The zero end of the tape is held in your left hand, positioned below the other part of the tape held in your right hand (see figure 5.4).

- Apply tension to the tape so that it fits snugly around the body part but does not indent the

Table 5.2 Standardized Sites for Bony Breadth Measurements

Site	Anatomical reference	Position	Measurement
Biacromial (shoulder)	Lateral borders of acromion processes of scapula	Horizontal	With client standing, arms hanging vertically and shoulders relaxed, downward and slightly forward, apply blades of anthropometer to lateral borders of acromion processes. Measurement is taken from the rear.
Chest	Sixth ribs on midaxillary line or fourth costosternal joints anteriorly	Horizontal	With client standing, arms slightly abducted, apply the large spreading caliper tips lightly on the sixth ribs on the midaxillary line. Take measurement at end of normal expiration.
Bi-iliac (bicristal)	Iliac crests	45° downward angle	With client standing, arms folded across the chest, apply anthropometer blades firmly at a 45° downward angle, at maximum breadth of iliac crest. Measurement is taken from rear.
Bitrochanteric	Greater trochanter of femur	Horizontal	With client standing, arms folded across the chest, apply anthropometer blade with considerable pressure to compress soft tissues. Measure maximum distance between the trochanters from the rear.
Knee	Femoral epicondyles	Diagonal or horizontal	With client sitting and knee flexed to 90°, apply caliper blades firmly on lateral and medial femoral epicondyles.
Ankle (bimalleolar)	Malleoli of tibia and fibula	Oblique	With client standing and weight evenly distributed, place the caliper blades on the most lateral part of lateral malleolus and most medial part of medial malleolus. Measurement is taken on an oblique plane from the rear.
Elbow	Epicondyles of humerus	Oblique	With client's elbow flexed 90°, arm raised to the horizontal, and forearm supinated, apply the caliper blades firmly to the medial and lateral humeral epicondyles at an angle that bisects the right angle at the elbow.
Wrist	Styloid process of radius and ulna, anatomical "snuff box"	Oblique	With client's elbow flexed 90°, upper arm vertical and close to torso, and forearm pronated, apply caliper tips firmly at an oblique angle to the styloid processes of the radius (at proximal part of anatomical snuff box) and ulna.

Data from Wilmore et al. (1988, pp. 28-38).

skin or compress the subcutaneous tissue. To apply the appropriate tension and to increase the reliability of your girth measurements, use an anthropometric tape measure with a spring-loaded handle.

• For some circumferences (e.g., waist, hip, and thigh), you should align the tape in a horizontal plane, parallel to the floor (see figure 5.4)

Standardized Procedures for Measuring Body Weight and Standing Height

To measure body weight and stature (standing height), you should follow standardized procedures as recommended in the *Anthropometric Standardization Reference Manual* (Gordon et al. 1988).

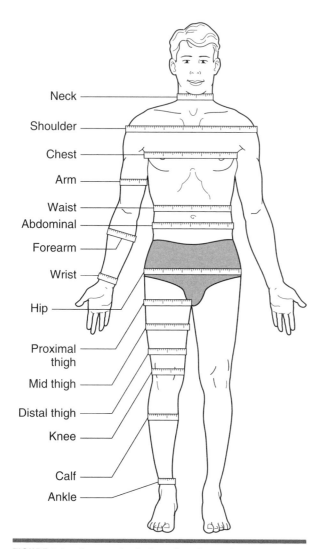

FIGURE 5.1   Anatomical sites for circumference measures.

Body Weight

1. Use a **beam scale** with movable weights or an **electronic digital scale** to measure body weight to the nearest 100 g. Spring scales are not recommended.

2. For measurement of body weight, the client stands on the platform of the scale with the body weight evenly distributed between the feet. Light indoor clothing, but no shoes, may be worn; however, a disposable paper gown is preferred to standardize this measure. One trial is usually sufficient to obtain an accurate measurement of body weight.

3. Check the accuracy of the scale periodically by placing standard calibration weights (see appendix C for manufacturer's address) on the scale. If the scale weight does not match the calibration weight, balance the scale by adjusting its calibration screw while the calibration weight is on the scale.

Standing Height

1. Use a **stadiometer** with a fixed or movable rod to measure standing height. Stature may be measured against a wall if the wall does not have a baseboard and the floor is not carpeted.

2. The client, barefoot, stands on a flat surface that is at a right angle to the vertical rod or board of the stadiometer. The weight is evenly distributed between the two feet, and the arms hang by the sides with palms facing the thighs. The heels are together, touching the vertical board of the stadiometer. The feet are spread at a 60° angle to each other. Whenever possible, the head, scapula, and buttocks should also be touching the vertical board. The head is erect with eyes focused straight ahead.

3. As the client inhales deeply, the horizontal board of the stadiometer is lowered to the most superior point on the head, compressing the hair. Standing height is measured to the nearest 0.1 cm.

Sources of Measurement Error

The accuracy and reliability of the anthropometric method for assessing body composition are potentially affected by *equipment, technician skill, client factors,* and the *prediction equation* selected to estimate body composition (Bray et al. 1978; Callaway et al. 1988). As with other body composition methods, the total prediction error is a function of the errors associated with both the anthropometric method and the reference method used to develop the prediction equation (i.e., law of propagation of errors). Acceptable errors for estimating %BF and FFM are, respectively, ≤3.5% BF (for men and women) and ≤2.8 kg (for women) and ≤3.5 kg (for men) (see table 2.1, p. 20). The following questions and responses address sources of measurement error for circumference and skeletal width measures.

Equipment

• **What equipment will I need to measure bony widths and body breadths or depths?** To measure bony widths, use a **skeletal anthropometer.** For body breadths and depths, use a **sliding** or **spreading caliper** (see figure 5.3). The precision characteristic (0.05-0.50 cm) and range of measurement (0-210 cm) depend on the type of skeletal anthropometer or caliper you are using (Wilmore et al. 1988). The instruments need to be carefully maintained and must be calibrated periodically so that their accuracy can be checked and restored if needed.

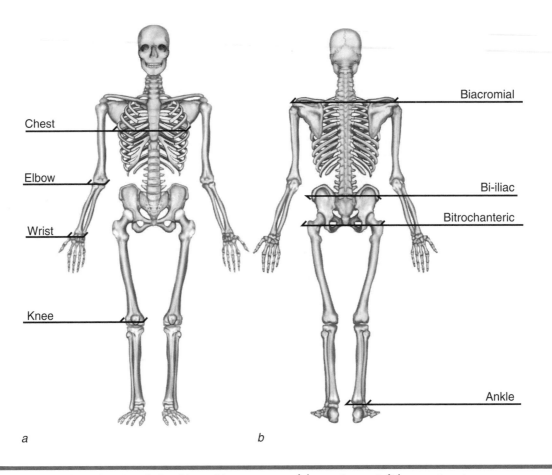

Chest

Elbow

Wrist

Knee

Biacromial

Bi-iliac

Bitrochanteric

Ankle

a   b

FIGURE 5.2   Anatomical sites for skeletal breadth measures: *(a)* anterior and *(b)* posterior view.

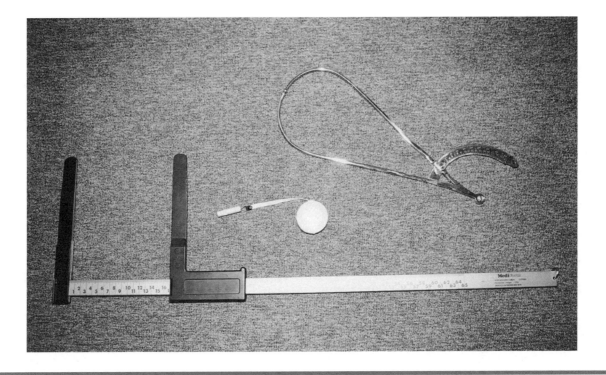

FIGURE 5.3   Skeletal anthropometers and anthropometric tape measure.

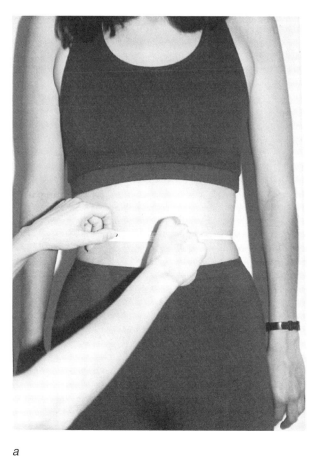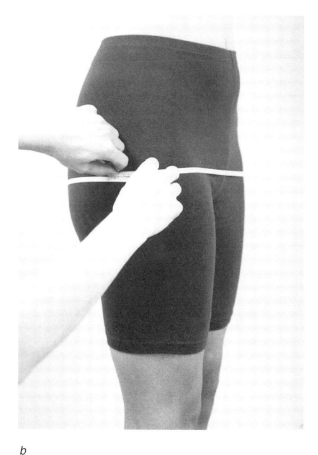

a   b

FIGURE 5.4   Measurement of *(a)* waist and *(b)* hip circumferences.

• **Can I use any type of tape measure to measure body circumferences?** Use an anthropometric tape measure (see figure 5.3) to measure circumferences. The tape measure should be made from a flexible material that does not stretch with use. Plastic-coated tape measures can be used if an anthropometric tape measure is not available. Some anthropometric tapes have a spring-loaded handle (i.e., Gulick handle) that allows a constant tension to be applied to the end of the tape during the measurement. Use of the spring-loaded handle is not recommended for circumference measurements requiring minimal tension (e.g., neck circumference) (Callaway et al. 1988).

Technician Skill

• **How much skill and practice are required to ensure accurate circumference and skeletal diameter measurements?** Compared with the SKF method, technician skill is not a major source of measurement error. However, you need to practice to perfect the identification of the measurement sites and your measurement techniques. Experts recommend practicing on at least 50 people and taking a minimum of three measurements for each site in rotational order (Callaway et al. 1988). Closely follow standardized procedures for locating measurement sites, positioning of the anthropometer or tape measure, and applying tension during the measurement.

• **Is there good agreement in circumference and skeletal diameter values when measurements are taken by two different technicians?** Variability in circumference measurements taken by different technicians is relatively small (0.2-1.0 cm), with some sites differing more than others (Callaway et al. 1988). Skilled technicians can obtain similar values even when measuring circumferences of individuals who are obese (Bray et al. 1978). However, practice is needed to perfect the identification of the measurement site and your measurement technique.

Client Factors

• **For clients who are obese, are circumferences more easily measured than SKFs?** As with the SKF method, it is more difficult to obtain consistent measurements of circumference for individuals who are obese compared to those who are lean (Bray & Gray 1988a). However, circumferences are preferable to SKFs for measuring clients who are obese, for several reasons:

- Measurement of circumferences requires less technician skill.

- You can measure circumferences of individuals who are obese regardless of their size, whereas the maximum aperture of the SKF caliper may not be large enough to allow measurement.

- Differences between technicians are less for circumference measurements (Bray & Gray 1988a).

• **Is it possible to measure bony widths of clients who are heavily muscled or obese?** Accurate measurement of bony diameters in persons who are heavily muscled or obese may be difficult because the underlying muscle and fat tissues must be firmly compressed. It may be difficult to identify and palpate bony anatomical landmarks, leading to error in locating the measurement site.

• **Does the menstrual cycle affect the size of circumference measures for women?** It is possible for the accuracy of circumference measurements to be affected by fluid retention and subcutaneous edema, particularly in women experiencing large weight gains during certain stages of their menstrual cycle.

Anthropometric Prediction Equations

• **What factors do I need to consider when selecting an anthropometric prediction equation for my client?** As mentioned earlier, anthropometric prediction equations should be selected based on the gender, age, ethnicity, and level of body fatness of your client. Using an inappropriate equation may produce systematic prediction error in estimating body composition. Also, we do not recommend using BMI to estimate body fatness of your clients because of the large prediction error associated with this method.

• **Do anthropometric prediction equations used to assess and monitor body composition of military personnel in the U.S. provide an accurate estimation of body fat?** These equations use various combinations of body weight, height, neck, abdomen, hip, thigh, arm, forearm, and wrist circumferences to predict either FFM or %BF of military personnel. Cross-validation of each equation developed for men and women in the Army, Navy, Air Force, and Marines yielded large, unacceptable prediction errors (SEE = 3.7-5.2% BF) (Hodgdon 1992). Therefore, the practice of dismissing personnel from the armed services because their %BF exceeds military standards may be questionable when their %BF is estimated from these anthropometric equations. Technicians should consider using a reference method, like densitometry or dual-energy X-ray absorptiometry, instead of field methods to assess body composition of military personnel facing this problem.

Using Anthropometric Indices for Classification of Disease Risk

Health and longevity are threatened when a person is either overweight or underweight. Individuals with body fat levels falling at or near the extremes of the body fat continuum are likely to have serious health problems that reduce life expectancy and threaten their quality of life. Persons who are overweight or obese have a higher risk of cardiovascular disease, dyslipidemia, hypertension, glucose intolerance, insulin resistance, diabetes mellitus, obstructive pulmonary disease, gallbladder disease, osteoarthritis, and certain types of cancer (U.S. Dept. of Health and Human Services 2000c). At the opposite extreme, underweight individuals with too little body fat tend to be malnourished. These people have a relatively higher risk of fluid-electrolyte imbalances, osteoporosis and osteopenia, bone fractures, muscle wasting, cardiac arrhythmias and sudden death, peripheral edema, and renal and reproductive disorders (Fohlin 1977; Mazess, Barden, & Ohlrich 1990; Vaisman et al. 1988).

In addition, research indicates that the total amount of body fat and the way in which fat is distributed in the body (i.e., **regional fat distribution**) are the principal causes for the high rate of chronic disease in individuals who are overweight or obese (Blair & Brodney 1999). The proliferation of research on body fat distribution and its relationship to disease has expanded at an exponential rate over the past 10 years, providing evidence that a link exists between increased abdominal fat and increased morbidity and mortality (Pi-Sunyer 1999). The impact of regional fat distribution on health is related to the amount of visceral fat in the abdominal cavity (i.e., **intra-abdominal fat**). **Visceral fat** is adipose tissue stored within and around the organs in the thoracic and abdominal cavities. Computed tomography (CT), dual-energy X-ray absorptiometry, and magnetic resonance imaging (MRI) provide direct measures of visceral fat and regional fat distribution. In large-scale epidemiological studies and clinical settings, however, indirect anthropometric indices are often used because of their simplicity and low cost. In these studies, BMI, WHR, waist circumference, sagittal abdominal diameter, and neck circumference have been used to assess body fat distribution and to identify at-risk individuals (Ben-Noun et al. 2001; Ohrvall et al. 2000; Pi-Sunyer 2000; Taylor et al. 1998; Turcato et al. 2000).

BMI

The BMI is commonly used to classify individuals as obese, overweight, and underweight; to identify individuals at risk for obesity-related diseases; and to monitor changes in body fatness of clinical populations (U.S. Dept. of Health and Human Services 2000c; World Health Organization 1998). It has been established that BMI is a significant predictor of cardiovascular disease and type 2 diabetes (Janssen et al. 2002). Because of this association and the fact that BMI is easily calculated (BMI = body weight/height squared), BMI is widely used in population-based and prospective studies to identify at-risk individuals.

BMI, however, is limited as an index of obesity (i.e., body fatness) because it does not take into account the composition of an individual's body weight. In addition, factors such as age, ethnicity, body build, and frame size affect the relationship between BMI and %BF. Thus, using BMI as an index of obesity may result in misclassifications of underweight, overweight, and obesity. Also, because BMI is a better measure of non-abdominal and abdominal subcutaneous fat than of visceral fat (Janssen et al. 2002), other anthropometric indices need to be used to assess fat distribution.

Table 5.3 provides standards for classifying BMI values. The World Health Organization (1998) defines **obesity** as a BMI of 30 kg/m² or more; **overweight** as a BMI between 25 and 29.9 kg/m²; and **underweight** as a BMI of less than 18.5 kg/m². These suggested cutoff values are based on the relationship between BMI and morbidity and mortality reported in observational studies in Europe and the United States. The use of BMI in health risk appraisals assumes that people who are disproportionately heavy are so because of excess FM. However, controversy exists concerning the most appropriate BMI cutoff value for designating obesity (Deurenberg 2001).

Table 5.3   Classification of Overweight and Obesity Based on Body Mass Index (BMI)

Classification	BMI value
Underweight	<18.5
Normal weight	18.5-24.9
Overweight	25.0-29.9
Obesity	
Class I	30.0-34.9
Class II	35.0-39.9
Class III	≥40.0

Data from WHO Report. 1998. *Obesity: preventing and managing the global epidemic.* Report of a WHO consultation on obesity. Geneva: World Health Organization.

As mentioned earlier, the relationship between BMI and %BF is affected by age, gender, ethnicity, and body build (Deurenberg et al. 1998, 1999; Snijder et al. 1999). For a given BMI value, older individuals have a greater %BF than their younger counterparts, and young adult males have lesser %BF than young adult females. Also, for a given level of %BF, age- and gender-matched Caucasians have a higher BMI (1.3-4.6 kg/m²) than other ethnic groups (e.g., African Americans, Chinese, Indonesians, Ethiopians, and Polynesians) (Deurenberg et al. 1998). These findings suggest that using a universal BMI cutoff value to define obesity (≥30 kg/m²) may not be appropriate. Ethnic-specific cutoff values need to be established based on the relationship between BMI and %BF and on morbidity and mortality risks in relation to BMI for specific ethnic groups (Deurenberg 2001).

As noted earlier, BMI is the ratio of body weight to height squared: BMI (kg/m²) = WT (in kilograms)/HT² (in meters). To calculate BMI, the body weight is measured in kilograms and the height is converted from centimeters to meters (cm/100). Alternatively, you can use a nomogram (see figure 5.5) to calculate your client's BMI (Bray 1978). To use the nomogram, plot your client's height and body weight in the appropriate columns and connect these two points with a ruler. Read the corresponding BMI at the point where the connecting line intersects the BMI column on the nomogram. You should be aware that the versions of Bray's nomogram appearing in *The Surgeon General's Report on Nutrition and Health* (U.S. Dept. of Health and Human Services 1988) and *Diet*

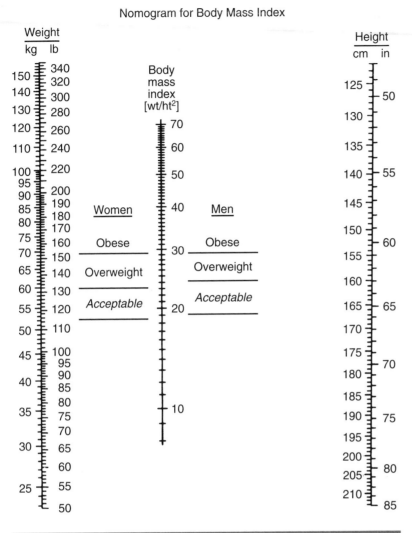

Nomogram for Body Mass Index

FIGURE 5.5 Nomogram for BMI.

Reprinted, by permission, from G.A. Bray, 1978, "Definitions, measurements, and classifications of the syndromes of obesity," *International Journal of Obesity 2 (2): 99-112.*

and Health (National Research Council 1989) are inaccurate because the orientation of the vertical columns of the original nomogram was mistakenly altered by the graphic artist (Kahn 1991).

WHR

In 1947, Vague introduced a system for differentiating types of obesity based on regional fat distribution. He coined the terms *android obesity* and *gynoid obesity* to refer to the localization of excess body fat mainly in the upper body (android) or lower body (gynoid). Android obesity is more typical of males; gynoid obesity is more characteristic of females. However, men and women who are obese can be, and often are, classified into either group. There are other terms used to describe types of obesity and regional fat distribution. **Android obesity** (apple-shaped) is frequently referred to as upper-body obesity or central adiposity, and **gynoid obesity** (pear-shaped) is often termed lower-body obesity. Individuals with **lower-body obesity** tend to deposit excess fat on the hips and thighs; those with **upper-body obesity** or **central adiposity** tend to deposit fat on the trunk and abdomen.

The WHR is commonly used as an indirect measure of lower- and upper-body fat distribution. Upper-body obesity or central adiposity, measured by the WHR, is moderately related (r = 0.48-0.61) to risk factors associated with cardiovascular and metabolic diseases in men and women (Ohrvall et al. 2000). Young adults with WHR values in excess of 0.94 for men and 0.82 for women are at high risk for adverse health consequences (Bray & Gray 1988b).

Although the WHR has been used as an anthropometric measure of central adiposity and visceral fat, this index has certain limitations:

- The WHR of women is affected by menopausal status (Svendsen et al. 1992; Weits et al. 1988). Postmenopausal women show more of a male pattern of fat distribution than premenopausal women do (Ferland et al. 1989).

- The WHR is not valid for evaluating fat distribution in prepubertal children (Peters et al. 1992).

- The accuracy of WHR in assessing visceral fat decreases with increasing levels of fatness.

- Hip circumference is influenced by subcutaneous fat deposition only, whereas waist circumference is affected by both visceral fat and subcutaneous fat deposition. Thus, the WHR may not accurately detect changes in visceral fat accumulation (Goran et al. 1995; van der Kooy et al. 1993).

The WHR is simply calculated by dividing waist circumference (measured in c by hip circumference (measured in c The measurement site for waist circ

Table 5.4   Waist-to-Hip Circumference Ratio Norms for Men and Women

	Age	RISK			
		Low	Moderate	High	Very high
MEN	20-29	<0.83	0.83-0.88	0.89-0.94	>0.94
	30-39	<0.84	0.84-0.91	0.92-0.96	>0.96
	40-49	<0.88	0.88-0.95	0.96-1.00	>1.00
	50-59	<0.90	0.90-0.96	0.97-1.02	>1.02
	60-69	<0.91	0.91-0.98	0.99-1.03	>1.03
WOMEN	20-29	<0.71	0.71-0.77	0.78-0.82	>0.82
	30-39	<0.72	0.72-0.78	0.79-0.84	>0.84
	40-49	<0.73	0.73-0.79	0.80-0.87	>0.87
	50-59	<0.74	0.74-0.81	0.82-0.88	>0.88
	60-69	<0.76	0.76-0.83	0.84-0.90	>0.90

Adapted, by permission, from Bray and Gray, 1988, "Obesity–Part 1–Pathogenesis," *Western Journal of Medicine* 149: 432. ©BMJ Publishing Group.

however, has not been universally standardized. The World Health Organization (1988) recommends measuring waist circumference midway between the lower rib margin and the iliac crest, and hip circumference at the widest point over the greater trochanters. In contrast, the *Anthropometric Standardization Reference Manual* (Callaway et al. 1988) recommends measuring the waist circumference at the narrowest part of the torso and hip circumference at the level of the maximum extension of the buttocks (see figure 5.4). The WHR norms (table 5.4) were established using the standardized measurement procedures described in the *Anthropometric Standardization Reference Manual*. You can use the WHR nomogram (figure 5.6), instead of hand calculations, to obtain WHR values for your clients. Plot the client's waist and hip circumferences in the corresponding columns of the nomogram and connect these points with a ruler. Read the WHR at the point where this line intersects the WHR column.

Waist Circumference

Waist circumference (WC) is gaining support as an alternative to WHR for assessing regional adiposity in field and clinical settings. Compared to the WHR, WC provides a more accurate indirect measure of visceral fat and is not greatly influenced by age, gender, standing height, and degree of overall adiposity (Han et al. 1997; Lemieux et al. 1996). WC is highly related (r = 0.76-0.88) to MRI and CT

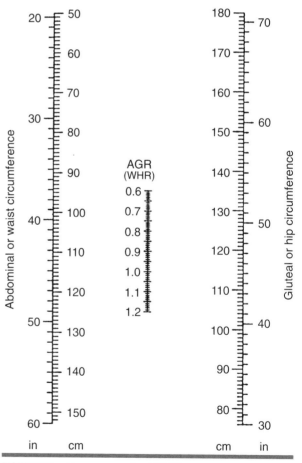

FIGURE 5.6   Nomogram for WHR.

Adapted, by permission, from Bray and Gray, 1988, "Obesity–Part 1–Pathogenesis," *Western Journal of Medicine* 149: 432, ©BMJ Publishing Group.

measures of intra-abdominal (visceral) fat in men and women (Han et al. 1997; Janssen et al. 2002; Zamboni et al. 1998) and to cardiovascular risk factors in older (67-78 yr of age) women (Turcato et al. 2000). Given WC and BMI independently contribute to the estimation of total non-abdominal fat, as well as abdominal subcutaneous and visceral fat, experts suggest using both of these anthropometric indices to assess total body and central adiposity in clinical settings (Janssen et al. 2002; Taylor et al. 1998). The National Cholesterol Education Program (NCEP 2001) recommends using WC cutoff values of >102 cm for men and >88 cm for women to evaluate obesity as a risk factor for coronary heart disease and metabolic diseases.

SAD

The **sagittal abdominal diameter (SAD)** is a measure of anteroposterior thickness of the abdomen at the umbilical level. Although the literature refers to this measure as a diameter, technically it is a depth measurement. In anthropometry, diameter refers to width or a side-to-side measurement, whereas a depth refers to a front-to-back measurement. Therefore, sagittal abdominal diameter should be called *sagittal abdominal depth*.

Research suggests that SAD is an excellent indirect measure of visceral fat (Ohrvall et al. 2000; Zamboni et al. 1998). Sagittal abdominal depth is strongly related to visceral adipose tissue in men (r = 0.82) and women (r = 0.76), even after adjustment for BMI (r = 0.66 and 0.63, respectively, for men and women) (Zamboni et al. 1998). However, the strength of this relationship was higher in persons who were lean to moderately overweight compared to those who were obese. Compared to WC, WHR, and BMI, the SAD was more strongly related to risk factors for cardiovascular disease and metabolic diseases in both women and men (Ohrvall et al. 2000). SAD is also associated with cardiovascular disease risk factors in older women, 67 to 78 yr of age (Turcato et al. 2000).

The procedures for measuring SAD have not been standardized. In most studies, SAD was measured with the client lying supine with the legs extended on an examination table. A sliding-beam anthropometer is used to measure the vertical distance (height to the nearest 0.1 cm) between the top of the table and the abdomen at the level of the umbilicus or iliac crests. In some studies, SAD was measured with the hips and legs flexed or with the client standing instead of lying supine.

Neck Circumference

Neck circumference is sometimes used in clinical settings as an indirect index of upper-body fat distribution and to identify individuals who are overweight or obese (Ben-Noun et al. 2001). This measure is related (r = 0.56-0.86) to BMI, WHR, WC, and hip circumference in men and women. Neck circumference norms for identifying men and women who are overweight (i.e., BMI = 25.0-29.9 kg/m^2) or obese (i.e., BMI $\geq$30 kg/m^2) are as follows: overweight men, 37 to 39.4 cm; overweight women, 34 to 36.4 cm; obese men, $\geq$39.5 cm; obese women, $\geq$36.5 cm (Ben-Noun et al. 2001). To measure neck circumference, follow the standardized procedure described in table 5.1.

Using Anthropometric Measures for Frame Size Classification and Anthropometric Profiles

Various anthropometric measures can be used to classify frame size, to create an anthropometric profile (somatogram), and to determine body type (somatotype) for your clients. You can use combinations of SKF, circumference, and skeletal diameter measures to visually assess your client's body profile (proportionality and symmetry), as well as regional changes resulting from weight management programs.

Frame Size

Skeletal diameters are used to classify frame size to improve the validity of using height-weight tables for evaluating body weights of clients. The rationale for including frame size is that skeletal breadths are important estimators of the bone and muscle components of FFM. Therefore, an estimate of frame size allows you to differentiate between persons who weigh more because of a large musculoskeletal mass and those who are overweight because of a large FM (Himes & Frisancho 1988). Since there are health implications for individuals who are overweight, critical evaluation of body weight is important.

There are numerous skeletal measurements that can be used to estimate frame size; however, the best estimators of frame size for evaluating body weight are those that are highly related to FFM (independent of stature) and poorly related to FM. Research indicates that wrist, ankle, and elbow breadths are valid measures of frame size

Table 5.5   Elbow Breadth Norms for Men and Women in the United States

| | Age (years) | FRAME SIZE | | |
		Small	Medium	Large
MEN	18-24	≤6.6	>6.6 and <7.7	≥7.7
	25-34	≤6.7	>6.7 and <7.9	≥7.9
	35-44	≤6.7	>6.7 and <8.0	≥8.0
	45-54	≤6.7	>6.7 and <8.1	≥8.1
	55-64	≤6.7	>6.7 and <8.1	≥8.1
	65-74	≤6.7	>6.7 and <8.1	≥8.1
WOMEN	18-24	≤5.6	>5.6 and <6.5	≥6.5
	25-34	≤5.7	>5.7 and <6.8	≥6.8
	35-44	≤5.7	>5.7 and <7.1	≥7.1
	45-54	≤5.7	>5.7 and <7.2	≥7.2
	55-64	≤5.8	>5.8 and <7.2	≥7.2
	65-74	≤5.8	>5.8 and <7.2	≥7.2

Reprinted from A.R. Frisancho, 1984, "New standard of weight and body composition by frame size and height for assessment of nutritional status of adults and the elderly." Reproduced with permission by the *American Journal of Clinical Nutrition* 40: 808-819. ©American Journal of Clinical Nutrition, American Society for Clinical Nutrition.

(Frisancho & Flegel 1983; Himes & Bouchard 1985). Frisancho and Flegel (1983) reported that elbow breadth is a good predictor of frame size because of its weak association with SKF (triceps and subscapular) measures. Given that reference data are available for elbow breadth, you can use this measure to classify your client's frame size (table 5.5). The anatomical landmarks for measuring elbow breadth are described in table 5.2 and illustrated in figure 5.2.

Anthropometric Profile

An **anthropometric profile** is a **somatogram** that graphically depicts your client's pattern of muscle and fat distribution with respect to reference values (figure 5.7). Circumferences for six muscular components (shoulders, chest, arm, forearm, thigh, and calf) and five nonmuscular components (umbilicus, hip, knee, wrist, and ankle) are measured. The relative deviation (%) of each circumference measurement from its reference value is calculated and graphed to visually display the individual's body proportionality. Somatograms may be especially useful for charting changes (pre- and posttest anthropometric profiles), monitoring progress of clients involved in weight management (diet/exercise) programs, and appraising the degree of proportionality and symmetry in the body. To assess your client's body profile, use the following steps:

Anthropometric Profile

1. Measure the 11 circumferences as described in table 5.1 to the nearest 0.1 cm using an anthropometric measuring tape. Measure the thigh circumference at the proximal location.

2. Calculate a reference value (R) by summing your client's 11 circumferences and dividing by 100. The value 100 is equal to the sum of the constant (k) values for a reference man and reference woman (see table 5.6): R = 11 circumferences/100. These constants were derived from data collected for 133 young men and 128 young women, 20 to 24 yr of age. For example, if the sum of the 11 circumferences equaled 734 cm, your client's reference value would be 7.34 cm (734/100).

3. Divide each circumference by its corresponding k value (see table 5.6) to obtain the d value: d = C/k. For example, if your male client's shoulder circumference measures 139.5 cm, d is equal to 139.5/18.47 or 7.55.

4. Subtract the reference value (R) obtained in step 2 from each d value calculated in step 3 to obtain the absolute difference (AD) score: AD = d − R. In our example for shoulder circumference, the AD is equal to 7.55 − 7.34 or 0.21. Note that when R is greater than d, a negative AD value will be obtained.

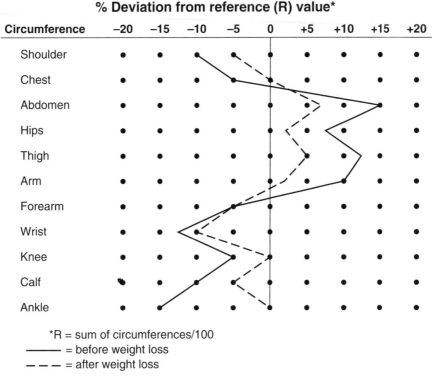

% Deviation from reference (R) value*

Circumference	−20	−15	−10	−5	0	+5	+10	+15	+20
Shoulder
Chest
Abdomen
Hips
Thigh
Arm
Forearm
Wrist
Knee
Calf
Ankle

*R = sum of circumferences/100
——— = before weight loss
— — — = after weight loss

FIGURE 5.7 Somatogram of a 45-yr-old woman before and after weight loss.

Table 5.6   Somatogram Constants (k) for a Reference Man and Woman

Circumference	Reference man	Reference woman
Shoulder	18.47	17.51
Chest	15.30	14.85
Abdomen	13.07	12.90
Hips	15.57	16.93
Thigh	9.13	10.03
Arm	5.29	4.80
Forearm	4.47	4.15
Wrist	2.88	2.73
Knee	6.10	6.27
Calf	5.97	6.13
Ankle	3.75	3.70
Total	100.00	100.00

Reference man and woman, age 20-24 yr (Behnke & Wilmore 1974).

5. Calculate the relative difference (%) by dividing each AD score by the R-value and multiplying by 100: % = AD/R × 100. For our example, % = 0.21/7.34 × 100 or 2.9%.

6. Plot the percentage obtained for each circumference on the somatogram (see figure 5.7).

Alternatively, you can use computer software (i.e., LifeSize) to obtain comprehensive anthropometric profiles for your clients (Olds & Norton 1999). This software allows you to calculate body type, percentage of body fat, BMI, and body surface area, as well as other anthropometric measures. For a full anthropometric profile, 9 SKFs, 13 circumferences, 6 skeletal diameters, and 10 height/length measurements are evaluated. The restricted profile includes 9 SKFs, 5 girths, and 2 skeletal breadth measures.

Somatotype

Anthropometric measures can be used to determine a client's body type or **somatotype.** The Heath-Carter anthropometric method (Carter 1982) is the most popular system for somatotyping. This method uses height, weight, SKFs, girths, and skeletal breadth measures to rate three components of body type for each client:

- **Endomorphy** is a measure of body fatness and is the first component, evaluated by measurement of the triceps, subscapular, and suprailiac SKF thicknesses.
- **Mesomorphy** is a measure of muscularity and is the second component, evaluated by measurement of height, bony widths (humerus and femur), and circumferences corrected for

SKF thicknesses (i.e., arm circumference adjusted for triceps SKF and calf circumference adjusted for calf SKF).

- **Ectomorphy** is a measure of proportionality between body weight and height and is the third component evaluated. The individual's body weight and the ratio of height to the cube root of body weight are used for this purpose.

You can use the Heath-Carter somatotype rating form (page 85) to obtain a rating for each of these three components by circling the value closest to your client's score and reading the corresponding component score from the column directly below that value. A somatotype score of 2-8-4, for example, would describe a lean, muscular individual who has below-average fatness (endomorphy rating = 2), a high degree of muscularity (mesomorphy rating = 8), and average proportionality (ectomorphy rating = 4). As an alternative to this rating form, you can obtain somatotype ratings for clients by using the LifeSize computer software developed by Olds and Norton (1999).

Key Points

- Circumference measures are affected by fat, muscle mass, and skeletal size. Skeletal size is directly related to LBM.
- Anthropometric prediction equations include combinations of various circumferences and bony diameter measures and are based on either population-specific or generalized models.
- Anthropometric prediction equations are better than SKF equations for estimating body composition of individuals who are obese.
- Anthropometric prediction equations estimate FFM, Db, or %BF.
- Technician skill, client factors, and the prediction equation used are potential sources of measurement error for the anthropometric method.
- A minimum of three measurements at each circumference or skeletal diameter site should be taken in rotational order.
- An anthropometric tape should be used to measure circumferences. Sliding and spreading calipers are used to measure bony widths.
- BMI is a crude index of overall body fatness and therefore should not be used to classify a client's level of body fatness.
- The BMI is commonly used in large-scale epidemiological surveys and clinical settings to classify individuals as obese, overweight, or underweight.

- Total body fat and regional fat distribution (upper- vs. lower-body obesity) are related to disease risk.
- WC and SAD are used in field and clinical settings to assess intra-abdominal (visceral) fat deposition.
- An anthropometric profile is a somatogram that visually depicts the regional distribution of fat and muscle in the body.
- The Heath-Carter anthropometric somatotyping method is widely used to evaluate body type.
- Endomorphy, mesomorphy, and ectomorphy are three components of an anthropometric somatotype.

Key Terms

Learn the definition for each of the following key terms. Definitions of terms can be found in the Glossary, page 227.

android obesity	regional fat distribution
anthropometric profile	sagittal abdominal diameter (SAD)
anthropometry	skeletal anthropometer
central adiposity	skeletal diameter
circumference	sliding or spreading caliper
ectomorphy	somatogram
endomorphy	somatotype
gynoid obesity	stadiometer
intra-abdominal fat	upper-body obesity
lower-body obesity	visceral fat
mesomorphy	waist circumference
neck circumference	waist-to-hip ratio (WHR)

Review Questions

1. Identify the basic assumptions and principles underlying the use of anthropometric methods for estimating body composition.

2. Why should you avoid using BMI to estimate your client's body composition? What factors affect the relationship between BMI and %BF?

3. What are the major sources of measurement error for the anthropometric method?

4. Briefly describe the standardized procedures for measuring circumferences and skeletal diameters.

5. What factors should you consider when selecting an anthropometric prediction equation for your client?

6. Why are anthropometric equations better suited than SKF equations for estimating body composition of clients who are obese?

7. Identify the recommended BMI cutoff values for classifying clients as underweight, overweight, and obese.

8. Explain why one universal BMI cutoff value should not be used to identify clients at risk due to obesity.

9. What are the limitations of using the WHR as a measure of central adiposity?

10. What indirect measures are commonly used to assess visceral fat?

11. What values of WHR, waist circumference, and neck circumference are used to identify individuals with a high risk of developing diseases due to central adiposity or obesity?

12. What anthropometric measures can be used to classify your client's frame size?

13. When assessing your client's anthropometric profile, what anthropometric measures can you use to describe the muscular and nonmuscular components of the profile?

14. Define the three components of an anthropometric somatotype and briefly describe the measures used to evaluate each of these components.

Heath-Carter Somatotype Rating Form

Name _____ Sex: M F No _____

Occupation _____ Age _____ Date _____

Project _____ Ethnic group _____ Measured by _____

Skinfolds (mm):

Triceps =

Subscapular =

Suprailiac =

Total skinfolds = ☐

Calf = ☐

Total skinfolds (mm)																								
Triceps	10.9	14.9	18.9	22.9	26.9	31.2	35.8	40.7	46.2	52.2	58.7	65.7	73.2	81.2	89.7	98.9	108.9	119.7	131.2	143.7	157.2	171.9	187.9	204.0
Subscapular	9.0	13.0	17.0	21.0	25.0	29.0	33.5	38.0	43.5	49.0	55.5	62.0	69.5	77.0	85.5	94.0	104.0	114.0	125.5	137.0	150.5	164.0	180.0	196.0
Suprailiac	7.0	11.0	15.0	19.0	23.0	27.0	31.3	35.9	40.8	46.3	52.3	58.8	65.8	73.3	81.3	89.8	99.0	109.0	119.8	131.3	143.8	157.3	172.0	188.0
First component	½	1	1½	2	2½	3	3½	4	4½	5	5½	6	6½	7	7½	8	8½	9	9½	10	10½	11	11½	12

Height (in.) = ☐

Bone: Humerus (cm) = ☐

Femur = ☐

Muscle: Biceps (cm) − (triceps skinfold) = ☐

Calf − (calf skinfold) = ☐

Height (in.)	55.0	56.5	58.0	59.5	61.0	62.5	64.0	65.5	67.0	68.5	70.0	71.5	73.0	74.5	76.0	77.5	79.0	80.5	82.0	83.5	85.0	86.5	88.0	89.5
Bone: Humerus (cm)	5.19	5.34	5.49	5.64	5.78	5.93	6.07	6.22	6.37	6.51	6.65	6.80	6.95	7.09	7.24	7.38	7.53	7.67	7.82	7.97	8.11	8.25	8.40	8.55
Femur	7.41	7.62	7.83	8.04	8.24	8.45	8.66	8.87	9.08	9.28	9.49	9.70	9.91	10.12	10.33	10.53	10.74	10.95	11.16	11.37	11.58	11.79	12.00	12.21
Muscle: Biceps (cm) − (triceps skinfold)	23.7	24.4	25.0	25.7	26.3	27.0	27.7	28.3	29.0	29.7	30.3	31.0	31.6	32.2	33.0	33.6	34.3	35.0	35.6	36.3	37.1	37.8	38.5	39.3
Calf − (calf skinfold)	27.7	28.5	29.3	30.1	30.8	31.6	32.4	33.2	33.9	34.7	35.5	36.3	37.1	37.8	38.6	39.4	40.2	41.0	41.8	42.6	43.4	44.2	45.0	45.8
Second component	½	1	1½	2	2½	3	3½	4	4½	5	5½	6	6½	7	7½	8	8½	9						

Weight (lb) = ☐

Ht/³√Wt. = ☐

Upper limit	12.32	11.99	12.53	12.74	12.95	13.15	13.36	13.56	13.77	13.98	14.19	14.39	14.59	14.80	15.01	15.22	15.42	15.63
Mid-point	and	12.16	12.43	12.64	12.85	13.05	13.26	13.46	13.67	13.88	14.01	14.29	14.50	14.70	14.91	15.12	15.33	15.53
Lower limit	below	12.00	12.33	12.54	12.75	12.96	13.16	13.37	13.51	13.78	13.99	14.20	14.40	14.60	14.81	15.02	15.23	15.43
Third component	½	1	1½	2	2½	3	3½	4	4½	5	5½	6	6½	7	7½	8	8½	9

	First component	Second component	Third component
Anthropometric somatotype			
Anthropometric plus photoscopic somatotype			

By _____

Rater _____

From Vivian H. Heyward and Dale R. Wagner, 2004. *Applied Body Composition Assessment, Second Edition.* (Champaign, IL: Human Kinetics). Reprinted, by permission, from T.G. Lohman, A.F. Roche, and R. Martorell, eds., 1988, *Anthropometric standardization reference manual* (Champaign, IL: Human Kinetics), 157.

CHAPTER 6

BIOELECTRICAL IMPEDANCE ANALYSIS METHOD

Key Questions

- What are the basic assumptions and principles underlying the use of the bioelectrical impedance analysis method?
- What is the traditional bioelectrical impedance analysis method and how has it been modified for clinical and home use?
- What are the standardized procedures for measuring whole-body impedance?
- What are the major sources of error for the bioelectrical impedance method? What pretesting guidelines do clients need to follow to minimize this error?

Bioelectrical impedance analysis (BIA) is a rapid, noninvasive, and relatively inexpensive method for evaluating body composition in field and clinical settings. Thomasett's (1962) pioneering work in the early 1960s established basic BIA principles. With this method, low-level electrical current is passed through the client's body, and the **impedance (Z)**, or opposition to the flow of current, is measured with a BIA analyzer. The individual's TBW can be estimated from the impedance measurement because the electrolytes in the body's water are excellent conductors of electrical current. When the volume of TBW is large, the current flows more easily through the body with less resistance. The resistance to current flow is greater in individuals with large amounts of body fat, given that adipose tissue is a poor conductor of electrical current due to its relatively small water content. Because the water content of the FFB is relatively large (~73% water), FFM can be predicted from TBW estimates. Individuals with a large FFM and TBW have less resistance to current flowing through their bodies than do those with a smaller FFM.

Hoffer et al. (1969) reported a strong relationship between total body impedance measures and TBW, suggesting that the BIA method may be a valuable tool for analyzing body composition and assessing TBW in the clinical setting. Indeed, since the work of Hoffer et al. (1969), many researchers have developed BIA prediction equations for TBW, FFM, and %BF. Houtkooper and colleagues (1996) summarized 55 BIA prediction equations (18 predicting TBW, 29 estimating FFM, and 8 for %BF or fat mass). They reported that the typical prediction errors (SEE) range from 0.9 to 1.8 kg TBW, 2.0 to 3.0 kg FFM, and 3.0% to 4.0% BF in adults.

Although the relative predictive accuracy of the BIA method is similar to that of the SKF method, BIA may be preferable in some settings because

- it does not require a high degree of technician skill,
- it is generally more comfortable and does not intrude as much on the client's privacy, and
- it can be used to estimate body composition of persons who are obese (Gray et al. 1989; Segal et al. 1988).

Recent technological advances have resulted in many changes and improvements in BIA methodology. The traditional BIA method, which is still most frequently used, involves a whole-body (wrist-to-ankle) measurement of impedance at a single (50-kHz) frequency. In an effort to eliminate some assumptions and to improve the clinical usefulness of BIA, many researchers have experimented with other approaches (e.g., series vs. parallel models, single vs. multiple frequencies, and whole-body vs. segmental measurement). This chapter addresses the basic assumptions and principles of the traditional BIA method and covers some of the variations along with BIA prediction models. For a comprehensive synthesis of the various BIA approaches, we recommend the reviews of Lukaski (1996) and Thomas et al. (1998) or the report of a National Institutes of Health (NIH) panel organized for the purpose of evaluating BIA technology (Ellis et al. 1999). Additionally, we summarize data from studies evaluating the accuracy of foot-to-foot (Tanita) and hand-to-hand (Omron) BIA analyzers for home use. Lastly, this chapter addresses standardized testing procedures and sources of measurement error for the **whole-body BIA method**.

Assumptions and Principles of the BIA Method

The volume of the body's TBW or FFM is indirectly estimated from bioelectrical impedance measures. Therefore, the principles of electrical conductivity apply, and certain basic assumptions about the geometric shape of the body and the relationship of impedance to the length and volume of the conductor are made.

BIA Assumptions

• **Assuming that the body is a perfect cylinder, at a fixed signal frequency (e.g., 50 kHz) the impedance (Z) to current flow through the body is directly related to the length (L) of the conductor (height) and inversely related to its cross-sectional area (A): $Z = \rho(L/A)$**, where ρ is the **specific resistivity** of the body's tissues and is assumed to be constant. To express this relationship in terms of Z and the body's volume, instead of its cross-sectional area, the equation is multiplied by L/L: $Z = \rho(L/A)(L/L)$. $A \times L$ is equal to volume (V), so rearranging this equation yields $V = \rho L^2/Z$. Thus, the volume of the FFM or TBW of the body is directly related to L^2, or height squared (HT^2), and indirectly related to Z.

However, the application of this equation may not be perfect because of the complex geometric

shape of the body (Van Loan 1990). Also, specific resistivity (ρ) is not constant and has been shown to vary among body segments because of differences in tissue composition, hydration levels, and electrolyte concentration (Kushner 1992). Chumlea et al. (1988) reported that the specific resistivity of the trunk is two to three times greater than that of the extremities. Also, the specific resistivity of the arms and legs is greater in adults compared to children and in those who are obese compared to those with normal weight (Chumlea et al. 1988; Fuller & Elia 1989).

• **The human body is shaped like a perfect cylinder with a uniform length and cross-sectional area.** The traditional, whole-body, tetrapolar BIA model relies on this assumption, but it is not entirely true. As illustrated in figure 6.1, the human body more closely resembles five cylinders (two arms, two legs, and trunk, excluding the head), connected in a series, instead of one large, perfect cylinder (Kushner 1992). Because the body seg-

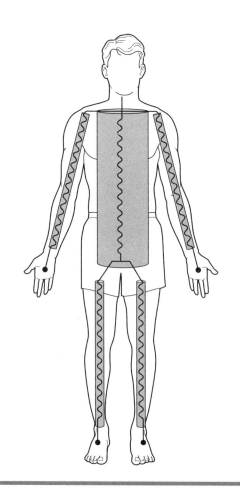

FIGURE 6.1 Five-cylinder model connected in electrical series.

Reprinted, by permission from R.F. Kushner, 1992, "Bioelectrical impedance analysis: A review of principles and applications," *Journal of the American College of Nutrition* 11: 201.

ments are not uniform in length or cross-sectional area, resistance to the flow of current through these body segments will differ. The segmental BIA approach addresses this assumption by summing the segments to obtain total body volume. Thus, instead of $V = \rho L^2/Z$, the segmental equation is $V = 2(\rho L^2/Z)_{leg} + 2(\rho L^2/Z)_{arm} + (\rho L^2/Z)_{trunk}$.

Principles

- **Biological tissues act as conductors or insulators, and the flow of current through the body will follow the path of least resistance.** The FFM contains large amounts of water (~73%) and electrolytes, making it a better conductor of electrical current than fat, which is anhydrous and a poor electrical conductor. At low frequencies (~1 kHz), the current passes through the extracellular water (ECW) only; at higher frequencies (500-800 kHz), it penetrates cell membranes and passes through the intracellular water (ICW), as well as the ECW (Lukaski 1987). The traditional single-frequency BIA method uses a low-level excitation current (500-800 μA) at 50 kHz to measure total body impedance. At a frequency of 50 kHz, BIA is really a measure of ECW, rather than TBW or FFM, but ECW is highly correlated with TBW and FFM in healthy individuals (Schoeller 2000). Thus, TBW and FFM are often estimated from this method.

- **Impedance is a function of resistance and reactance. Resistance (R)** is a measure of pure opposition to current flow through the body, while **reactance (Xc)** is the opposition to current flow caused by **capacitance** (voltage storage) produced by the cell membrane (Kushner 1992). In the traditional BIA model, it is assumed that resistors and capacitors act in a series; that is, $Z^2 = R^2 + Xc^2$ or $Z = \sqrt{(R^2 + Xc^2)}$. Since R is much larger than Xc (at a 50-kHz frequency) during measurement of whole-body impedance, R is a better predictor of FFM and TBW than Z (Lohman 1989b). For these reasons, the **resistance index (HT²/R)**, instead of HT²/Z, is used in many BIA models to predict FFM or TBW (Lohman 1989b). However, in a parallel BIA model (i.e., $Z^{-2} = R^{-2} + Xc^{-2}$), Xc, R, and Z purportedly reflect ICW, ECW, and TBW, respectively (Lukaski 1996).

BIA Models and Approaches

The traditional BIA method involves measurement of whole-body resistance using a tetrapolar, wrist-to-ankle electrode configuration at a single frequency for the purpose of estimating TBW or FFM. However, technological advances and changes in theoretical modeling have led to a number of variations in the traditional BIA method. These varia-

tions use sophisticated models to assess segmental body composition and fluid subcompartments, thereby improving the clinical usefulness of BIA. Also, user-friendly BIA analyzers, designed for home use and individual monitoring of health and fitness, use upper-body or lower-body impedance measures to estimate body composition. This section describes prediction models for the traditional BIA method, variations of the traditional model, and **upper- and lower-body BIA methods**.

Traditional BIA Prediction Model

The assumptions and principles of the traditional BIA method, also known as the electrophysical model, were stated earlier in this chapter (see pp. 88-89). This model assumes that there is only one conducting path and that the body consists of a series of resistors. An electrical current, injected at a single frequency of 50 kHz, is used to measure whole-body impedance (i.e., wrist-to-ankle) for the purpose of estimating TBW and FFM. This whole-body, single-frequency series model is equal or superior to alternative BIA methods (e.g., segmental and dual- or multifrequency BIA) for predicting TBW in healthy clients who have a normal ratio of ECW to ICW (Gudivaka et al. 1999; Kyle & Pichard 2000; Organ, Bradham et al. 1994; Simpson et al. 2001; Wotton, Thomas et al. 2000). One should not use the traditional approach, however, to estimate TBW or fluid compartments (i.e., ECW and ICW) of clients who have an altered hydration status (e.g., in some clinical populations).

Whole-body bioimpedance measures (Z, R, and Xc) are used in BIA prediction equations to estimate TBW and FFM. These prediction equations are based on either population-specific or generalized models (see chapter 2). The population-specific equations are valid for and should be applied only to individuals whose physical characteristics match the sample from which the equation was derived. Researchers have developed equations that are age-specific (Deurenberg, van der Kooy et al. 1990; Lohman 1992), ethnic-specific (Rising et al. 1991; Stolarczyk et al. 1994), fatness-specific (Gray et al. 1989; Segal et al. 1988), and physical activity level-specific (Houtkooper et al. 1989). Alternatively, generalized BIA equations have been developed for heterogeneous populations varying in age, gender, and body fatness (Deurenberg, van der Kooy et al. 1990; Gray et al. 1989; Kushner & Schoeller 1986; Kyle, Genton, Karsegard et al. 2001; Lukaski & Bolonchuk 1988; Van Loan & Mayclin 1987).

Regardless of the approach (population specific or generalized), the predictive accuracy of BIA equations typically is improved by the inclusion of body weight, along with HT² and R, in the BIA regression model. The human body is not a perfect cylinder with a uniform cross-sectional area, and the specific resistivity of tissues is not constant. Thus, including body weight in the equation may be one way of accounting for the complex geometric shape of the body, as well as individual differences in trunk size (Kushner 1992). Generally, the resistance index, HT^2/R, is a relatively stronger predictor of FFM than body weight is (Lohman 1992). Most BIA equations use either HT^2/R or HT^2 and R separately as predictors of FFM.

Although Xc is not typically included as a predictor in most BIA equations, it is moderately related to relative body fatness. Xc may reflect changes in fluid distribution and hydration of adipose tissue and FFM associated with increased fatness (Baumgartner et al. 1988). Several researchers have noted that Xc accounted for a significant proportion of the variance in FFM and have included Xc in their prediction equations (Kyle, Genton, Karsegard et al. 2001; Stolarczyk et al. 1994). Kyle, Genton, Karsegard et al. (2001) commented that Xc was a better indicator of decreases in FFM with aging than age itself.

The predictive accuracy of the BIA method for estimating FFM may be improved when a multicomponent model is used as the reference measure in the derivation of equations, particularly for population subgroups whose FFB composition differs from the assumed values of the 2-C model (e.g., American Indians, those who are elderly, and children). Several researchers have developed BIA equations based on multicomponent reference measures (Baumgartner et al. 1991; Guo et al. 1989; Houtkooper et al. 1989; Lohman 1992; Stolarczyk et al. 1994; Van Loan et al. 1990; Williams et al. 1995).

Evans, Arngrimsson, and coworkers (2001) reported excellent agreement between TBW estimates from BIA and deuterium oxide (D_2O) dilution. Because BIA is a much faster and easier method than the reference standard of isotope dilution (see chapter 3), it is tempting to use BIA instead of the dilution method to estimate TBW for multicomponent models. Using TBW estimates from BIA and D_2O in a 4-C model resulted in similar mean %BF values, but the individual differences between $\%BF_{4C-D2O}$ and $\%BF_{4C-BIA}$ were large (−5.6 to 5.5% BF) (Evans et al. 2001). Also, compared to 2-C model estimates of %BF from densitometry, use of the $4-C_{BIA}$ model did not improve the accuracy of estimating %BF; thus, even though BIA may provide a good estimate of TBW, we do not recommend using BIA as a substitute for isotope dilution in multicomponent models.

ations in the Traditional BIA Model

are several variations to the whole-body, -frequency, serial BIA model. This section describes some of the more common variations and their uses.

Parallel Model

As mentioned previously, the traditional BIA model assumes that the body's resistors are arranged in series. Furthermore, impedance instruments are configured such that a resistor and a capacitor are wired in series; thus, the Z, R, or Xc displayed by BIA analyzers is based on this series model. However, in the human body, resistors and capacitors are oriented both in series and in parallel (Lukaski 1996). It has been suggested that the parallel model is more consistent with human physiology (Ellis et al. 1999; Gudivaka et al. 1999). The **parallel model** ($Z^{-2} = R^{-2} + Xc^{-2}$) is simply the reciprocal of the **series model** ($Z^2 = R^2 + Xc^2$). Lukaski (1996) gives the following formulas for transforming series R and Xc values obtained from the BIA analyzer into parallel model equivalents:

$$R_p = R_s + [(Xc_s)^2/(R_s)]$$
$$Xc_p = Xc_s + [(R_s)^2/(Xc_s)]$$

where the subscript letters p and s indicate parallel and series, respectively.

The parallel model is thought to be most useful in estimating ICW or body cell mass (BCM) (Ellis et al. 1999). This model is preferred especially for measuring patients who are malnourished or have a fluid imbalance (Kotler et al. 1996; Talluri et al. 1999). Also, Gudivaka et al. (1999) reported that the parallel model was accurate for measuring changes in ICW.

Segmental BIA Model

Body parts with small cross-sectional areas have the greatest effect on impedance. For example, the forearm, which accounts for only slightly more than 1% of body weight, contributes 25% to whole-body impedance (Fuller & Elia 1989). Likewise, although the trunk region represents most of the FFM, it contributes relatively little to whole-body resistance. The theoretical upper limb-trunk-lower limb resistance ratio is 13.8:1: 11.8 (Organ, Bradham et al. 1994). Thus whole-body BIA is relatively insensitive to changes in the trunk region. For this reason, **segmental BIA (SBIA)**, which involves measuring the resistance for each segment (i.e., upper limb, trunk, and lower limb) separately, may be more useful in patients with altered fluid distribution.

Organ, Bradham et al. (1994) described the theory of SBIA and the electrode placement for this method. The procedures for SBIA have since been simplified and standardized (Cornish et al. 1999). The SBIA approach offers no clear advantage over the traditional whole-body model for estimating TBW in healthy clients (Organ, Bradham et al. 1994; Wotton, Thomas et al. 2000). However, research suggests that SBIA is the preferred approach to assess regional fluid changes and to monitor ECW in patients with abnormal fluid distribution such as those undergoing hemodialysis (Song et al. 1999; Zhu et al. 1998, 1999, 2000).

Multifrequency BIA

Single-frequency BIA devices operate at a frequency of 50 kHz. At this frequency, BIA is primarily reflecting the ECW compartment, but some of the current also passes through the cells. Given the principle that low-frequency (~1 kHz) current does not penetrate the cells and that complete penetration occurs only at a very high frequency (~1 mHz), **multifrequency BIA (MFBIA)** or **bioelectrical impedance spectroscopy (BIS)** devices are designed to scan a wide range of frequencies. Theoretically, this approach provides estimates of ECW, ICW, and TBW.

With multiple frequencies there are many data points, making MFBIA data difficult to comprehend. Thomas et al. (1998) summarized six approaches or models used to analyze MFBIA data. The **Cole model** is a parallel model that uses resistance values at zero and infinite frequencies. This model is recommended for analyzing MFBIA data (Cornish & Ward 1998; Gudivaka et al. 1999; Schoeller 2000) and for assessing TBW, ICW, and ECW of patients with abnormal fluid distribution. It was also reported to be the best model for predicting changes in ECW, ICW, and TBW (Gudivaka et al. 1999).

Lower-Body and Upper-Body BIA Analyzers

Inexpensive lower-body (i.e., foot-to-foot) and upper-body (i.e., hand-to-hand) BIA devices are

available and have been marketed for home use (see figure 6.2). The Tanita analyzers measure lower-body impedance between the right and left legs as the individual stands on the analyzer's electrode plates. The handheld Omron Body Logic® analyzer measures upper-body impedance between the right and left arms (see figure 6.2). These analyzers provide estimates of %BF and FFM using proprietary equations developed by the manufacturers. Typically, it is not possible to obtain impedance (i.e., resistance and reactance) data from these analyzers. However, they do provide the general public with a low-cost, simple, and reasonably accurate means of self-assessing body fat.

Lower-Body Analyzers

The development of foot-to-foot BIA analyzers for home use began in 1992, and the Tanita Corporation now markets about 20 different models of lower-body analyzers that vary in weight capacity, software and memory, and data output. With this method, the client stands barefoot on a footpad similar to a bathroom scale, and lower-body impedance is measured with pressure contact electrodes. Each footpad has two electrode plates—one for the ball of the foot and the other for the heel. The current is applied by the anterior electrodes (balls of feet), and the impedance is measured by the posterior electrodes (heels). Research shows good agreement between impedance values measured by pressure contact and traditional gel electrodes (Nuñez et al. 1997; Tan et al. 1997).

Using a multicomponent model to compare the group and individual predictive accuracy of foot-to-foot (Tanita-305) and whole-body (Bodystat-1500) BIA analyzers, Jebb and colleagues (2000) reported similar CEs for the Tanita (0.9% BF) and Bodystat (−1.5% BF) analyzers. However, the individual predictive accuracy of the Tanita (±10.2% BF) was slightly greater than that of the Bodystat (±9.3% BF).

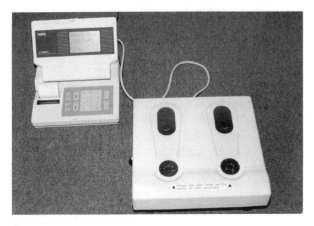

a

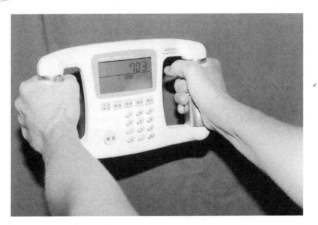

b

FIGURE 6.2 *(a)* Lower-body and *(b)* upper-body bioimpedance analyzers.

Compared to 2-C model estimates of FFM obtained from underwater weighing, the average FFM of heterogeneous samples of adults is estimated reasonably well using Tanita analyzers (SEE = 3.5-3.7 kg) (Cable et al. 2001; Utter et al. 1999). There has also been good agreement with SKF estimates of %BF in collegiate wrestlers (Utter et al. 2001) and with dual-energy X-ray absorptiometry model estimates of FFM in children (Sung et al. 2001; Tyrrell et al. 2001). Tyrrell et al. (2001) derived a FFM prediction equation for children using foot-to-foot impedance measures obtained from a Tanita analyzer. Also, the lower-body BIA method is similar in accuracy to the hydrodensitometry (HD) method for assessing changes in %BF over time (Utter et al. 1999).

Upper-Body Analyzers

In the late 1990s, Omron Healthcare developed a low-cost hand-to-hand BIA analyzer for home use. The plate electrodes are in the handles of the device, and the client simply holds it in both hands with the arms outstretched parallel to the ground while standing erect. The manufacturer's proprietary equation was developed and cross-validated on a large and diverse, heterogeneous sample from three laboratories using HD to obtain 2-C model reference measures of %BF and FFM (Loy et al. 1998). The group predictive accuracy (SEE) for estimating FFM was 3.9 kg for men and 2.9 kg for women. In an independent cross-validation of the Omron analyzer, Gibson and colleagues (2000) reported slightly smaller prediction errors (SEE = 2.9 kg for men and 2.2 kg for women). Loy et al. (1998) also noted that the average FFM estimates from the Omron device were similar to values obtained using whole-body (RJL and Valhalla) analyzers. Lastly, in a study of Japanese men, the accuracy of upper-body (Omron, HBF-300), lower-body (Tanita, TBF-102), and whole-body (Selco, SIF-891) analyzers was compared to 2-C model reference measures of %BF obtained from HD. The average difference between reference and predicted %BF values was slightly smaller for the Omron (2.2% BF) compared to the whole-body (3.3% BF) and lower-body (3.2% BF) analyzers (Demura, Yamaji, Goshi, Kobayashi et al. 2002). However, estimation errors tended to be greater at the lower and upper extremes of the %BF distribution using the Omron and Tanita devices.

Recently, Omron developed BIA prediction equations to estimate the body composition of physically active adults. These equations are programmed in the new Omron analyzer (model HBF-306), along with prediction equations for nonactive adults and children. The predictor variables in the manufacturer's equation for this unit are upper-body impedance, age, gender, height, weight, and physical activity level (i.e., "athlete" or "nonathlete"). The prediction errors for athletes (SEE = 3.8% BF and 3.6% BF for male and female athletes, respectively) were somewhat less than that for nonathletes (SEE = 4.5% BF) (Yamanoto 2002).

This device has been tested on ethnically diverse samples of European and Asian populations. Generally the group predictive accuracy is good for these population subgroups, but individual prediction errors can be high (Deurenberg et al. 2001; Deurenberg & Deurenberg-Yap 2002). Deurenberg et al. (2001) noted that 24% of the obese females and 44% of the obese males in their study would have been misclassified (false negatives) based on the Omron data. Compared to that for a 4-C model, the SEE was 4.5% BF; the error in estimating %BF using this analyzer was also related to the age, level of body fatness, and arm span-to-height ratio of the subjects (Deurenberg & Deurenberg-Yap 2002).

Other manufacturers have developed hand-to-hand BIA analyzers, such as the Body Comp Scale by American Weights and Measures (Rancho Santa Fe, CA) and the palm-sized SlimStep® by Ya-Man (San Jose, CA). However, to our knowledge, there is no research assessing the predictive accuracy and validity of these devices.

Using the BIA Method

The whole-body tetrapolar method uses four electrodes applied to the hand, wrist, foot, and ankle (figure 6.3). An excitation current (200-800 μA) at 50 kHz is applied at the source or drive (distal) electrodes on the hand and foot, and the voltage drop due to impedance is detected by the sensor (proximal) electrodes on the wrist and ankle.

Commonly used population-specific and generalized BIA equations are presented in later chapters and are summarized in appendixes B.1 through B.4. To use these equations, you obtain R and Xc directly from your BIA analyzer. You will estimate the %BF of your client by determining the fat mass (FM = BW − FFM) and dividing FM by the client's body weight [%BF = (FM/BW) × 100].

At this time, we do not recommend using the FFM and %BF estimates obtained directly from your BIA analyzer unless you (a) know which equations are programmed in the analyzer's computer, (b) obtain information from the manufacturer regarding the validity and accuracy of these equations, and (c) determine if these equations are generalizable and applicable to your clients.

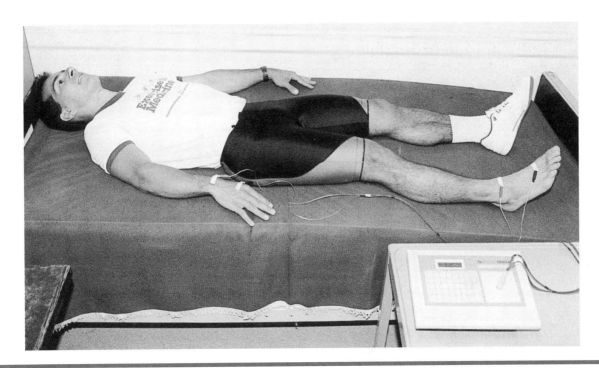

FIGURE 6.3 Tetrapolar electrode placement and client positioning for whole-body BIA.

Table 6.1   Bioelectrical Impedance Analysis Approaches and Uses

Approach	Model	Recommended use
Traditional BIA (whole body, tetrapolar, single frequency)	Series	To estimate TBW and FFM in healthy clients with normal hydration status and normal fluid distribution
	Parallel	To estimate ICW and BCM
Segmental BIA	Series	To measure fluid distribution or regional fluid accumulation in clinical populations
	Parallel	To measure regional or segmental ICW
Multifrequency BIA	Cole	To estimate ECW, ICW, and TBW; to monitor changes in the ECW/BCM and ECW/TBW ratios in clinical populations
Upper-body (hand-to-hand) BIA Lower-body (leg-to-leg) BIA	NA	To estimate %BF in healthy clients with normal hydration status and normal fluid distribution

Key: BCM = body cell mass; FFM = fat-free mass; ECW = extracellular water; ICW = intracellular water; TBW = total body water; NA = not applicable (these analyzers are based on series model but do not provide impedance or resistance data).

As mentioned previously, for healthy clients with normal fluid balance, the traditional whole-body, single-frequency tetrapolar BIA method is a valid approach for estimating TBW. In certain clinical populations with abnormal fluid distribution, it is preferable to use a different BIA approach to assess ECW, ICW, and BCM. Table 6.1 summarizes the different BIA approaches and their recommended uses.

BIA Technique

The accuracy of the BIA method is highly dependent on the control of factors that may increase the measurement error (Gonzalez et al. 1999, 2002). Your client must adhere to the "BIA Pretesting Client Guidelines" (page 94), which are designed to control for fluctuations in hydration status.

BIA Pretesting Client Guidelines

- No eating or drinking within 4 hr of the test
- No exercise within 12 hr of the test
- Client should urinate within 30 min of the test
- No alcohol consumption within 48 hr of the test
- No diuretic medications within seven days of the test
- No testing of female clients who perceive they are retaining water during that stage of their menstrual cycle

Use the following standardized testing procedures to minimize error when using the BIA method:

Standardized Procedures for the Whole-Body BIA Method

- Take bioimpedance measures on the right side of the body with the client lying supine on a nonconductive surface in a room with normal ambient temperature (~25° C).
- Clean the skin at the electrode sites with an alcohol pad.
- Place the sensor (proximal) electrodes (figure 6.4) on (a) the dorsal surface of the wrist so that the upper border of the electrode bisects the styloid processes of the ulna and radius, and (b) the dorsal surface of the ankle so that the upper border of the electrode bisects the medial and lateral malleoli. You can use a measuring tape and surgical marking pen to mark these points for electrode placement.
- Place the source (distal) electrodes at the base of the second or third metacarpal-phalangeal joints of the hand and foot (figure 6.4). Make certain that the distance between the proximal and distal electrodes is at least 5 cm.
- Attach the lead wires to the appropriate electrodes. Red leads are attached to the wrist and ankle, and black leads are attached to the hand and foot.
- Make certain that the client's legs and arms are comfortably abducted (NIH [1994] recommendation is a 30-45° angle from the trunk). There should be no contact between the thighs or between the arms and the trunk as this will "short-circuit" the electrical path, dramatically ⁻ting the impedance value.

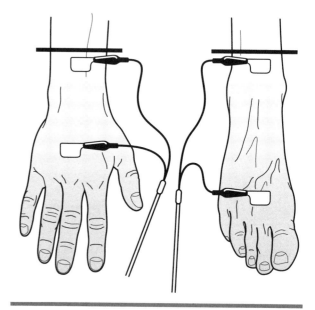

FIGURE 6.4 Proximal and distal electrode placement for whole-body BIA.

Organ and colleagues developed a protocol for the SBIA method using a six-electrode configuration (Organ, Bradham et al. 1994). This technique was later simplified by removal of some of the additional electrodes, as the contralateral wrist and ankle were found to be the optimum electrode sites for measuring resistance of the arm-trunk and leg-trunk segments (Cornish et al. 1999). Standardized electrode placement for SBIA method is described elsewhere (see Cornish et al. 1999).

Sources of Measurement Error

Reviews of BIA studies have reported good reliability for this method (Kushner 1992; Kushner et al. 1996; Van Loan 1990). The reliability (CV) for multiple resistance measurements is generally about 1% to 2% for same-day assessments and 2% to 3.5% for multiple-day measures. Vehrs et al. (1998) concluded that the within-day, between-day, and intermachine reliabilities of resistance measurements were equal (r > .97). The within-day (CV = 0.4-1.5%) and between-day (CV = 1.0-3.6%) variability of the foot-to-foot BIA analyzers is similar to that of the whole-body devices (Nuñez et al. 1997), and test-retest reliability was similar (r = .99) for whole-body, upper-body, and lower-body analyzers (Demura, Yamaji, Goshi, Kobayashi et al. 2002).

The accuracy and precision of the BIA method are affected by instrumentation, client factors, technician skill, environmental factors, and the prediction equation used to estimate FFM (Kushner 1992; Lohman 1989b; Van Loan 1990). The theoretical error is estimated to be ~1.8 kg if the reference method (e.g., HD) is error free (Lohman 1992). Unfortunately, this is not the case; so part (approximately 20-50%) of the total prediction error associated with the BIA method and the equations can be attributed to error in the reference method. Therefore, an SEE of ≤3.5 kg for men and ≤2.8 kg for women is acceptable (see table 2.1, p. 20).

Instrumentation

There are a number of manufacturers of BIA analyzers. The NIH panel of BIA experts emphasized the need for standardization of equipment and recommended that all BIA instruments should report resistance, reactance, and prediction equations, as well as calculated body composition values (Ellis et al. 1999). The panel also noted that raw data such as source current, frequency, range, and accuracy should be made available. This section addresses questions you may have regarding BIA analyzers.

- **Can different types of whole-body BIA analyzers be used interchangeably?** Research demonstrates significant differences in whole-body resistance when different brands of single-frequency analyzers are used (Graves et al. 1989; Oldham 1996; Smye et al. 1993). For example, Smye et al. (1993) reported lower resistance values (6% or 32-36 Ω) for the Holtain device compared to the Bodystat, RJL, and EZ Comp analyzers. Graves et al. (1989) noted that the correlation between resistance values measured with the Valhalla and Bioelectrical Sciences (BES) analyzers was only r = .59; the average %BF estimated for men from one BIA equation when using the resistance from these two instruments differed by 6.3% BF.

Although there is a high correlation (r = .99) between resistance values measured with the Valhalla and RJL analyzers, the Valhalla analyzer produced significantly higher resistances than the RJL analyzer for men (~16 Ω) and women (~19 Ω), corresponding to a systematic underestimation of FFM in men (~1.3 kg) and women (~1.0 kg) (Graves et al. 1989). Also, differences may exist among the same model of BIA instrument. The Z values from three RJL (model 101) analyzers differed by 7 to 16 Ω, which corresponded to a difference in FFM of 2.1 kg for some individuals (Deurenberg, van der Kooy et al. 1989).

- **Can MFBIA analyzers be used interchangeably?** The interinstrument and interoperator error for measuring impedance among three devices (SEAC SFB2 and MFBIA) was reported to be less than 1% (Ward et al. 1997). However, significant differences in impedance values between two different makes of MFBIA analyzers (SEAC SFB3 and Xitron 4000B) were found during testing on intensive care patients (Bolton et al. 1998).

Instrumentation is a substantial source of measurement error and one limitation of the BIA method. To control for this error, you should use the same instrument when monitoring changes in your client's body composition. Additionally, it is important to calibrate the BIA analyzer prior to testing and to follow the manufacturer's instructions for its calibration.

Client Factors

How Does My Client's Hydration Level Affect the Accuracy of Bioimpedance Measures?

A major source of error with the BIA method is intraindividual variability in whole-body resistance due to factors that alter the client's state of hydration. Between 3.1% and 3.9% of the variance in resistance may be attributed to day-to-day fluctuations in body water (Jackson et al. 1988). Factors such as eating, drinking, dehydrating, and exercising alter the individual's hydration state, thereby affecting total body resistance and the estimate of FFM (table 6.2). Taking resistance measures 2 to 4 hr after a meal decreases R (Deurenberg et al. 1988). Likewise, Gallagher and colleagues (1998) found a significant decrease in impedance 2 hr after consumption of breakfast, and this effect lasted for 5 hr postconsumption. In contrast to these studies, others found greater individual variability and smaller changes in R when taking a measurement at only 1 hr postmeal (Fogelholm et

Table 6.2   Summary of Factors Affecting Bioelectrical Impedance Measures

Factor	Effect on resistance (Ω)	Effect on fat-free mass (kg)	Reference
Type of analyzer (Vahalla vs. RJL)	↑16-18[a]	↓1.0-1.3	Graves et al. 1989
Eating or drinking within 4 hr	↓13-17	↑1.5	Deurenberg et al. 1988
Dehydration	↑40	↓5.0	Lukaski 1986
Aerobic exercise (low intensity)	NC	NC	Deurenberg et al. 1988
Aerobic exercise (moderate-high intensity)	↓50-70	↑12.0	Khaled et al. 1988; Lukaski 1986
Menstrual cycle (follicular vs. premenstrual)	↓5-8[b]	NC	Gleichauf & Rose 1989
Menstrual cycle (menses vs. follicular)	↑7[c]	NC	
Electrode placement	↑10	NC	Elsen et al. 1987
	↑70	↓11	Lukaski 1986
Electrode configuration (ipsilateral vs. contralateral)	NC	NC	Lukaski et al. 1985
Electrode configuration (right side vs. left side)	NC	NC	Graves et al. 1989
Room temperature (14° C vs. 35° C)	↑35[d]	↓2.2	Caton et al. 1988

[a]Valhalla > RJL; [b]Follicular > premenstrual; [c]Menses > follicular; [d]14° C > 35° C.
NC = no change.

al. 1993; Rising et al. 1991). Kushner et al. (1996) concluded that eating or drinking has a minimal influence on whole-body Z during the first hour but is likely to cause a decrease in Z (<3%) at 2 to 4 hr postconsumption.

How Does Exercise Affect Bioimpedance Measures?

Kushner et al. (1996) suggested three ways in which exercise may influence BIA measurements:

- Increased blood flow and warming of skeletal muscle tissue reduce Z and the specific resistivity (ρ) of muscle.
- Increased cutaneous blood flow, skin temperature, and sweating can also lower Z values.
- Fluid loss due to exercise, however, increases Z.

The effect of aerobic exercise on resistance measurements is partially dependent on exercise intensity and duration. Jogging and cycling at moderate intensities (~70% $\dot{V}O_2$max) for 90 to 120 min produced substantial decreases in R, resulting in a large overestimation of FFM (Khaled et al. 1988; Lukaski 1986). In contrast, cycling at relatively lower intensities (100 and 175 W) for 90 min had a much smaller effect (1-9 Ω) (Deurenberg

et al. 1988). Liang and Norris (1993) reported a decreased R of about 3% immediately after 30 min of moderate-intensity exercise, but R returned to normal 1 hr postexercise with water ad libitum. The decrease in R after strenuous exercise most likely reflects the relatively greater loss of body water in the sweat and expired air as compared to the loss of electrolytes. This leads to a higher electrolyte concentration in the body's fluids, thereby lowering R values (Deurenberg et al. 1988).

Another study indicated that the BIA method adequately predicted changes in TBW after heat-induced dehydration and glycerol-induced hyperhydration but not following exercise-induced dehydration; thus, factors other than just total fluid volume affect BIA measures following exercise (Koulmann et al. 2000). The researchers hypothesized that the redistribution of body fluid volumes to active muscles during exercise, inducing a relative increase of hydration in these segments (legs), might partially conceal the decrease of fluid volumes in the other, less active segments (trunk and arms).

Can I Take Bioimpedance Measurements at Any Time During My Client's Menstrual Cycle?

Although the menstrual cycle alters ICW, TBW, the ratio of ECW to ICW, and body weight (Mitchell et al. 1993), there are only small changes in bioimpedance measures (Z and R) between the follicular and premenstrual stages and between menses and the follicular stage (Deurenberg et al. 1988; Gleichauf & Rose 1989). However, the average body weight of the women studied was stable (less than 0.2 kg) during the menstrual cycle. In women experiencing relatively large body weight gains (2-4 kg) during the menstrual cycle, a large part of the weight gain is due to an increase (1.5 kg on average) in TBW (Bunt et al. 1989).

Until we have more conclusive data dealing with this issue, we recommend taking BIA measurements at a time during the menstrual cycle when the client perceives that she is not experiencing a large weight gain. This practice should minimize error and yield a more accurate estimate of FFM for your clients.

Technician Skill

• **Is there high agreement between bioimpedance values measured by two different technicians?** Technician skill is not a major source of measurement error. There is virtually no difference in R measurements taken by different technicians, provided that standardized procedures for electrode placement and client positioning are closely followed (Jackson et al. 1988). The proximal sensor electrodes, in particular, need to be correctly positioned at the wrist and ankle. A 1-cm displacement of the sensor electrodes may result in a 2% error in R (Elsen et al. 1987). Lukaski (1986) reported a 16% increase in R (~79 Ω) due to improper electrode placement. Also, technicians must take care when removing and replacing electrodes. Segmental Z values may vary as much as 3.6% with replacement of just one electrode (Lozano et al. 1995).

• **How does body position affect bioimpedance measures?** Deviations in body position from the standard supine, abducted position alter Z values as much as 12% (Lozano et al. 1995); moving from a standing to a supine position results in an immediate increase in Z of about 3% because of fluid shifts (Kushner et al. 1996). Also, it is necessary to standardize the amount of time that a client has been supine before Z is recorded; Z gradually increases for several hours when a person is in the supine position (Kushner et al. 1996). Experts recommend having

your client lie supine for at least 10 min before you take BIA measurements (Ellis et al. 1999).

• **Should I measure whole-body bioimpedance on the right or left side of the body?** The differences between R measurements using ipsilateral (right arm-right leg or left arm-left leg) and contralateral (right arm-left leg or left arm-right leg) electrode placements are generally small (Graves et al. 1989; Lukaski et al. 1985; Organ, Bradham et al. 1994).

Environmental Factors

How does temperature affect bioimpedance values? Researchers have demonstrated that ambient temperature affects skin temperature, and in turn, R varies inversely with skin temperature (Caton et al. 1988; Gudivaka et al. 1996; Liang et al. 2000). Cool ambient temperatures cause a drop in skin temperature (24° C compared to 33° C under normal conditions), resulting in a significant increase in total body R and a decrease in estimated FFM (Caton et al. 1988). Liang et al. (2000) reported a slightly greater difference in R (46 Ω) between cold (17° C ambient temperature and 28.7° C skin temperature) and hot (35° C ambient and 35.8° C skin temperature) conditions. The inverse relationship between impedance and skin temperature occurs across all frequencies (Gudivaka et al. 1996). Gudivaka et al. (1996) noted that the error in predicted TBW is <1% when the ambient temperature is between 22.3° and 27.7° C. We recommend a room temperature of about 25° C (77° F) for BIA testing.

Factors Affecting BIA Prediction Equation Selection

BIA prediction equations should be selected based on the client's age, gender, ethnicity, physical activity level, and level of body fatness. Use of inappropriate equations can lead to systematic prediction errors in estimating FFM. This is a major potential source of error for the BIA method. The BIA equations recommended in later chapters were carefully selected based on research substantiating their applicability and generalizability to specific population subgroups.

Key Points

- BIA is a rapid, noninvasive, and nonintrusive method for measuring body composition.
- Impedance to current flow through the body is directly related to the square of the individual's height and indirectly related to cross-sectional area.
- FFM, with its water and electrolytes, is a good conductor of electrical current; fat is a poor conductor. Therefore, BIA equations can be used to estimate FFM and TBW.
- The traditional BIA approach includes a tetrapolar (wrist-to-ankle) electrode configuration, serial model at a single fixed frequency. This approach is recommended for estimating TBW in healthy clients with normal fluid distribution.

- The parallel model is simply the reciprocal of the serial model and is recommended for estimating ICW and BCM.
- SBIA is useful for patients with altered fluid distribution.
- MFBIA provides a way of estimating all fluid compartments (TBW, ICW, and ECW) and monitoring the changes in these compartments.
- Low-cost lower-body and upper-body BIA analyzers designed for home use provide reasonable group estimates of %BF, but the individual prediction error for these analyzers is greater than for whole-body BIA analyzers.
- The BIA method may be more suitable than the SKF method for measuring body composition of clients who are obese.
- Population-specific BIA equations are applicable only to individuals from a specific group, for example, children, persons who are elderly, and those who are obese.
- Eating, drinking, dehydration, exercise, and menstrual cycle stage may affect bioimpedance measures. Therefore, it is necessary to follow client pretesting guidelines to ensure accuracy when using the BIA method.
- The same analyzer should be used for monitoring change in body composition over time because there is no standardization of BIA analyzers and different analyzers may not produce the same resistance measure.
- Although technician skill is not a major source of error with the BIA method, the clinician must pay attention to correct electrode placement and client positioning for an accurate measurement.

Key Terms

Learn the definition for each of the following key terms. Definitions of terms can be found in the Glossary, page 227.

bioelectrical impedance analysis (BIA)

bioelectrical impedance spectroscopy (BIS)

capacitance

Cole model

impedance (Z)

lower-body BIA method

multifrequency BIA (MFBIA)

parallel model

reactance (Xc)

resistance (R)

resistance index (HT²/R)

segmental BIA (SBIA)

series model

specific resistivity (ρ)

upper-body BIA method

whole-body BIA method

Review Questions

1. How does the predictive accuracy of the BIA method compare to that of the SKF method, and why may the BIA method be preferable in some situations?
2. Describe the underlying principles and basic assumptions of the BIA method. How is this method used to estimate FFM?
3. What variables are typically used in BIA prediction equations for estimating TBW and FFM?
4. Describe how the parallel, segmental, and MFBIA approaches differ from the traditional whole-body, single-frequency serial model. What are the uses for each approach?
5. Compare and contrast the "home-use" BIA devices with traditional tetrapolar BIA.
6. Briefly describe the standardized procedures for measuring whole-body impedance.
7. What is the precision of the BIA method? Identify the major sources of measurement error for the BIA method.
8. What steps should be taken to control the hydration status of your client prior to BIA measurement?

NEAR-INFRARED INTERACTANCE METHOD

Key Questions

- What are the basic assumptions and principles of the near-infrared interactance method?
- What variables are included in near-infrared interactance prediction models?
- What is the procedure for taking measurements with Futrex analyzers, and how do you obtain optical density values from these devices?
- What are the major sources of measurement error for the near-infrared interactance method?
- Can optical density be used to accurately estimate body composition?
- Are the manufacturer's Futrex equations valid for estimating body fat percentage?

Near-infrared spectroscopy has been used since 1968 to measure the protein, fat, and water content of agricultural products (Norris 1983). Conway and colleagues (1984) applied this technology to study human body composition using a high-precision (6-nm), expensive, computerized spectrophotometer. Scanning at wavelengths from 700 nm to 1100 nm, they developed a calibration (prediction) equation that had good agreement with %BF derived from deuterium oxide dilution (D_2O). However, this equation was developed using 36 subjects and was cross-validated on only 17 subjects.

Using computer simulation to predict the accuracy of a low-cost, portable, wide-slit spectrophotometer, Conway and Norris (1987) suggested that a lesser-precision (50 nm instead of 6 nm) instrument, suitable for fieldwork, could be designed. However, the prediction accuracy of the **near-infrared interactance (NIR)** method, which estimates %BF from the reflectance of near-infrared light off the underlying tissue, would be compromised using this device. Shortly thereafter, a less expensive, commercial NIR analyzer (Futrex-

5000) based on the results of this research was marketed (Conway et al. 1984; Conway & Norris 1987). The Futrex NIR analyzers estimate %BF from optical density measured at only one site (biceps brachii). This is an attractive field method because it is rapid, painless, and easy to perform, and the client does not have to disrobe. However, much skepticism surrounds the use of this instrument and the manufacturer's NIR equations to estimate body fatness. This chapter addresses the basic assumptions and principles of the NIR method, NIR prediction models, testing procedures, and sources of measurement error for the NIR method.

Assumptions and Principles of the NIR Method

The NIR analyzer indirectly measures the tissue composition (fat and water) at various sites. Therefore, certain relationships are assumed and principles of prediction applied with use of the NIR method to estimate total body fatness.

Assumptions

• **The degree of infrared light absorbed and reflected is related both to the composition of the tissues (water, fat, and protein) through which the light is passing and to the specific wavelength of the near-infrared light.** Conway et al. (1984) demonstrated that the peak absorption wavelengths for pure fat and pure water are 930 nm and 970 nm, respectively. The shape of the interactance curve at these two wavelengths is a function of the amount of fat and water present in the sample being measured (figure 7.1). The Futrex-5000 measures the **optical density (OD)**, or the amount of light reflected by the underlying tissues, at two wavelengths, 940 nm (OD1) and 950 nm (OD2) (Futrex 1988). It is not clear why the manufacturer selected 940-nm and 950-nm wavelengths instead of the 930-nm and 970-nm wavelengths that were identified as the absorption peaks for fat and water.

• **The NIR light penetrates the tissues to a depth of up to 4 cm and is reflected off the bone back to the detector.** Therefore, the OD measures both subcutaneous and intramuscular fat. Although the manufacturer makes this claim, there are no research data to substantiate it. In fact, we found that the slope of the relationship between OD and SKF measures at the biceps site was significantly stronger for leaner women (22% BF) compared to

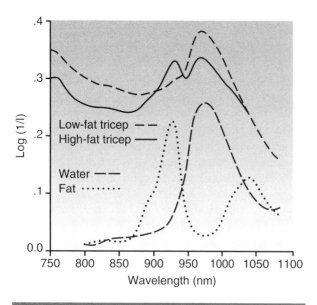

FIGURE 7.1   Near-infrared spectra for pure fat and distilled water and for low-fat and high-fat triceps sites.

Reprinted from J.M. Conway, K.H. Norris, and C.E. Bodwell, 1984, "A new approach for the estimation of body composition: Infrared interactance," *American Journal of Clinical Nutrition* 40:1125. Reproduced with permission by the American Journal of Clinical Nutrition. © Am J Clin Nutr. American Society for Clinical Nutrition.

fatter women (39% BF) (Quatrochi et al. 1992). This finding suggests that deep penetration (up to 4 cm) of the NIR light beam may be disrupted by fat layering and irregularities at the fat-muscle junction, especially in persons who are obese and have large amounts of subcutaneous fat.

Principles

• **There is an inverse and linear relationship between OD measures and subcutaneous fat at the biceps site and total body fatness.** The less NIR light re-emitted (i.e., more absorbed), the greater the amount of subcutaneous fat. The relationship between ODs, subcutaneous fat (SKF thickness), and %BF depends on the site being measured. Research demonstrates a stronger relationship between OD and SKF measures at the biceps site (r = –.66 to –.79) compared to other commonly used SKF sites (r = –.01 to –.48) (Brooke-Wavell et al. 1995; Hortobagyi, Israel, Houmard, McCammon et al. 1992; McLean & Skinner 1992; Quatrochi et al. 1992). Furthermore, the OD at the biceps site is a better predictor of %BF estimated by hydrodensitometry than the ODs measured at other common SKF sites or any combination of these sites (Elia et al. 1990; Heyward, Jenkins et al. 1992; Hortobagyi, Israel, Houmard, McCammon et al. 1992; McLean & Skinner 1992; Quatrochi et al. 1992; Wilmore et al. 1994). These findings are surprising given that the biceps site is not commonly found in SKF equations. However, the correlations are only moderate between ODs and SKFs measured at a given site, demonstrating that the OD measure is affected not only by subcutaneous fat thickness but also by the composition (fat and water content) of other tissues (skin, muscle, and bone). Conway and Norris (1987) postulated that the combination of skin thickness and subcutaneous fat thickness at the biceps allows sufficient penetration of the near-infrared beam. Thus, the OD at the biceps may be a fairly good indicator of total body fatness (Conway & Norris 1987).

• **Gender and age are independent predictors of total Db and %BF.** Data from the first NIR study suggested that the slope and intercept of the regression line depicting the relationship between near-infrared measures and %BF may be different for women and men (Conway et al. 1984). The manufacturer's NIR models include gender as a predictor in the equations. The Futrex-5000/XL model (for adults only) does not include age, but the other NIR models (5000A series and 6100/XL) use age as a predictor. Although one study reported that age accounted for less than 1% of the variance in %BF (Schreiner et al. 1995), other researchers have noted that age is a significant predictor, accounting for an additional 2% to 13% of the variance in total Db (Heyward, Jenkins et al. 1992; Hicks et al. 2000; Hortobagyi, Israel, Houmard, O'Brien et al. 1992). These findings suggested that both age and gender should be included in NIR prediction models, especially if the equations are to be applied to heterogeneous populations of males and females varying greatly in age.

NIR Prediction Models

Table 7.1 summarizes prediction models for various types of Futrex NIR analyzers. The manufacturer's NIR model for the Futrex-5000/XL is a single-site model using ODs at the biceps (OD1 and OD2), body weight, height, gender, and exercise level to predict %BF (Futrex 1988): %BF = C_0 + C_1 (biceps OD2) + C_2 (gender, 0.01 for males and –0.01 for females) + C_3 (body weight in pounds/100) + C_4 (height in inches/100) + C_5 (biceps OD1) + C_6 (exercise level: <15 min/day = 0.00; 15-30 min/day = 0.02; 30-60 min/day = 0.05; and >60 min/day = 0.08). According to the manufacturer, the values of the coefficients, C_0 to C_6, vary among instruments because of small differences in the original optical standard installed in each Futrex-5000 unit (Futrex 1988). As mentioned previously, age is not a predictor in the Futrex-5000/XL model but is included in the Futrex-5000A and 6100 models. Age should be used in NIR models developed for samples varying greatly in age (Heyward, Jenkins et al. 1992; Hicks et al. 2000; Israel et al. 1989).

Researchers have reported that some of the variables in the Futrex-5000 manufacturer's model are not significant predictors of Db and %BF. For example, because there is an extremely strong relationship (r = .93-.99) between OD1 and OD2 at the biceps site and only a small difference between average OD1 and OD2 measurements, only one of these ODs or the average of the two enters the prediction model (Fornetti et al. 1999; Heyward, Jenkins et al. 1992; Hicks et al. 2000; Israel et al. 1989; Oppliger et al. 2000). NIR research consistently shows that body weight and height are significant predictors of Db, FFM, or %BF. However, it is important to point out that biceps OD measures (either OD1 or OD2) explain an additional proportion of the variability in Db, FFM, or %BF (Fornetti et al. 1999; Heyward, Jenkins et al. 1992; Hicks et al. 2000; Hortobagyi, Israel, Houmard, McCammon et al. 1992; Houmard et al. 1991; McLean & Skinner 1992; Oppliger et al. 2000; Vehrs et al. 1998; Wilmore et al. 1994).

There is some controversy over the use of exercise or physical activity level as a predictor variable. Some researchers have suggested that future NIR equations eliminate exercise level as a predictor variable because it is difficult to quantify and has the potential of dramatically affecting the estimate of %BF (Israel et al. 1989; McLean & Skinner 1992; Wilmore et al. 1994). In fact, exercise level is not included as a predictor in the manufacturer's equation for the latest Futrex NIR device (Futrex-6100/XL).

Table 7.1   Futrex Near-Infrared Interactance (NIR) Prediction Models

Model[a]	Calibration and marketed use	Predictors in model equations
1100	For home use; for adults only; track changes	BW, HT, OD
5000/XL	For adults only	BW, HT, OD, gender, exercise level, (frame size)[b]
5000A/ZL	For children (5-17 yr) and adults	BW, HT, OD, gender, exercise level, age
5000A/WL	For high school wrestlers or lean adolescent males (13-18 yr); for adults, children, and high school male athletes	BW, HT, OD, gender, age, athletic status
6100/XL	For adults only	BW, HT, OD[c], gender, age

[a]These are the latest versions of the Futrex NIR. The 1100, 5000/XL, and 6100/XL models take the place of the 1000, 5000, and 6000 models, respectively.

[b]Body frame size is an additional predictor in the Futrex-5000 series when this device is used to estimate total body water.

[c]The 6100/XL measures optical density at six wavelengths: 810, 910, 932, 944, 976, and 1,023 nm; the 5000 series uses NIR wavelengths of 940 and 950 nm.

Key: BW = body weight; HT = height; OD = optical density at the biceps.

Although the sum of SKFs (ΣSKF) from multiple sites is a better indicator of total body fatness than SKF measurements from just one site (Jackson & Pollock 1985), measuring different sites or combining ODs from multiple sites does not appear to improve the prediction of %BF compared to using the biceps site alone (Conway & Norris 1987; Elia et al. 1990; Heyward, Jenkins et al. 1992; Hortobagyi, Israel, Houmard, McCammon et al. 1992; McLean & Skinner 1992; Quatrochi et al. 1992; Williams et al. 1995; Wilmore et al. 1994). Although Oppliger and colleagues (2000) found that the best equation for estimating %BF among high school wrestlers used OD measures from several sites, this equation was only slightly better than the equation using only the biceps site. While OD measures at the biceps appear to be the best predictor of body fat in adolescents and adults, Fuller et al. (2001) noted that the biceps OD2 and SKF measurements at the subscapular site are more highly related to %BF$_{4\text{-}C}$ for children (8-12 yr of age).

No NIR equations have been developed using multicomponent estimates of %BF. Several researchers, however, have used either hydrostatic weighing with a 2-C model or dual X-ray absorptiometry (DXA) as reference methods to develop NIR prediction equations (Fornetti et al. 1999; Heyward, Jenkins et al. 1992; Hicks et al. 2000; Oppliger et al. 2000). The Oppliger et al. (2000) equation estimates %BF, and the Fornetti et al. (1999) equation predicts FFM. The other equations (Heyward, Jenkins et al. 1992; Hicks et al. 2000) estimate Db, allowing you to use the appropriate population-specific conversion formula to estimate %BF from Db (see table 1.4, p. 9).

Using the NIR Method

Numerous researchers have reported unacceptable prediction errors for Futrex model equations; therefore, we do not recommend using the %BF estimates obtained directly from your Futrex analyzer. Instead, you can use OD values and NIR prediction equations developed for various population subgroups (see appendixes B.1-B.4). To obtain OD values from the Futrex analyzers, access the OD measurement mode using the following steps (Futrex 1988).

Accessing the Optical Density Measurement Mode on the Futrex-5000 NIR Analyzer

1. Turn the unit on and let it count down for 15 sec.
2. Enter "Clear 881." The analyzer will display OD1.

3. Place the NIR light wand into the optical standard (Teflon® block, 1 cm thick). Press "Enter" twice. (Your analyzer is programmed by the manufacturer to take two measurements at each site. Therefore, you must press the Enter key twice.)
4. OD1 and OD2 values for the standard will be alternately displayed. Record these values. You can set the number of measurements taken at the site from one to eight by following the manufacturer's instructions (Futrex 1988).
5. Place the NIR light probe perpendicular to the measurement site, and press the Enter key the designated number of times (one to eight). OD1 and OD2 values for your client will be alternately displayed. Record these values.
6. When assessing many clients during one test session, periodically measure and record the OD values for the optical standard (step 3) every 10 minutes. This corrects for electronic drift in your analyzer (Futrex 1988).
7. Calculate ΔOD2 by subtracting the client's OD value from the corresponding standard OD value (e.g., biceps ΔOD2 = OD2 standard – OD2 client).

NIR Technique

Figure 7.2 illustrates the NIR measurement technique. Following the standardized procedures listed next will increase the accuracy and precision of your OD measurements:

Standardized Procedures for Optical Density Measurements

1. Before measuring your client, calibrate your analyzer using the optical standard (Teflon block) supplied by the manufacturer. Record the OD1 and OD2 standard values.
2. Carefully identify and mark the NIR site, which should always be on the right side of the body. The biceps OD is measured on the anterior midline over the belly of the biceps brachii muscle, midway between the acromion process of the scapula and the antecubital fossa of the elbow (Futrex 1988).
3. Firmly place the NIR probe, with its light shield, perpendicular to the measurement site. Use pressure equivalent to a "firm" handshake (Futrex 1988).
4. Hold the probe firmly in place while your client pushes the Enter key on the analyzer a designated number of times (usually two). Make certain the client uses the left hand when pressing the keypad.

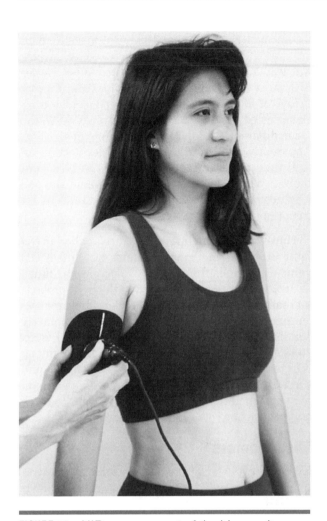

FIGURE 7.2   NIR measurement of the biceps site.

5. Read the digital display that alternately flashes the client's OD1 and OD2 values. Record these numbers.

Sources of Measurement Error

Many researchers have reported high within- and between-day reliability for ODs at the biceps site (r = .95-.99) and %BF estimates (r = .91-.98) obtained from the Futrex-5000 (Cassady et al. 1993; Eaton et al. 1993; Fornetti et al. 1999; Heyward, Jenkins et al. 1992; Hortobagyi, Israel, Houmard, McCammon et al. 1992; Hortobagyi, Israel, Houmard, O'Brien et al. 1992; Quatrochi et al. 1992; Vehrs et al. 1998; Wilmore et al. 1994). Schreiner et al. (1995) reported the between-day reliability of %BF estimated from NIR (95.3%) to be higher than that of waist girth (93.4%) and waist-to-hip ratio (82.4%). They also noted only small differences in OD values measured by two trained technicians. Additionally, research has shown high reliability (r > .97) and close agreement between %BF values measured

with two Futrex-5000 analyzers (Vehrs et al. 1998). The following sections address potential sources of measurement error for the NIR method, such as instrumentation, technician skill, client factors, and prediction equations.

Instrumentation

Have the Futrex NIR Analyzers Been Validated Against Chemical Analysis or More Sophisticated Spectrophotometers?

A highly sophisticated and precise (6 nm) computerized spectrophotometer (Neotec Instruments, Pacific Scientific, Silver Springs, MD) was used to pioneer the application of the NIR method for human body composition assessment and to explore the feasibility of designing a low-cost, less precise (50 nm) NIR analyzer for this purpose. Apparently, the Futrex-5000 NIR analyzer (figure 7.3) was developed based on Conway and Norris' (1987) computer-simulated analysis of the expected accuracy for a less expensive, wide-slit (50 nm) spectrophotometer. To our knowledge, OD values from the Futrex analyzers have not been compared to those obtained from the Neotec computerized spectrophotometer. Unlike the Neotec instrument, which was validated through measurement of the water, fat, and protein content of beef and pork carcasses, the Futrex analyzers have never undergone chemical validation (Williams et al. 1995).

Are Different NIR Analyzers Available for Body Composition Analysis?

Futrex, Inc. is the only manufacturer of a commercial NIR device that obtains OD values for the estimation of %BF. In the late 1990s, Futrex developed a new line of NIR analyzers. As mentioned earlier, table 7.1 summarizes various prediction models for Futrex analyzers. Most of the upgrades in the Futrex models (i.e., the 1100, 5000/XL, and

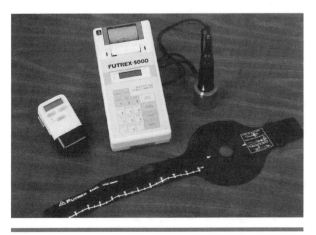

FIGURE 7.3   Futrex-5000 and Futrex-1000 analyzers.

6100/XL have replaced the 1000, 5000, and 6000, respectively) are designed to make this product more user-friendly (e.g., color print and ability to download data to computers) rather than change the way the ODs are measured. All of these analyzers, with the exception of the Futrex-6100/XL, measure NIR light at two different wavelengths. The Futrex-6100/XL measures NIR light at up to six different wavelengths. To date, we are unaware of any cross-validation studies for these newer models.

The vast majority of NIR research has been done using the Futrex-5000 analyzer, but a few studies have used other models. The Futrex-1000 is a handheld, battery-operated device designed for home use. It estimates %BF from OD measurements at the biceps site, body weight, and height; age and gender are not included as predictors (see table 7.1). Eckerson et al. (1998) reported large errors in the estimation of %BF for both men (CE = 5.0, SEE = 5.1, TE = 7.4) and women (CE = 4.4, SEE = 4.6, TE = 6.3) using the Futrex-1000. In studies that have included both the Futrex-5000 and the Futrex-1000, lower validity coefficients and higher prediction errors have been reported for the Futrex-1000 (Smith et al. 1997; Stout, Eckerson, Housh, & Johnson 1994; Stout, Eckerson, Housh, Johnson, & Betts 1994; Stout et al. 1996). Even though the Futrex-5000A is marketed for use with children, Smith et al. (1997) observed that the Futrex-5000 model provides a more accurate estimate of %BF for female gymnasts (13-17 yr of age) than the 5000A model. Likewise, Cassady et al. (1993) reported unacceptable %BF prediction errors when using the Futrex-5000A manufacturer's equation to assess body composition of children. We found only one study that compared %BF estimates from the Futrex-6000 and DXA. In this study of females who were obese, the mean difference between %BF$_{\text{Futrex-6000}}$ and %BF$_{\text{DXA}}$ was small (1.4% BF); however, the individual predictive accuracy (95% limits of agreement = −8.0 to 10.7% BF) for the Futrex-6000 were large (Panotopoulos et al. 2001). Based on these limited data, it appears that these models of Futrex analyzers are no better than the Futrex-5000 and in some cases are even worse at estimating %BF.

Technician Skill

• **Is the NIR method more difficult for the technician to perform than the SKF method?** Technician skill is not a major source of measurement error for the NIR method. There is little difference in biceps OD values when the same individual is measured independently by two NIR technicians (Heyward et al. 1993). The measurement error associated with differences between technicians was less (2.2-2.4%) than that reported (2.8-8.8%) for SKF measures (Lohman, Pollock et al. 1984; Morrow, Fridye et al. 1986). When two independent technicians tested 39 subjects with the Futrex-5000, the mean difference was 0.5% BF (Schreiner et al. 1995). Apparently, the NIR method requires less technician skill than the SKF method because the NIR light probe is placed firmly on the measurement site and it is not necessary to isolate the subcutaneous fat from the underlying muscle tissue.

However, the amount of pressure applied to the light probe during measurement may affect OD values. OD measurements may decrease as much as 10% when pressure applied to the light probe is increased (Elia et al. 1990). The manufacturer recommends applying "firm pressure, approximately equal to the pressure of a firm handshake" (Futrex 1988, p. 14). This may be difficult to control among NIR technicians. In addition, the technician should use a light shield (15 cm, doughnut-shaped foam pad) attached to the end of the NIR probe to block out extraneous room light during testing.

Client Factors

Exercise level, variations in skin color, and hydration status may be potential sources of error for the NIR method.

• **How does my client's hydration status affect the OD measurement?** To date, no one has studied the effects of factors that may influence hydration level (i.e., eating, drinking, exercise, and menstrual cycle stages) on OD measurements. For the bioelectrical impedance analysis (BIA) method, the within-individual variability measured on different days is 3.1% to 3.9%; much of this variability is attributed to changes in TBW over test days (Jackson et al. 1988). Assuming that the OD is a measure of the tissue composition (fat and water) at the biceps site, it is feasible that a change in OD values across days may reflect fluctuations in the water content of muscle tissue similar to those with BIA. However, research shows that day-to-day variability of the OD measures at the biceps account for <1% of the total variance of the Futrex measurements (Heyward, Jenkins et al. 1992; Vehrs et al. 1998). Also, the variance attributed to the subject × day interaction was less for the OD measures (OD1 = 0.9%, OD2 = 1.5%) compared to that with BIA resistance (2.0%) (Vehrs et al. 1998). Thus, it appears that the Futrex-5000 is either less affected by or less sensitive to fluctuations in TBW than BIA.

• **Does the client's skin tone affect the amount of infrared light absorbed and reflected?** Skin color or skin tone explains a significant proportion (12-16%) of the variability in OD measures at the biceps site, even after controlling for the SKF thickness (Wilson & Heyward 1993). We developed a skin-tone wheel and numerically coded various shades of black, brown, red, yellow, pink, and white to assess skin tones of a sample of American Indian, African American, Hispanic, and Caucasian men. The subject's skin color at the biceps site was matched to a color on the skin-tone wheel. Subjects with darker skin tones tended to have higher OD values. A higher OD results in a greater underestimation of %BF. In fact, in a study of African American and Caucasian football players, Houmard et al. (1991) reported that the %BF of African American players was underestimated (–7.3%) to a greater extent than that of Caucasian players (–3.3%). However, Hortobagyi and colleagues reported that the correlation between OD and SKF measures at the biceps site was virtually identical for African Americans (r = –.65) and Caucasians (r = –.67) and therefore concluded that skin color does not appear to affect OD values (Hortobagyi, Israel, Houmard, O'Brien et al. 1992). Additional research is needed to determine whether or not skin tone should be included in NIR prediction models.

NIR Prediction Equations

• **Are the manufacturer's equations that are programmed into my Futrex analyzer valid for estimating the body fat percentage of my clients?** There are over 20 cross-validation studies of the manufacturer's equations for the Futrex NIR analyzers. With few exceptions, the prediction errors have been large (SEE and TE > 3.5% BF). Also, we found similar prediction errors (CE = –1.93%, SEE = 4.06%, TE = 4.47%) when the Futrex-5000 manufacturer's equation was compared to 4-C model estimates of %BF in African American men (unpublished observations). In most cases, the Futrex-5000 equation underestimates %BF by as much as 2% to 10% BF. The degree of underestimation appears to be directly related to the level of body fatness (Elia et al. 1990; Heyward, Cook et al. 1992). Some studies noted large underestimations of average %BF when the Futrex-5000 equation was compared to multicomponent model estimates of %BF for women with obesity (Fuller et al. 1994) and for patients with arthritis who were overweight (Heitmann et al. 1994). Thus, the %BF of fatter clients is likely to be more grossly underestimated than that of leaner clients when one uses the Futrex-5000 manufacturer's equation.

Because many research studies have reported unacceptable prediction errors, we do not recommend using the Futrex manufacturer's equations to assess %BF for your clients. Some researchers have used OD values obtained from Futrex analyzers to develop prediction equations that accurately assess Db, FFM, or %BF. Population-specific equations have been developed and cross-validated for women aged 20 to 72 yr (Heyward, Jenkins et al. 1992), American Indian women 18 to 60 yr of age (Hicks et al. 2000), college female athletes (Fornetti et al. 1999), and high school wrestlers (Oppliger et al. 2000). These equations are presented in later chapters dealing with these population subgroups.

Key Points

- The OD is a measure of the fat and water composition of the tissues at the measurement site.
- The amount of infrared light reflected back to the analyzer is inversely and linearly related to subcutaneous and total body fat.
- OD measures have good within- and between-day reliability.
- OD explains a significant portion of the variability in %BF beyond that accounted for by body weight and height.
- The biceps is the best single site for estimating body fat of adults using the NIR method.
- OD values obtained from Futrex NIR analyzers can be used in population-specific NIR prediction equations to estimate body composition of certain population subgroups.
- The manufacturer's NIR equations programmed into the Futrex-1000, Futrex-5000, and Futrex-5000A analyzers do not accurately estimate %BF and should not be used.
- Cross-validation research for the newer Futrex NIR analyzers is lacking (e.g., Futrex-6100/XL).

Key Terms

Learn the definition for each of the following key terms. Definitions of terms can be found in the Glossary, page 227.

near-infrared interactance (NIR)

optical density (OD)

Review Questions

1. How do the Futrex NIR analyzers differ from the NIR technology that was used to pioneer this method of body composition assessment?

2. What are the peak absorption wavelengths for fat and water? What wavelengths are used in Futrex NIR analyzers to measure OD?

3. How does the level of body fatness affect the amount of light reflected back to the NIR analyzer?

4. What measurement site has the strongest relationship to SKF thickness and total body fat?

5. Identify the variables included in the Futrex manufacturer's prediction equations. Do all of these predictor variables contribute significantly to the estimation of %BF?

6. Describe how you can access the OD mode using a Futrex NIR analyzer. Describe the testing procedure.

7. How does the reliability of the NIR method compare to that for other field methods of body composition assessment?

8. Differentiate among the various models of Futrex NIR analyzers. Which model has been studied the most?

9. What client factors potentially affect OD measurements?

10. How can you use the NIR method to obtain an estimate of body composition for your clients?

11. Should you use the manufacturer's equations that are programmed into the Futrex analyzers?

Body Composition Methods and Equations for Healthy Populations

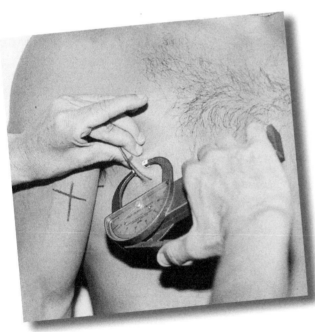

This section addresses suitable reference and field methods to assess the body composition of healthy populations. In each chapter, you will find information about the fat-free composition, suitable reference methods and models, as well as recommended field methods and prediction equations for assessing the body composition for each of the following groups:

- Children (chapter 8)
- Older adults (chapter 9)
- Ethnic groups including African-Americans, American Indians, Asians, Caucasians, and Hispanics (chapter 10)
- Athletes and physically-active individuals (chapter 11)

BODY COMPOSITION
AND CHILDREN

Key Questions

- How does the fat-free body composition of children differ from that of adults?
- How do the components and overall density of the fat-free body change during childhood and adolescence?
- What body composition models and methods can be used to obtain reference measures of body composition for children?
- Is dual-energy X-ray absorptiometry a viable reference method for children?
- Is air displacement plethysmography as good as hydrostatic weighing for measuring body density of children?
- What field methods and prediction equations may be used to estimate body composition of children and adolescents?
- How accurate are skinfold and bioelectrical impedance equations in estimating body composition for groups of children and for individual children?

Over the last two decades, the prevalence of childhood overweight and obesity has increased at an alarming rate (National Center for Health Statistics 2001; Ogden et al. 2002; Strauss & Pollack 2001). The prevalence of obesity (10.9%) has doubled and prevalence of overweight (15.5%) has increased for children in the United States (Ogden et al. 2002; Styne 2001). A major concern is that children who are obese tend to become obese adults who have a relatively high risk of developing diseases and disorders associated with excess body weight and body fatness. Because of these public health implications, the epidemic increase in childhood overweight and obesity has stimulated much interest in identifying accurate ways to assess the body composition of children in school and clinical settings.

The accurate assessment of body composition in children and youth is complicated and challenging. Children are chemically immature; changes in the proportions and densities of the FFB components due to growth and maturation directly affect overall FFB density. Changes in FFB density are caused by a decrease in TBW and an increase in bone mineral during growth and development. Therefore, multicomponent body composition models must be used to establish reference data and to develop field method prediction equations for children (Lohman 1992). To date the amount of research using this approach is limited.

As mentioned in chapter 1, molecular-level multicomponent models require the measurement of total Db, bone mineral, and/or TBW. Traditionally, hydrostatic weighing has been used to measure Db; however, some children, especially those who are young or overweight, have difficulty complying with the testing procedures. Therefore, researchers are testing the validity of alternative methods (e.g., air displacement plethysmography and dual-energy X-ray absorptiometry) that potentially could be used to obtain reference measures of body composition for children. However, much more research is needed before any one of these methods can be universally accepted as a reference method for children. The lack of a suitable reference method is a major issue that must be resolved before accurate prediction equations can be developed to estimate the body composition of children in field and clinical settings. In light of this limitation, previously published prediction equations, even those based on 4-C molecular model reference measures in children, should be viewed with caution.

This chapter describes the FFB composition of children, compares body composition models and reference methods for children, and presents field method prediction equations based on multicomponent model estimates of body composition for children. The prediction equations were selected using the evaluation criteria outlined in chapter 2 but with the following modifications:

- The prediction error (SEE) for estimating FFM of children should be less (2.2 kg) than the maximum values for adults (SEE = 2.8 kg for women and 3.5 kg for men) because of the relatively smaller FFM in children.

- All prediction equations selected were derived from either
  - multicomponent models that account for interindividual variability in the mineral or water components of the FFB, or
  - 2-C model age- and gender-specific conversion formulas that adjust for changes in FFB density.

- Some prediction equations use combinations of variables from different body composition methods (e.g., SKF, bioelectrical impedance, and circumferences) in the same equation. These equations may not be practical in clinical or school settings because the practitioner needs to have a bioelectrical impedance analyzer, SKF caliper, and anthropometric tape measure to measure all of these variables. Therefore, whenever possible, we selected only "pure" prediction equations for each field method (i.e., SKF measures only for SKF equations, bioimpedance measures only for BIA equations, optical density measures only for near-infrared interactance equations, and circumference and/or skeletal breadth measures only for anthropometric equations).

Fat-Free Body Composition of Children

The changes in the water, mineral, and protein components of the FFB due to growth and matu-

ration influence the density of the FFB in children (Boileau et al. 1984; Fomon et al. 1982; Haschke 1983; Roemmich et al. 1997). From birth to 22 yr of age, the FFB density (FFBd) steadily increases in males from ~1.063 to 1.102 g/cc and in females from ~1.064 to 1.096 g/cc (Lohman 1986). Using a 4-C model approach, Roemmich and colleagues (1997) calculated the FFBd of prepubertal and pubertal children (8-14 yr). They reported that the average FFBd of prepubertal children (1.084 g/cc for boys, 1.086 g/cc for girls) is significantly less than that of pubertal children (1.087 g/cc for boys, 1.091 g/cc for girls). Wells et al. (1999) reported an average FFBd of 1.0864 ± 0.0074 g/cc for 8- to 12-yr-old boys and girls. The average FFBd of boys and girls did not differ significantly.

As the child ages, the hydration of the FFB decreases while the mineral content of the FFB increases, resulting in an overall increase in FFBd (Roemmich et al. 1997; Wells et al. 1999). The hydration of the FFB decreases from 79% at 1 yr of age to 74% at 20 yr of age (Lohman 1989a). In contrast, the bone mineral content of the FFB increases from 3.7% in infants to ~7.0% in adulthood (Fomon et al. 1982). Furthermore, for any given value of %BF, the overall total Db of children is less than that of adults (Lohman, Slaughter et al. 1984), reflecting differences in the FFB composition of children and adults.

Body Composition Models and Reference Methods

This section addresses the validity of 2-C, multicomponent (3-C and 4-C), and dual-energy X-ray absorptiometry (DXA) models and methods for assessing body composition of children.

Two-Component Models

Chapter 1 dealt with the limitations of using a 2-C model to estimate relative body fatness of individuals whose FFB composition differs from the assumed constants (i.e., 73.8% water, 19.4% protein, and 6.8% mineral). The classic 2-C models of Siri and Brozek are limited especially in children and adolescents because the proportions or densities of the FFB components and their average FFBd differ significantly from values for the adult reference body. Compared to 4-C models, these 2-C models systematically overestimate the %BF of children by 3% to 5% (Lohman 1992; Nielsen et al. 1993; Roemmich et al. 1997). Therefore, one should not use these 2-C models, or field method prediction equations based on these models, to obtain reference measures or to estimate the body composition of children and adolescents.

To address this problem, researchers have developed 2-C model population-specific conversion formulas (see table 1.4, p. 9) to estimate %BF from Db in children. The constant values in these age- and gender-specific equations were theoretically or empirically derived using average values for the water, mineral, and protein components of the FFB to estimate FFBd of children and adolescents (Lohman 1992; Wells et al. 1999; Weststrate & Deurenberg 1989). These conversion formulas have been used to derive reference body composition measures for the development and cross-validation of field method prediction equations for children and adolescents (Reilly et al. 1996), as well as for comparison of reference methods and models (Fields & Goran 2000; Nicholson et al. 2001; Nielsen et al. 1993; Roemmich et al. 1997; Wells et al. 1999). Compared to 4-C models, the age- and gender-specific conversion formulas provide fairly reasonable estimates of average %BF (CE = –2.7% to 1.6% BF; prediction error = 1.9-3.4% BF) for groups of children. However, the 95% limits of agreement (i.e., ±2 SD of mean difference between the 4-C and 2-C models) for estimating %BF of individuals within a group are as much as –7.5% to 5% BF (Roemmich et al. 1997).

Traditionally, Db has been assessed using the hydrostatic weighing (HW) method. As previously mentioned, this method may not be suitable for children who have difficulty remaining still and expiring all of the air from lungs while under water. Recently, researchers have examined the suitability of using air displacement plethysmography (ADP) as a practical alternative to HW for measuring Db and %BF of children (Fields et al. 2002). The reliability (precision) of ADP is excellent (0.8-1.0% BF). Compared to HW, the ADP method tends to overestimate $\%BF_{2-C}$ by 0.8% to 1.7% BF (Demerath et al. 2002; Dewit et al. 2000; Fields & Goran 2000). The 95% limits of agreement between methods were wide (–8.7 to 10.4% BF), reflecting a large degree of variation in the difference between ADP and HW methods for individuals within a group (Demerath et al. 2002). Also, both of these methods significantly underestimated $\%BF_{4-C}$ by 2.7% BF (ADP) and 3.9% BF (HW) in children (Fields & Goran 2000). Although ADP is an attractive alternative to HW for children, additional research is needed to assess the validity of this method using 4-C models as the reference standard. If you are restricted to using only a 2-C model, be certain to select an appropriate age- and gender-specific conversion formula to estimate %BF from Db measured by either ADP or HW.

Three-Component Models

Three-component model equations account for interindividual variability in either TBW or TBM

(see chapter 1). The Siri (1961) and Boileau et al. (1984) 3-C equations adjust Db for the relative proportion of water in the body, whereas the Lohman (1986) 3-C equation adjusts Db for the relative amount of TBM in the body. In one study of children and adolescents (8-20 yr of age), the Boileau 3-C model equation significantly overestimated $\%BF_{4-C}$ by ~2% BF (TE = 2.2% BF) in boys and girls (Nielsen et al. 1993). However, other researchers reported excellent agreement (CE < 1% BF; TE < 1% BF; 95% limits of agreement < 1% BF) between 3-C (water) and 4-C model estimates of %BF in children (Roemmich et al. 1997; Wells et al. 1999). These data suggest that the 3-C water model provides accurate and valid reference measures of body composition for children and adolescents.

In comparison to the 4-C model, the Lohman 3-C mineral equation is less accurate than the 3-C water equation in children. The TE (~3.0% BF) of the 3-C mineral model was more than two times greater than that of the 3-C water model, and the 95% limits of agreement were substantially larger (±6.1% BF), suggesting that this 3-C model should not be used to obtain reference measures for children and adolescents (Roemmich et al. 1997).

DXA Model

In recent years, DXA has been used to derive reference body composition measures for the development and cross-validation of field method prediction equations for children and adolescents (Morrison et al. 2001; Nicholson et al. 2001; Okasora et al. 1999; Tyrrell et al. 2001). However, experts suggest that further research and standardization of this technology (e.g., type of equipment, data collection mode, and software version) are needed before DXA can be firmly established as a reference method for assessing body composition (Fields et al. 2002; Kohrt 1995, 1998; Lohman 1996).

A number of researchers have investigated the validity of DXA for assessing reference measures of body composition in children and youth by comparing %BF from DXA ($\%BF_{DXA}$) to 4-C model estimates ($\%BF_{4-C}$). When pediatric software versions and correction factors that account for the relatively smaller body size of children are used, DXA (Hologic and Lunar) and 4-C models yield similar average estimates of %BF for groups of children (Fields & Goran 2000; Wells et al. 1999). However, the accuracy of DXA in estimating %BF of individuals is variable, with fairly large 95% limits of agreement between these models (–6.5 to 8.3% BF) (Roemmich et al. 1997; Wells et al. 1999).

In summary, the 4-C or 3-C water model should be used to obtain reference measures of body composition for children and adolescents. Although DXA and the 2-C model age- and gender-specific conversion formulas yield fairly good %BF estimates for groups, there may be considerable error (as much as ±9% BF) with use of these models to estimate the %BF of an individual child.

Field Methods and Prediction Equations

Field methods are commonly used in school and clinical settings to estimate body composition of children for the purpose of monitoring changes during growth and development and classifying their levels of body fatness. In school settings, health and physical educators need to interpret body composition results for children and parents. Children should be taught how to achieve and maintain a healthy body weight through lifestyle modifications (e.g., physical activity and nutrition). Information about changes in body composition and body fatness due to maturation should be provided so that the children, especially girls, can understand that these changes in their bodies during puberty are normal.

Tips for Assessing Body Composition of Children

On the basis of work by Thomas and Whitehead (1993), we suggest the following approach for incorporating body composition into the health and physical education curriculums:

Tips for Assessing Body Composition of Children

1. Prior to body composition testing, inform the parents so they will understand the purpose and procedures of this assessment.

2. Instruct students regarding concepts and procedures for measuring body composition.

3. Maintain records of these measures over time to assess the interaction effects of growth, maturation, diet, and physical activity on body composition changes.

4. Measure only standardized sites and follow established procedures.

5. If you feel it is necessary, ask a teacher, nurse, or the child's parent to be present during body composition testing.

6. Ensure confidentiality by sharing test results only with the child and the parents.

7. Provide personal feedback and "group" interpretations of the results.

8. Do not use body composition test results for grading purposes.

9. Be sure to make the body composition assessment a positive experience for each child. Do not label, criticize, or ridicule children during any phase of this assessment.

Practitioners also need to consider sensitivity and ethical issues regarding the touching of children by adults, particularly when measuring certain SKF sites. For modest and self-conscious children, especially young adolescent females, we recommend that you avoid measuring the subscapular and suprailiac SKF sites.

SKF Equations for Children

Table 8.1 presents SKF equations for children and adolescents. The overwhelming majority of these equations estimate Db instead of %BF. Use of these equations is limited by the fact that it is necessary to employ 2-C model population-specific conversion formulas to estimate %BF from Db. The Slaughter (1988) equations, however, are based on 4-C model reference measures of %BF. These equations are age-, gender-, and fatness-specific and use the sum (Σ) of two SKFs to predict %BF (see table 8.1). The prediction error for these equations ranged from 3.6% to 3.8% BF. These equations may be used to assess the body composition of African American and Caucasian boys and girls 8 to 17 yr of age. Fatness-specific equations (triceps + subscapular SKF) were developed for children whose $\Sigma2SKF$ is less than or greater than 35 mm. The intercept of the triceps + subscapular equation ($\Sigma2SKF \leq 35$ mm) for boys varies depending on maturation stage (see table 8.3).

We reviewed studies that cross-validated previously published SKF equations for children (Goran et al. 1996; Gutin et al. 1996; Janz et al. 1993; Lohman, Caballero et al. 2000; Nicholson et al. 2001; Reilly et al. 1995; Roemmich et al. 1997; Wells et al. 1999; Wong et al. 2000). The predictive accuracy of these equations varies because the studies used different models (i.e., 2-C, 4-C, or DXA models), reference methods (i.e., HW, TBW, or DXA), ages (i.e., 4-17 yr of age) and ethnic groups (i.e., American Indian, African American, or Caucasian), and SKF calipers (i.e., Harpenden, Holtain, or Lange). Moreover, studies using DXA for reference measures or bone mineral values for 4-C models employed different manufacturers, software versions, and scanning modes. Also, the type of SKF caliper did not always match that used in the original validation studies (see table 8.1). All of these factors potentially influence the cross-validation results and make interpretation of these findings difficult.

Generally, the cross-validation of the Slaughter SKF equations against 4-C model reference

Table 8.1 Comparison of Skinfold Equations for Children

Name of equation	Reference model, methods, and caliper type	Validation sample	Equation and validation statistics for original study*
Brook 1971	2-C; TBW_{D2O}; Harpenden	23 children, 1-11 yr; ethnicity NR	Boys: $Db = 1.1690 - 0.0788$ (log Σ4SKF)[b] $R^2 = $ NR; SEE = NR Girls: $Db = 1.2063 - 0.0999$ (log Σ4SKF)[b] $R^2 = $ NR; SEE = NR
Deurenberg, Pieters et al. 1990	2-C; HW; Holtain	114 boys and 98 girls, 7-12 yr; ethnicity NR	Boys: $Db = 1.133 - 0.0561$ (log Σ4SKF)[b] $+ 1.7$ (age $\times 10^{-3}$) $R^2 = 0.43$; SEE = 0.0095 g/cc Girls: $Db = 1.1187 - 0.0630$ (log Σ4SKF)[b] $+ 1.9$ (age $\times 10^{-3}$) $R^2 = 0.47$; SEE = 0.0095 g/cc
Durnin & Rahaman 1967	2-C; HW; Harpenden	48 boys, 12-16 yr, and 38 girls, 13-16 yr; ethnicity NR	Boys: $Db = 1.1533 - 0.0643$ (log Σ4SKF)[b] $R^2 = 0.58$; SEE = 0.0083 g/cc Girls: $Db = 1.1369 - 0.0598$ (log Σ4SKF)[b] $R^2 = 0.61$; SEE = 0.0081 g/cc
Johnston et al. 1988	2-C; HW; Harpenden	140 Canadian boys and 168 Canadian girls, 8-14 yr; ethnicity NR	Boys: $Db = 1.1660 - 0.0070$ (log Σ4SKF)[b] $R^2 = 0.49$; SEE = 0.0060 g/cc Girls: $Db = 1.1440 - 0.0600$ (log Σ4SKF)[b] $R^2 = 0.45$; SEE = 0.0050 g/cc
Lohman, Caballero et al. 2000	2-C; TBW_{D2O}; Lange	98 American Indian children, 8-11 yr	%BF $= 17.66 - 0.08$ (age) $+ 2.4$ (sex)[a] $+ 0.21$ (BW) $+ 0.38$ (tri SKF) $+ 0.20$ (SI SKF) $R^2 = 0.83$; RMSE = 3.36% BF
Slaughter et al. 1988	4-C; HW, D_2O, SPA; Harpenden	242 AA and CA children, 8-17 yr	Boys: %BF $= 0.735$ (Σ2SKF)[c] $+ 1.0$ $R^2 = 0.77$; SEE = 3.8% BF Boys: %BF $= 0.783$ (Σ2SKF)[d] $+ 1.6$ (if Σ2SKF > 35 mm) $R^2 = $ NR; SEE = NR Boys: %BF $= 1.21$ (Σ2SKF)[d] $- 0.008$ (Σ2SKF)$^2 + I$[e] $R^2 = 0.78$; SEE = 3.6% BF Girls: %BF $= 0.610$ (Σ2SKF)[c] $+ 5.1$ $R^2 = 0.77$; SEE = 3.8% BF Girls: %BF $= 0.546$ (Σ2SKF)[d] $+ 9.7$ (if Σ2SKF > 35 mm) $R^2 = $ NR; SEE = NR Girls: %BF $= 1.33$ (Σ2SKF)[d] $- 0.013$ (Σ2SKF)$^2 - 2.5$ $R^2 = 0.78$; SEE = 3.8% BF

*Only the Lohman, Caballero et al. (2000) equation was cross-validated in the original study. The bootstrap cross-validation analysis for this equation yielded an $R^2 = 0.83$ and RMSE = 3.4% BF.

[a]Sex = 0 for boys; 1 for girls; [b]Σ4SKF = biceps + triceps + subscapular + suprailiac; [c]Σ2SKF = triceps + calf; [d]Σ2SKF = triceps + subscapular; [e]I = intercept varies (see table 8.3).

Key: HW = hydrostatic weighing; Db = body density; TBW = total body water; D_2O = deuterium oxide dilution technique; SPA = single-photon absorptiometry; tri = triceps; SI = suprailiac; R^2 = multiple correlation coefficient squared; RMSE = root mean square error; SEE = standard error of estimate; NR = not reported; AA = African American; CA = Caucasian.

measures indicates that these equations provide a fairly reasonable estimate of average %BF (CE = –0.3 to 1.3% BF) for groups of African American and Caucasian children and adolescents (Roemmich et al. 1997; Wong et al. 2000) but grossly underestimate the average %BF of American Indian children (8-11 yr) by ~14% BF (Lohman, Caballero et al. 2000). The prediction errors (SEE) for these equations generally range from fair (4.5% BF) to good (3.5% BF), but the 95% limits of agreement are wide (±8-10% BF) and variable (Roemmich et al. 1997; Wells et al. 1999; Wong et al. 2000). This means that these equations may over- or underestimate the %BF of some individuals by as much as 8% to 10% BF.

Using 2-C model reference measures, Janz et al. (1993) cross-validated the Slaughter SKF equations for girls and boys. HW was used to determine total Db, and total Db was converted to %BF using Lohman's (1992) age-gender conversion formulas. For girls, both equations had acceptable prediction errors (SEE = 3.5-3.6% BF). However, the Σ triceps + calf equation slightly overestimated (+1.7% BF) the average %BF of the girls. For boys, the prediction error for the Σ triceps + calf SKF equation (SEE = 4.6% BF) was unacceptable and varied with maturation level. The 95% limits of agreement were not reported. Overall, these studies suggest that the Slaughter SKF equations need further refinement to improve their predictive accuracy for groups and individuals.

Until these equations are refined or new SKF equations based on 4-C models and standardized equipment and measurement protocols are developed and cross-validated, we recommend using the Slaughter SKF equations for estimating %BF of African American and Caucasian children and adolescents (see tables 8.1 and 8.3). Keep in mind, however, that the error in estimating %BF may be fairly large (±10% BF) for some individuals. To minimize prediction error, you should use the same type of SKF caliper that Slaughter used (i.e., Harpenden) to measure the sites. Also, as already noted, the triceps + calf SKF equations may be more suitable than the triceps + subscapular equations for children who are not comfortable having the subscapular site measured.

To date the Lohman, Caballero et al. (2000) equation is the only SKF equation developed specifically for American Indian children, but it is based on only 2-C model reference measures of %BF estimated from TBW. The bootstrap cross-validation procedure, described in chapter 2, was used to internally cross-validate this equation. The prediction error for this 2-C model (i.e., %BF estimated from TBW) equation was good (root mean square error [RMSE] = 3.4% BF) for estimating the average %BF for a group of American Indian children who were 8 to 11 yr of age and had 23% to 60% BF. The estimated prediction error for individuals within the group, however, was not reported. This equation needs to be cross-validated against 4-C model reference measures for independent samples from this ethnic group.

Use table 8.3 to select an appropriate SKF equation based on your client's age, gender, ethnicity, and ΣSKF. You can use either table 1.2 (see p. 6) or percent fat charts (see figure 8.1, a and b) to classify levels of body fatness for children and adolescents (Lohman 1987).

Bioelectrical Impedance Equations for Children

Bioelectrical impedance (BIA) equations for children and adolescents are presented in table 8.2. With the exception of the ethnic-specific equation developed for American Indian children, all of these BIA equations estimate FFM instead of %BF. Only one (i.e., the Houtkooper equation) is based on multicomponent reference measures of FFM. Houtkooper and colleagues (1992) used the Boileau et al. (1984) 3-C model that adjusted Db for TBW. The equation was cross-validated on samples from three different laboratories, and its prediction error was 2.1 kg. This BIA equation can be used to estimate FFM of children and adolescents ranging in age from 10 to 19 yr and in body fatness from 6.5% to 36% BF. Although Deurenberg, Kusters et al. (1990) found that age was significantly related to impedance measures in children, Houtkooper et al. (1992) reported that including age as a predictor did not significantly improve the predictive accuracy of their BIA equation.

We reviewed studies that cross-validated previously published BIA equations for children (Goran et al. 1996; Kim et al. 1994; Lohman, Caballero et al. 2000; Nicholson et al. 2001; Reilly et al. 1996; Roemmich et al. 1997). The predictive accuracy of these equations varies because the studies used different models (i.e., 2-C, 4-C, or DXA models), reference methods (i.e., HW, TBW, or DXA), bioimpedance analyzers (i.e., Bodystat, RJL, and Valhalla analyzers), and age (i.e., 6-14 yr of age) or ethnic (i.e., American Indian, African American, Asian, and Caucasian) groups. Moreover, only one of the studies evaluated the accuracy of this equation against 4-C model reference measures (Roemmich et al. 1992). In all but one (i.e., Nicholson et al. 2001), the type of bioimpedance analyzer did not match that used in the original validation studies.

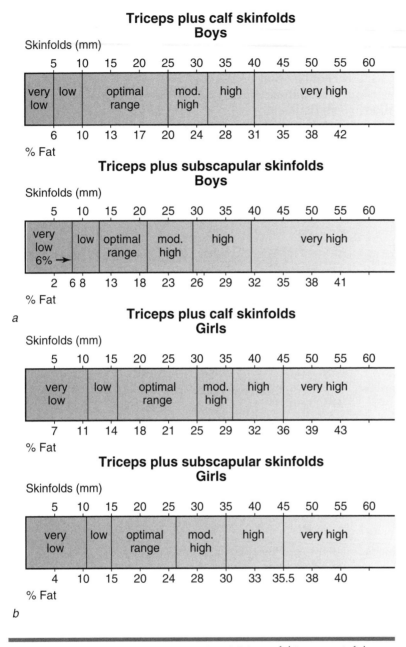

FIGURE 8.1 Percent body fat charts for children: *(a)* boys and *(b)* girls.

Reprinted, by permission, from T.G. Lohman, 1987, *Measuring body fat using skinfolds* [videotape] (Champaign, IL: Human Kinetics).

All of these factors make interpretation of the findings difficult.

Despite these differences, the Houtkooper equation appears to provide a fairly reasonable estimate of body composition for independent groups of Caucasian children. Compared to 4-C model reference measures, this equation accurately estimated average %BF for a group of children 8 to 14 yr of age; however, the error of the equation for individuals within the group was large (95% limits of agreement

= ±11% BF) (Roemmich et al. 1997). Compared to 2-C model reference measures, obtained by using age- and gender-specific conversion formulas to derive %BF from Db, the Houtkooper equation estimated average FFM with negligible bias (–0.3 kg or 1.4% BF); and individual errors (95% limits of agreement) in estimation (0-2.9% BF) were much less than those (–5 to 10% BF) for other BIA equations (Reilly et al. 1996). Compared to DXA and TBW reference measures, however, the Houtkooper equation systematically underestimated the average %BF for groups of African American, American Indian, and Caucasian children (–7.4 to –11.1% BF): The 95% limits of agreement also were fairly large (±8.3-9.6% BF) (Lohman, Cabellero et al. 2000; Nicholson et al. 2001). Some of this error may be due to differences in reference methods as well as in the age and ethnicity of samples.

Until new BIA equations, based on 4-C models and standardized measurement protocols for cross-validation studies, are developed, we recommend using the Houtkooper BIA equation for estimating body composition of Caucasian children and adolescents 10 to 19 yr of age (see table 8.3). Keep in mind, however, that the error in estimating %BF may be fairly large (±11% BF) for some individuals. To minimize prediction error, you should use the same type of bioimpedance analyzer that Houtkooper used (i.e., RJL) to measure whole-body resistance.

To assess the body composition of children younger than 10 yr, we recommend using the equations of Kushner and colleagues (1992) (see table 8.3). The Kushner equation estimates the TBW, instead of FFM, of prepubertal children ages 6 to 10 yr. To develop this equation, the deuterium (D$_2$O) dilution method was used to obtain reference values for TBW. The prediction error for this equation was 1.41 L. To convert TBW estimates to FFM, one may use the Lohman, Boileau et al. (1984) age-gender constants for hydration of the FFB (see table 8.3).

Some ethnic-specific BIA equations have been developed for American Indian and Japanese Native children, as well as for children from four ethnic

Table 8.2   Comparison of Bioelectrical Impedance Analysis Equations for Children

Name of equation	Reference model, methods, and analyzer	Validation sample	Equation and validation statistics for original study	Cross-validation analysis		
				Sample	Method	Statistics
Boileau 1996	4-C; HW, DXA (model and SV NR), D_2O; Valhalla 1990B	129 CA boys and girls, 8-16 yr	FFM = 4.138 + 0.657 (HT^2/R) + 0.16 (BW) − 0.131 (sex)[a] R^2 = 0.97; SEE = 1.8 kg	*	*	*
Deurenberg, van der Kooy et al. 1991	2-C; HW; RJL 101	166 boys and girls, 7-15 yr; ethnicity NR	FFM = 0.360 (BW) + 0.406 (HT^2/R) + 5.58 (HT) + 0.56 (sex)[b] − 6.48 R^2 = 0.97; SEE = 1.7 kg	166 boys and girls, 7-15 yr; ethnicity NR	Internal XV (sample split into 2 groups)	d = 0.1 to 0.2 kg; LA = ±3.4 kg
Houtkooper et al. 1992	3-C; HW, D_2O; RJL 101	94 CA boys and girls, 10-14 yr	FFM = 0.61 (HT^2/R) + 0.25 (BW) + 1.31 R^2 = 0.95; SEE = 2.1 kg	63 CA boys and girls, 10-19 yr	External XV on 2 independent samples	d = 1.4 to 1.7 kg; SEE = 1.5 to 2.5 kg; LA = NR
Kim et al. 1994	2-C; HW, Selco SIF-891 analyzer	84 Japanese Native boys, 9-14 yr	FFM = 0.56 (HT^2/Z) + 0.20 (BW) + 1.66 R^2 = 0.92; SEE = 1.3 kg Combined samples: R^2 = 0.97; SEE = 1.6 kg	57 Japanese Native boys, 9-14 yr	External XV	d = 0.3 kg; SEE = 1.4 kg; LA = ±3.2 kg
Kushner et al. 1992	TBW; $H_2{}^{18}O$; RJL 101	24 CA boys and girls, 5-10 yr	TBW = 0.593 (HT^2/R) + 0.065 (BW) + 0.04 R^2 = 0.99; SEE = 1.41 L	13 CA boys and girls, 6-10 yr	External XV	d = 0.05 L; LA = NR
Lewy et al. 1999	DXA; Lunar (model and SV NR); RJL (model NR)	34 AA boys and girls	FFM = 0.84 (HT^2/R) + 1.10 R^2 = 0.97; SEE = 1.5 kg	**	**	**

Table 8.2

Name of equation	Reference model, methods, and analyzer	Validation sample	Equation and validation statistics for original study	Cross-validation analysis		
				Sample	Method	Statistics
Lohman, Caballero et al. 2000	2-C; D_2O; Valhalla 1990B	98 American Indian children, 8-11 yr	%BF = 23.64 − 0.71 (age) + 0.83 (sex)[c] + 0.83 (BW) − 0.60 (HT^2/R) + 0.04 (Xc) R^2 = 0.78; RMSE = 3.8% BF	98 American Indian children, 8-11 yr	Bootstrap XV	R^2 = 0.78; RMSE = 3.8% BF
Tyrrell et al. 2001	DXA; Lunar (DPX-L, SV and scan mode NR); Tanita model NR	82 European, Maori, and Pacific Island children, 5-11 yr	FFM = 0.31 (HT^2/Z) + 0.17 (HT) + 0.11 (BW) + 0.942 (sex)[d] − 14.96 R^2 = 0.97; SEE = NR; LA = −2.9 to 1.4 kg	**	**	**
Watanabe et al. 1993	2-C; HW; analyzer NR	163 Japanese Native girls, 9-15 yr	FFM = 0.42 (HT^2/Z) + 0.60 (BW) − 0.75 (arm C) + 7.72 R^2 = 0.94; SEE = 1.9 kg	**	**	**

[a]Sex = −1 for males; 1 for females; [b]Sex = 1 for males; 0 for females; [c]Sex = 0 for males; 1 for females; [d]Sex = 2 for males; 1 for females.

** Equation not cross-validated in original study.

Key: HW = hydrostatic weighing; BW = body weight; HT = height; TBW = total body water; D_2O = deuterium oxide dilution technique; $H_2^{18}O$ = oxygen isotope dilution technique; HT^2/R = resistance index; HT^2/Z = impedance index; C = circumference; d = absolute mean difference between BIA-predicted and reference values; LA = 95% limits of agreement; R^2 = multiple correlation coefficient squared; RMSE = root mean square error; SV = software version; SEE = standard error of estimate; NR = not reported; AA = African American; CA = Caucasian.

Table 8.3 Summary of Recommended Equations for Children

	Method/Equipment	Ethnicity/Gender	Equation	Reference
SKF (Σ TRICEPS + CALF)	Harpenden caliper	AA and CA boys (8-17 yr)	1. $\%BF_{4\text{-}C} = 0.735 (\Sigma SKF) + 1.0$	Slaughter et al. 1988
		AA and CA girls (8-17 yr)	2. $\%BF_{4\text{-}C} = 0.610 (\Sigma SKF) + 5.1$	Slaughter et al. 1988
SKF (Σ TRICEPS + SUBSCAPULAR)	[$\Sigma SKF > 35$ mm] Harpenden caliper	AA and CA boys (8-17 yr)	3. $\%BF_{4\text{-}C} = 0.783 (\Sigma SKF) + 1.6$	Slaughter et al. 1988
		AA and CA girls (8-17 yr)	4. $\%BF_{4\text{-}C} = 0.546 (\Sigma SKF) + 9.7$	Slaughter et al. 1988
	[$\Sigma SKF < 35$ mm] Harpenden caliper	AA and CA boys (8-17 yr)	5. $\%BF_{4\text{-}C} = 1.21 (\Sigma SKF) - 0.008 (\Sigma SKF)^2 + I*$	Slaughter et al. 1988
		AA and CA girls (8-17 yr)	6. $\%BF_{4\text{-}C} = 1.33 (\Sigma SKF) - 0.013 (\Sigma SKF)^2 - 2.5$	Slaughter et al. 1988
BIA	RJL 101 analyzer	CA boys and girls (5-10 yr)	7. $TBW (L)^a = 0.593 (HT^2/R) + 0.065 (BW) + 0.04$	Kushner et al. 1992
	RJL 101 analyzer	CA boys and girls (10-19 yr)	8. $FFM_{3\text{-}C} (kg) = 0.61 (HT^2/R) + 0.25 (BW) + 1.31$	Houtkooper et al. 1992
	Selco SIF-891 analyzer	Japanese Native boys (9-14 yr)	9. $FFM_{2\text{-}C} (kg) = 0.56 (HT^2/Z) + 0.20 (BW) + 1.66$	Kim et al. 1994
	Analyzer NR	Japanese Native girls (9-15 yr)	10. $FFM_{2\text{-}C} (kg) = 0.42 (HT^2/Z) + 0.60 (BW) - 0.75 (arm C) + 7.72$	Watanabe et al. 1993

Key: ΣSKF = sum of skinfolds (mm); HT = height (cm); BW = body weight (kg); R = resistance (Ω); Xc = reactance (Ω); Z = impedance (Ω); arm C = arm circumference (cm); TBW = total body water (L); 4-C, 3-C, 2-C = four-component, three-component, or two-component model; NR = not reported; AA = African American; CA = Caucasian.

*I = intercept substitutions based on maturation and ethnicity for boys:

Age	African American	Caucasian
Prepubescent	−3.2	−1.7
Pubescent	−5.2	−3.4
Postpubescent	−6.8	−5.5

aTo convert TBW to FFM, use the following age-gender hydration constants:

Boys: 5-6 yr FFM (kg) = TBW / 0.77   Girls: 5-6 yr FFM (kg) = TBW / 0.78
      7-8 yr FFM (kg) = TBW / 0.768   7-8 yr FFM (kg) = TBW / 0.776
      9-10 yr FFM (kg) = TBW / 0.762   9-10 yr FFM (kg) = TBW / 0.77

groups residing in New Zealand (see table 8.2). Note that these equations are based on either 2-C or DXA model reference measures only and therefore may not provide the best estimates of body composition for children in field and clinical settings.

The equation for American Indian boys and girls (see table 8.2; Lohman, Caballero et al. 2000) residing in the southwestern region of the United States was internally cross-validated using the bootstrap cross-validation procedure described in chapter 2. The prediction error for this 2-C model (i.e., FFM and %BF estimated from TBW) equation was fairly good (RMSE = 3.8% BF) for estimating the average %BF for a group of American Indian children 8 to 11 yr of age and having 23% to 60% BF. The estimated prediction error for individuals within the group, however, was not reported. The predictive accuracy of the equation needs to be tested on independent samples of children from this ethnic group.

The 2-C model BIA equations developed for Japanese Native children have acceptable prediction errors (1.3-1.9 kg of FFM) for estimating FFM_{2-C} for groups of children (see table 8.2; Kim et al. 1994; Watanabe et al. 1993); individual predictive accuracy was not assessed. Although the equation for Japanese Native girls includes a circumference measure (upper arm girth), it has been included because it is the only equation that we are aware of for this population subgroup.

The generalized equation for New Zealand children (5-11 yr of age) was developed for European, Maori, and Pacific Island boys and girls, as well as children from mixed descent (see table 8.2; Tyrrell et al. 2001). DXA was used to obtain references measures of FFM; and leg-to-leg, instead of whole-body, impedance was measured with a Tanita analyzer (see chapter 6). For the total validation sample, this equation slightly underestimated (−0.75 kg) average FFM_{DXA} and overestimated $%BF_{DXA}$ by 2.5% BF. For individuals, the 95% limits of agreement for FFM and %BF were fairly good (see table 8.2). This equation needs to be cross-validated to assess its validity and applicability to other groups and individuals in this population subgroup.

Using DXA to obtain reference measures of FFM, Lewy and colleagues (1999) developed a BIA equation for African American children (see table 8.2). This equation is based on only a small sample of 19 boys and 15 girls and has not been cross-validated. Therefore, at this time, we cannot recommend using it.

The bioimpedance equations programmed in the Omron handheld analyzer (model HBF-306) were developed and cross-validated for African American, Hispanic, and Caucasian children and adolescents 10 to 18 yr of age. These equations are based on 2-C model reference measures (Db by HW) using the Brozek et al. (1963) formula to convert Db to %BF. The predictor variables in the gender-specific equations for boys and girls are height, body weight, age, and arm-to-arm impedance. The group prediction errors for these equations were 2.8 kg and 2.5 kg, respectively, for boys and girls. The individual prediction errors or 95% limits of agreement were ±2.8 kg for boys and ±2.4 kg for girls (Omron Institute of Life Science 2002). Although the individual and group prediction errors are reasonable, these equations are limited in that they provide only a 2-C model estimate of body fat for children.

Near-Infrared Interactance Equations for Children

At the present time, there is limited research on use of the near-infrared interactance (NIR) method to estimate body composition of children. To evaluate the potential of using NIR optical density (OD) measures instead of SKFs for assessing body composition in children (8-12 yr), one study compared the strength of the relationships between 4-C model estimates of body composition (%BF and FFM) and OD and SKF measures at four sites (Fuller et al. 2001). For both SKF and OD measures, the subscapular site had the strongest correlation with $%BF_{4-C}$. The relationship between subscapular SKF and $%BF_{4-C}$ (r = 0.84 and 0.76 for boys and girls, respectively) was only slightly better than that between $%BF_{4-C}$ and subscapular OD2 measures (r = −0.78 and −0.74, respectively, for boys and girls), suggesting that NIR has potential for estimating body composition in children.

The Futrex-5000A NIR analyzer includes equations for children (5-12 yr) and adolescents (13-18 yr). Cross-validation of the manufacturer's equations indicated systematic overestimation of the average $%BF_{2-C}$ of children and adolescents by 2.5% to 4.1% BF. Also, the prediction errors for these equations (SEE = 4.9-5.5% BF) were large (Cassady et al. 1993; Klimis-Tavantzis et al. 1992). Therefore, we do not recommend using these equations to assess body composition of children. More research is needed to develop and cross-validate new NIR prediction equations for children using 4-C model reference measures of body fat. These studies should explore the potential of using OD measures, in combination with other variables such as body weight, height, age and ethnicity, to estimate $%BF_{4-C}$.

Anthropometric Equations for Children

Boileau et al. (1981) developed anthropometric prediction equations using two different samples of boys ages 8 to 11 yr. These equations include combinations of circumferences (arm, wrist, and thigh) and skeletal diameters (wrist and biacromial) to predict Db. The prediction errors for the equations were acceptable (SEE = 0.0072-0.0075 g/cc); however, these equations significantly under- or overestimated the average Db in the two samples of children by as much as −0.013 to 0.010 g/cc. This difference was attributed to biological variability between the samples and to differences in the methods used to measure residual lung volume. The authors also noted that SKFs were better predictors of Db than circumferences and skeletal widths for both samples. On the basis of these results, we do not recommend using these anthropometric equations to assess body composition of children.

Key Points

- During childhood and adolescence, FFB density progressively increases.
- As children age, the hydration of the FFB decreases and the mineral content of the FFB increases.
- The classic 2-C models of Siri and Brozek systematically overestimate %BF of children and adolescents.
- Either the 4-C or 3-C (water) models should be used to obtain reference measures of body composition for children and adolescents.
- Compared to 4-C models, the 2-C age- and gender-specific conversion formulas provide a fairly reasonable estimate of %BF for groups of children; however, the accuracy of these estimates for individual children varies between −7.5% and 5.0% BF.
- Compared to HW, ADP overestimates the average Db of children. The 95% limits of agreement for individuals are −4.4 to 9.6% BF.
- Compared to 4-C models, DXA yields similar estimates of average %BF for groups of children, but the individual accuracy ranges between −6.5 and 8.3% BF.
- The SKF and BIA methods can be used to estimate the body composition of prepubescent, pubescent, and postpubescent children using the Slaughter SKF equation or the Houtkooper BIA equation.
- Ethnic-specific SKF equations have been developed for American Indian, African American, and Caucasian children. Additional cross-validation studies are needed to confirm the accuracy of these equations for estimating %BF$_{4-C}$.
- The Slaughter SKF equations provide a fairly reasonable estimate of average %BF$_{4-C}$ for groups of African American and Caucasian children and adolescents; however, for individuals the prediction error is ±8% to 10% BF.
- Ethnic-specific BIA equations have been developed for American Indian, African American, Japanese Native, and New Zealand (European, Maori, and Pacific Island) children. More cross-validation studies are needed to establish the predictive accuracy and validity of these equations.
- The Houtkooper BIA equation provides a fairly good estimate of body composition for groups of Caucasian children and adolescents; the error for individuals, however, may be as large as ±11% BF.
- The NIR and anthropometric methods should not be used to assess body composition of children.
- Information about body composition should be incorporated in the health and physical education school curriculums.
- Practitioners need to be concerned about children's privacy and to interpret their body composition results in a positive manner.
- Newly developed prediction equations for children should be based on multicomponent models to account for interindividual variability in mineral and water components of the FFB.

Key Terms

All key terms used in this chapter have been identified and defined in previous chapters.

Review Questions

1. Describe the changes in the water and mineral components of the FFB during childhood and adolescence. How do these changes affect the overall FFB density of children?

2. Why do the classic 2-C model conversion formulas overestimate the %BF of children and adolescents?

3. Identify two molecular models that can be used to obtain reference measures of body composition for children and adolescents.

4. Identify the limitations of using HW and ADP for measuring Db of children. Which 2-C model population-specific formulas may be used to convert Db to %BF for children?

5. How accurate is DXA for estimating body composition of children? Is DXA a suitable alternative method for obtaining reference measures of body composition for children? Explain.

6. What steps should you take to ensure that body composition testing is a positive experience for each child?

7. Identify two field methods and the respective prediction equations that you can use to estimate body composition of children and adolescents. Specify the age and ethnic groups for whom these equations are applicable.

8. Compare the predictive accuracy of field methods for estimating body composition of individual children. Which field method is best?

CHAPTER 9

BODY COMPOSITION AND OLDER ADULTS

Key Questions

- How does the fat-free body composition of older adults differ from the reference body?
- What are the effects of aging on the components and overall density of the fat-free body?
- What body composition models and methods can be used to obtain reference measures of body composition for older adults?
- Is dual-energy X-ray absorptiometry a viable reference method for older adults?
- Is air displacement plethysmography as good as hydrostatic weighing for measuring body density of older adults?
- What field methods and prediction equations may be used to estimate body composition of older men and women?
- How accurate are skinfold, bioelectrical impedance, and anthropometric equations in estimating body composition for groups of older adults and for older individuals?

With aging come significant changes in body composition that affect the overall health status of older individuals. As the body ages, FM increases and relatively more fat is deposited internally as visceral fat, thereby increasing one's risk of diseases associated with a centralized pattern of fat deposition. The marked decline in FFM, primarily caused by a loss of skeletal muscle mass and bone mineral, impairs muscular strength and increases the risk of disability and frailty in the aging population.

Although the general pattern of changes in body composition during aging is known, researchers are trying to confirm the amount, direction, and variability of changes in the relative proportions of the water, protein, and mineral components of the FFB due to aging. Much of what we know about these factors is speculative because it is based on cross-sectional, rather than longitudinal, studies.

So far, it appears that the chemical composition of the FFB does not influence the average FFB density in older adults as much as it does in children (see chapter 8). However, there is large interindividual variation in the FFB composition of older individuals. Therefore, experts agree that multicomponent body composition models need to be used to establish reference data and to develop field method prediction equations for older adults (Baumgartner et al. 1991; Heymsfield et al. 1989; Lohman 1992).

This chapter describes FFB composition of older individuals, compares body composition models and reference methods, and presents field method prediction equations based on multicomponent model estimates of body composition for older men and women. We evaluated prediction equations applicable to individuals 60 yr of age or older using the criteria outlined in chapter 2.

Fat-Free Body Composition of Older Adults

Changes in the water, mineral, and protein components of the FFB due to aging affect the density of the FFB (FFBd) in older adults. Based on 4-C model measures of the relative water, mineral, and protein components of the FFB, the average FFBd of older (>65 yr) women and men ranges between 1.093 and 1.099 g/cc (Baumgartner et al. 1991; Heymsfield et al. 1989; Visser et al. 1997; Yee et al. 2001). The variability in average FFBd most likely reflects differences among samples, as well as differences in methods used to measure FFB components (e.g., manufacturers, models, and software versions for dual-energy X-ray absorptiometry [DXA] measures of bone ash). Also, within a given sample, there is a large amount of interindividual variability (SD = 0.010-0.012 g/cc) in FFBd (Heymsfield et al. 1989; Visser et al. 1997). This points to the importance of using multicomponent models to obtain reference measures of body composition for older adults.

Between the ages of 25 and 65 yr, there is a substantial decrease in LBM (10-16%) due to losses in bone mass **(osteopenia)** and skeletal muscle mass **(sarcopenia)** with aging (Heymsfield et al. 1989; Kuczmarski 1989; Roubenoff 2000). Typically, the average relative TBM in the FFB of older Caucasian adults is less (6.0-6.6% FFB) than that assumed (6.8% FFB) for the reference body (Baumgartner et al. 1991; Clasey et al. 1999; Visser et al. 1997; Yee et al. 2001). The average relative hydration of the FFB of older women (73.6-75.6% FFB) and older men (72.4-74.4% FFB) varies depending on the age, ethnicity, and level of body fatness of the sample (Baumgartner et al. 1991; Clasey et al. 1999; Goran et al. 1998; Visser et al. 1997). Based on these cross-sectional studies, it is not entirely clear if the relative hydration of the FFB decreases, stays the same, or increases with aging. Longitudinal studies are needed to clarify this point. Thus, the 2-C model conversion formulas reported in chapter 1 (see table 1.4, p. 9) for estimating %BF from Db for older men and women should be used only when it is impractical to obtain 4-C model measures of body composition. These conversion formulas were calculated using average values reported in the literature for FFBd of the older adults derived from 4-C models. It is highly likely that these formulas will need to be modified when the discrepancy about the relative hydration of the FFB in elderly persons is resolved.

Body Composition Models and Reference Methods

Changes in the proportions and densities of the FFB components due to aging directly affect overall FFBd. Because the relative mineral content of the FFB decreases with aging, the density of the FFB in older adults is typically less than 1.10 g/cc. Generally, you should not use the 2-C models of Siri (1961) and Brozek et al. (1963) for estimating the relative body fatness of older individuals.

Molecular-level 4-C models are used to assess the bone mineral, TBW, and Db of older individuals (Baumgartner et al. 1991; Heymsfield et al. 1990). In the past, hydrostatic weighing (HW) has been widely used to measure Db. For many older adults, however, this method is extremely stressful. Some older individuals are not be able to get in and out of the HW tank, to bend over far enough to be completely submerged, and to maximally expire all of the air from their lungs for estimation of residual volume and underwater weight. These difficulties may substantially increase the measurement error (10-15% BF) of the HW method (Chumlea & Baumgartner 1989). Therefore, researchers are testing the validity of alternative methods (e.g., air displacement plethysmography and DXA) for obtaining reference measures of body composition for older adults.

Two-Component Models

The use of the classic 2-C models of Siri and Brozek is limited in older adults because the proportions or densities of the FFB components and average FFBd may differ from those for the adult reference body. On average, these 2-C models typically systematically overestimate the %BF for groups of older adults by 2% to 5% BF (Baumgartner et al. 1991; Clasey et al. 1999; Williams, Going, Lohman, Hewitt et al. 1992; Yee et al. 2001). Although Visser et al. (1997) reported no significant differences between 2-C and 4-C model estimates of average %BF (<1% BF) for groups of older Caucasian or African American men and women, they noted substantial individual errors when the 2-C model was used to estimate %BF of older individuals within their sample. Baumgartner et al. (1991) reported 95% limits of agreement between %BF$_{4-C}$ and %BF$_{2-C}$ of ±6.2% and ±8.3% BF, respectively, for women and men 65 to 94 yr of age. Likewise, Clasey et al. (1999) reported wide 95% limits of agreement between 2-C and 4-C model estimates of %BF for older women (±12.0% BF) and older men (±10.7% BF). Because there is a large degree of interindividual variability in the FFB of elderly individuals, experts advocate using multicomponent models to establish reference measures of body composition in this population (Baumgartner et al. 1991; Clasey et al. 1999; Visser et al. 1997).

HW has been widely used to measure Db and estimate %BF from 2-C or multicomponent molecular models in older adults. As mentioned previously, this method may not be suitable for elderly individuals who have difficulty complying with HW testing procedures. Recently, Yee and colleagues (2001) compared Db measures from HW and air displacement plethysmography (ADP) for older (70-79 yr) adults. They reported that the ADP and HW methods yielded similar average Db values. Both methods also provided similar 2-C, 3-C, or 4-C model estimates of %BF in older men and women. These results suggest that ADP may be a practical alternative to HW for assessing body composition of older adults in research and clinical settings. However, more research is needed to confirm these findings.

Three-Component Models

One can take into account interindividual variability in either TBW or TBM by using 3-C models. The Siri (1961) 3-C formula adjusts Db for the relative proportion of TBW; the Lohman (1986) 3-C formula adjusts Db for the relative amount of TBM. The average difference between Siri 3-C and 4-C model estimates of %BF in groups of older adults is ≤1.0% BF for older women and ≤1.9% BF for older men (Clasey et al. 1999; Yee et al. 2001). For individuals, the 95% limits of agreement between these models are relatively small (−0.24% to 1.68% BF) (Clasey et al. 1999). These data suggest that the 3-C water model provides valid and accurate reference measures of body composition in older adults.

In contrast, the Lohman (1986) 3-C mineral model is less accurate than the 3-C water model for estimating %BF of older adults. Yee et al. (2001) compared this 3-C model to 4-C model estimates of %BF for groups of older men and women (70-79 yr). The Lohman 3-C model systematically underestimated average %BF$_{4-C}$ by 7.3% BF in men and 2.5% BF women. The limits of agreement between these two models for individual estimates of %BF were not reported. Although more research comparing these models is needed, these data suggest that this 3-C model should not be used to obtain reference measures for older adults.

DXA Model

Recently, DXA has been used as a reference method for the development and cross-validation of field method prediction equations for older adults (Genton et al. 2001; Kyle, Genton, Karsegard et al. 2001; Ravaglia et al. 1999; Roubenoff et al. 1997). To test the validity of using DXA for assessing reference measures of body composition in older populations, studies have compared DXA and 4-C model estimates of %BF (Clasey et al. 1997, 1999; Snead et al. 1993). Clasey and colleagues (1997) reported no significant difference between average %BF$_{DXA}$ and %BF$_{4-C}$ (<1% BF) for older men and women (59-79 yr); however, the TE (5.0% BF) and 95% limits of agreement (±10.1% BF) between these models were unacceptably large, suggesting that DXA should not be used to obtain reference body composition measures for older adults. The authors also noted that the scan mode (i.e., pencil beam vs. array) affected the accuracy of DXA measures. The array scan mode produced significantly greater %BF and bone mineral content values than the pencil-beam mode. Clasey et al. concluded that the pencil-beam mode should be used in research and clinical settings. In a more recent study, Clasey et al. (1999) reported that DXA overestimated average %BF$_{4-C}$ in women by 2.2% BF. In men, %BF$_{4-C}$ was underestimated by 1.7% BF on average. The TE (5.1-5.3% BF) and 95% limits of agreement (±9.4% BF for women; ±10.3% BF for men) were unacceptable.

Snead et al. (1993) noted that DXA significantly underestimates average %BF$_{4-C}$ by approximately 4% to 5% BF in older men and women (60-81 yr). When FM was artificially manipulated by placement of lard over the trunk or thigh region, DXA markedly underestimated FM overlying the trunk compared to the thigh. Given that older men and women tend to accumulate excess fat in the upper body, the authors concluded that the lower %BF$_{DXA}$ was due to an underestimation of truncal fat in older adults.

Until DXA technology is improved and standardized (e.g., type of equipment, scan mode, and software version), experts agree that DXA should not be used alone as a reference standard to compare body composition models and methods or to develop and cross-validate field method prediction equations for older adults (Clasey et al. 1997; Kohrt 1995, 1998; Lohman 1996). Although the DXA model may yield fairly good estimates of %BF for groups of older adults, the individual estimation errors are substantial (as much as ±10% BF). Instead, DXA should be used in conjunction with 4-C models. Alternatively, the 3-C water model may be used to obtain reference body composition measures for older adults in research and clinical settings.

Field Methods and Prediction Equations

Unfortunately, relatively few field method prediction equations for older adults are based on 4-C model reference measures of body composition. Of the field method prediction equations listed in tables 9.1 and 9.2, only the SKF and BIA (bioelectrical impedance) equations of Williams (Williams, Going, Lohman, Hewitt et al. 1992; Williams et al. 1995) and the BIA equation of Baumgartner et al. (1991) were developed using 4-C models. All of the other SKF, anthropometric, and BIA equations are limited by the fact that they use either 2-C (HW) or DXA models to obtain reference measures of body composition in older adults.

SKF Equations for Older Adults

Table 9.1 presents SKF equations for older adults. The Durnin and Womersley (1974) equations are age and gender specific. These equations predict Db of women and men ranging in age from 16 to 72 yr. However, the number of males (N = 24) and females (N = 37) in the older (50+ yr) category was limited. Cross-validation studies show that these equations tend to either over- or underestimate average $\%BF_{2-C}$ or $\%BF_{4-C}$ for groups of older men by –4.8% to 2.3% BF and for groups of older women by –6.0% to 7.0% BF (Brodowicz et al. 1994; Ravaglia et al. 1999). For individuals, the 95% limits of agreement for older men (±8.4-13.0% BF) and women (±16.4% BF) are large (Ravaglia et al. 1999; Reilly et al. 1994).

The Visser et al. (1994) equation is a generalized equation that estimates the Db of older men and women (see table 9.1). This equation was cross-validated in the original study. The *internal cross-validation* (i.e., equations developed for each of two subgroups of the sample were applied to the other subgroup) indicated that this equation gave a fairly reasonable estimate of average Db for each subgroup, but the prediction errors were only fair (SEE = 0.0100-0.0117 g/cc). For *external cross-validation,* the equation was applied to an independent group of older adults. Results indicated that this equation overestimated average Db by 0.0072 g/cc for women and 0.0011 g/cc for men.

Unlike the other SKF equations developed for older adults, the Williams, Going, Lohman, Hewitt et al. (1992) and Kwok et al. (2001) equations estimate %BF instead of Db. For the Williams et al. equation, a 4-C model was used to obtain reference measures of %BF. The prediction errors were good to excellent (see table 9.1). When these equations were cross-validated against $\%BF_{4-C}$ for an independent sample of older adults (Clasey et al. 1999), the average $\%BF_{4-C}$ was underestimated by 1.9% BF for men and by 5.8% BF for women, and the prediction errors were large (TE = 7.9-9.7% BF). Also, for individuals within each gender, the 95% limits of agreement were ±15.6% BF.

The Kwok et al. (2001) equation estimates %BF of older Chinese women and men using DXA to obtain reference measures of %BF. The prediction error for this generalized equation was fair, but the internal and external cross-validation analyses indicated that this equation underestimates $\%BF_{DXA}$ by less than 2.0% BF on average (see table 9.1). For individuals, the 95% limits of agreement were fairly large (–11.0% to 7.5% BF).

In summary, it appears that the group and individual predictive accuracy of the SKF equations developed for older adults is not adequate. In fact, some experts discourage using the SKF method to assess body composition of elderly clients. With aging, adipose tissue is redistributed, with relatively more subcutaneous and internal fat stored on the trunk than the extremities (Chumlea & Baumgartner 1989). Also, age-related decreases in the elasticity and hydration of the skin, as well as shrinkage in the size of fat cells, may increase the compressibility of subcutaneous adipose and connective tissues (Kuczmarski 1989). Experts therefore recommend using alternative methods, such as circumference measures or BIA, to estimate body composition in older adults.

BIA Equations for Elderly Clients

BIA equations for older adults are presented in table 9.2. As previously mentioned, only two of these equations (i.e., those of Baumgartner et al. [1991] and Williams et al. [1995]) are based on 4-C model estimates of %BF and FFM. The gender-specific equations developed by Williams et al. (1995) had small prediction errors (SEE = 1.5 kg); however, these equations have not been cross-validated, and the sample size was relatively small (see table 9.2). Thus, the generalized BIA equation developed by Baumgartner et al. (1991) is most likely a better alternative for assessing the body composition of both older women and men. This equation was developed using a 4-C model, correcting Db for TBW and TBM, and a jackknife regression procedure (see chapter 2). Using this approach, the reported prediction error was very good (SEE = 2.5 kg).

Cross-validation of the Baumgartner equation for additional independent samples of older women and men produced inconsistent results. In one study that cross-validated this equation with a 4-C model, Goran et al. (1998) estimated FM instead of

Table 9.1 Comparison of Skinfold Equations for Older Adults

Name of equation	Reference model, methods, and caliper type	Validation sample	Equation and validation statistics for original study	Cross-validation analysis		
				Sample	Method	Statistics
Deurenberg, van der Kooy, Hulshop et al. 1989	2-C; HW; Harpenden	35 men and 37 women, 60-83 yr; ethnicity NR	Men: $Db = 1.1193 - 0.0525 (\log_{10} \Sigma 4SKF)^a$ $R^2 = 0.55$; SEE = 0.0082 g/cc Women: $Db = 1.0494 - 0.0253 (\log_{10} \Sigma 4SKF)^a$ $R^2 = 0.37$; SEE = 0.0082 g/cc	**	**	**
Durnin & Womersley 1974	2-C; HW; Harpenden and Lange	24 men and 37 women, 50-72 yr; ethnicity NR	Men: $Db = 1.1715 - 0.0779 (\log_{10} \Sigma 4SKF)^a$ $R^2 = NR$; SEE = 0.0092 g/cc Women: $Db = 1.1339 - 0.0645 (\log_{10} \Sigma 4SKF)^a$ $R^2 = NR$; SEE = 0.0082 g/cc	**	**	**
Kwok et al. 2001	2-C; DXA (Hologic QDR-2000; SV 5.67A); Holtain	533 Chinese women and men, 69-82 yr	$\%BF = 1.120 (BMI) + 17.308 (\log_{10} \Sigma 2SKF)^b + 6.137 (sex)^c - 27.149$ $R^2 = 0.81$; SEE = 4.1% BF	80 Chinese adults, 69-82 yr	Internal XV	$d = 0.6\%$ BF; LA = -8.7% to 7.4% BF for women $d = 1.8\%$ BF; LA = -10.4% to 6.9% BF for men
				78 Chinese adults, 65-86 yr	External XV	$d = 2.1\%$ BF; LA = -11.0% to 6.8% BF for women $d = 0.1\%$ BF; LA = -7.6% to 7.5% BF for men
Visser et al. 1994	2-C; HW; Harpenden	76 men and 128 women, 60-87 yr; ethnicity NR	$Db = 1.0481 - 0.0300 (\log_{10} \Sigma 2SKF)^b + 0.0186 (sex)^d$ $R^2 = 0.55$; SEE = 0.0117 g/cc	204 adults, 60-87 yr	Double XV	$d = 0.0003$ g/cc; LA = NR for group 1 $d = 0.0005$ g/cc; LA = NR for group 2
				23 adults, 62-82 yr	External XV	$d = 0.0072$ g/cc; LA = NR for women $d = 0.0011$ g/cc; LA = NR for men

(continued)

Table 9.1 *(continued)*

Name of equation	Reference model, methods, and caliper type	Validation sample	Equation and validation statistics for original study	Cross-validation analysis		
				Sample	Method	Statistics
Williams, Going, Lohman, Hewitt et al. 1992	4-C; HW, D₂O, DPA (Lunar DP3); Harpenden	91 CA men and 116 CA women, 34-84 yr	Men: %BF = 0.486 (Σ4SKF)[e] – 0.0015 (Σ4SKF)^2 + 0.067 (age) – 3.83 R^2 = 0.80; SEE = 2.9% BF Women: %BF = 0.428 (Σ4SKF)[f] – 0.0011 (Σ4SKF)^2 + 0.127 (age) – 3.01 R^2 = 0.75; SEE = 3.8% BF	**	**	**

[a]Σ4SKF = biceps + triceps + subscapular + suprailiac; [b]Σ2SKF = biceps + triceps; [c]Sex = 1 for men; 2 for women; [d]Sex = 0 for women; 1 for men; [e]Σ4SKF = chest + subscapular + midaxillary + thigh; [f]Σ4SKF = triceps + subscapular + abdomen + calf.
** Equation not cross-validated in original study.

Key: HW = hydrostatic weighing; Db = body density; D₂O = deuterium oxide dilution technique; DPA = dual-photon absorptiometry; DXA = dual-energy X-ray absorptiometry; BMI = body mass index; XV = cross-validation; *d* = absolute mean difference between SKF-predicted and reference values; LA = 95% limits of agreement; R^2 = multiple correlation coefficient squared; SEE = standard error of estimate; SV = software version; NR = not reported; CA = Caucasian.

Table 9.2 Comparison of Bioelectrical Impedance Analysis Equations for Older Adults

Name of equation	Reference model, methods, and analyzer	Validation sample	Equation and validation statistics for original study	Cross-validation analysis		
				Sample	Method	Statistics
Baumgartner et al. 1991	4-C; HW, ^3H_2O, DPA (Lunar DP-4); RJL model NR	98 CA men and women, 65-94 yr	FFM = 0.28 (HT^2/R) + 0.27 (BW) + 4.50 (sex)[a] − 1.732 R^2 = 0.91; SEE = 2.5 kg	98 CA adults, 65-94 yr	Jackknife	R^2 = 0.91; SEE = 2.5 kg
Deurenberg, van der Kooy et al. 1990	2-C; HW; RJL 101	75 men and women, 60-83 yr; ethnicity NR	FFM = 0.671 (HT^2/R) + 3.1 (sex)[a] + 3.9 R^2 = 0.88; SEE = 3.1 kg	**	**	**
Kyle, Genton, Karsegard et al. 2001; Geneva equation	2C; DXA (Hologic QDR-4500; SV 8.26a); Xitron 4000B	343 CA men and women, 22-94 yr	FFM = 0.58 (HT^2/R) + 0.231 (BW) + 0.130 (Xc) + 4.429 (sex)[a] − 4.104 R^2 = 0.97; SEE = 1.7 kg; LA = ±3.5 kg	343 CA adults, 22-94 yr	Double XV	d = −0.9 to 0.7 kg; TE = 1.7 kg; LA = ±3.5 kg

(continued)

Table 9.2 *(continued)*

Name of equation	Reference model, methods, and analyzer	Validation sample	Equation and validation statistics for original study	Cross-validation analysis		
				Sample	Method	Statistics
Roubenoff et al. 1997	2-C; DXA (Lunar DPX-L, SV 1.3); RJL 101	455 CA men and women, mean age 78 yr	Men: FFM = 0.4273 (HT²/R) + 0.1926 (BW) + 0.0667 (Xc) + 9.1536 $R^2 = 0.72$; SEE = 3.4 kg Women: FFM = 0.4542 (HT²/R) + 0.1190 (BW) + 0.0455 (Xc) + 7.7435 $R^2 = 0.77$; SEE = 2.1 kg	455 CA adults, mean age 78 yr	Jackknife	SEE = 3.4 kg for men and 2.1 kg for women
				96 CA and Hispanic adults, 60+ yr	External XV	d = 2.5 kg; SEE = 2.3 kg for men and 1.6 kg for women; LA = NR
Williams et al. 1995	4-C; HW, DXA (Lunar DPX, SV 3.6), D₂O; Valhalla 1990B	23 CA women and 25 CA men, 49-80 yr	Men: FFM = 0.54 (HT²/R) + 0.13 (BW) + 0.13 (Xc) − 0.11 (age) + 8.71 $R^2 = 0.92$; SEE = 1.5 kg Women: FFM = 0.37 (HT²/R) + 0.16 (BW) + 11.94 $R^2 = 0.76$; SEE = 1.5 kg	**	**	**

** Equation not cross-validated in original study.

[a]Sex = 0 for women; 1 for men.

Key: BW = body weight; DPA = dual-photon absorptiometry; DXA = dual-energy X-ray absorptiometry; HW = hydrostatic weighing; Db = body density; TBW = total body water; ³H₂O = tritium dilution technique; D₂O = deuterium oxide dilution technique; HT²/R = resistance index; Xc = reactance; R² = multiple correlation coefficient squared; d = absolute mean difference between BIA-predicted and reference values; LA = 95% limits of agreement; SEE = standard error of estimate; SV = software version; NR = not reported; CA = Caucasian.

FFM. They reported that the Baumgartner equation slightly overestimated the average FM of older men by <1.0 kg, but grossly overestimated the average FM of older women (7.7 kg). The prediction errors, expressed in kilograms of FM, were fairly large for both men (4.0 kg) and women (4.2 kg); and limits of agreement for individuals were not reported.

Two other studies cross-validated the Baumgartner equation using DXA to obtain reference measures of %BF from which FFM was calculated (Genton et al. 2001; Kyle, Genton, Karsegard et al. 2001). Genton et al. (2001) reported that the Baumgartner equation overestimated average FFM_{DXA} of older women (4.3 kg) and men (1.4 kg), but the prediction errors (SEE = 2.4-2.8 kg; TE = 2.0-3.6 kg) were good. For individuals, the 95% limits of agreement were better for men (–2.7 to 5.6 kg) than for women (–1.5 to 10.1 kg). In addition, Kyle, Genton, Karsegard et al. (2001) noted that the Baumgartner equation underestimated the average FFM_{DXA} by 2.9 kg for a combined group of older men and women, but the overall prediction error was good (SEE = 2.8 kg).

Recently, Kyle, Genton, Karsegard et al. (2001) developed and cross-validated a single BIA prediction equation for assessing FFM of Caucasian adults 22 to 94 yr of age. This equation is also referred to as the Geneva equation. Approximately one-third of the total sample (N = 343) was at least 60 yr of age. One drawback, however, is that reference measures of FFM were obtained using DXA. Cross-validation of this equation yielded excellent prediction errors and narrow 95% limits of agreement (see table 9.2). Additional cross-validation of the Geneva equation for a sample of 206 Caucasian women and men, 65 to 94 yr of age, produced similar results. These findings suggest that the Geneva equation may be very useful in assessing body composition of older adults; however, additional studies are needed to test the predictive accuracy of this equation against 4-C model reference measures of FFM in older adults.

To assess body composition of older adults in field and clinical settings, we suggest using either the Baumgartner or the Geneva equation (see table 9.3). Keep in mind, however, that the Geneva equation is based on reference measures of FFM obtained using DXA and that DXA has not yet been universally accepted as the single best reference method for assessing body composition of older adults. To minimize the prediction error for these equations, you should measure whole-body resistance and reactance with the same type of bioimpedance analyzer that Baumgartner (RJL) and Kyle (Xitron 4000B) used to develop their equations.

Near-Infrared Interactance Equations for Older Adults

As discussed in chapter 7, much more research needs to be done before the near-infrared interactance (NIR) method can be applied to population subgroups such as older women and men. We advise against using the Futrex-5000 manufacturer's NIR equation because of its poor predictive accuracy. Therefore, at this time, we recommend using alternative field methods like BIA or anthropometric equations to assess body composition of your older clients.

Anthropometric Equations for Older Adults

Generalized anthropometric equations have been developed that use a combination of circumference measures to estimate Db in women (15-79 yr) and to estimate $\%BF_{2-C}$ in men (22-78 yr) (Tran & Weltman 1988, 1989). Large samples of women (N = 400) and men (N = 462) were used to generate these equations (see table 9.3).

Cross-validation of these equations indicated that the predictive accuracy for the female equation (TE = 0.0082 g/cc or 3.6% BF) was good, whereas the TE for the male equation was somewhat larger (TE = 4.4% BF). These anthropometric equations use the average of two abdominal circumferences, along with other predictor variables, to estimate either Db or %BF. The two abdominal circumferences are measured (a) midway between the xiphoid process of the sternum and the umbilicus and (b) at the level of the umbilicus. Measure the iliac circumference at the level of the anterior superior iliac spines; measure hip circumference at the level of the symphysis pubis anteriorly and the maximal protrusion of the buttocks posteriorly. To obtain %BF estimates from Db for older women, use the appropriate age-gender conversion formula (see table 9.3).

Clasey et al. (1999) cross-validated the Tran and Weltman equations against 4-C model reference measures of %BF for older adults. The average $\%BF_{4-C}$ of men and women was accurately estimated using these equations, with less than a ±1% BF difference between measured and predicted %BF. However, the prediction errors were fairly large (TE = 4.8-6.4% BF). For individuals, the 95% limits of agreement were ±9.7% BF for women and ±12.9% BF for men. Although the individual prediction errors are unacceptable, the group and individual predictive accuracy of the Tran and Weltman anthropometric equations was better than that of selected SKF equations (i.e., Jackson and Pollock Σ7SKF and Williams equations) for estimating %BF for older men and women (Clasey et al. 1999).

Table 9.3 Summary of Recommended Equations for Older Adults

	Method/Equipment	Ethnicity/Gender	Equation	Reference
BIA	RJL model NR	CA women and men (65-94 yr)	1. FFM_{4-C} (kg) = 0.28 (HT^2/R) + 0.27 (BW) + 4.50 (sex)[a] − 1.732	Baumgartner et al. 1991
	Xitron 4000B	CA women and men (22-94 yr)	2. FFM_{DXA} (kg) = 0.58 (HT^2/R) + 0.231 (BW) + 0.130 (Xc) + 4.429 (sex)[a] − 4.104	Kyle, Genton, Karsegard et al. 2001 Geneva equation
ANTHROPOMETRY		CA women (15-79 yr)	3. Db (g/cc)[b] = 1.168297 − [0.002824 × abdominal C] + [0.0000122098 × (abdominal C)^2] − [0.000733128 × hip C] + [0.000510477 × HT] − [0.000216161 × age]	Tran & Weltman 1989
		CA men (15-78 yr)	4. $\%BF_{2-C}$ = −47.371817 + 0.57914807 (abdominal C) + 0.25189114 (hip C) + 0.21366088 (iliac C) − 0.35595404 (BW)	Tran & Weltman 1988

[a]Sex = 0 for women; 1 for men; [b]Use the following age-gender formula to convert Db to %BF for older women (60-90 yr): %BF = [(5.02/Db) − 4.57] × 100.

Key: HT = height (cm); BW = body weight (kg); R = resistance (Ω); Xc = reactance (Ω); C = circumference (cm); abdominal C (cm) = average of two abdominal circumferences measured (1) anteriorly midway between the xiphoid process of the sternum and the umbilicus and laterally between the lower end of the rib cage and iliac crests and (2) at the umbilical level; 4-C, 2-C = four-component or two-component model; CA = Caucasian.

Key Points

- The overall density of the FFB tends to decrease with aging primarily due to the loss of bone mineral and skeletal muscle mass.

- There is a wide degree of interindividual variability in the FFB composition of older men and women.

- The classic 2-C model of Siri systematically overestimates the average %BF for groups of older adults. For individuals, the predictive accuracy of the 2-C model ranges from ±6.2% to 12.0% BF.

- In research settings, either the 4-C or 3-C (water) models should be used to obtain reference measures of body composition for older adults.

- Compared to 4-C models, DXA typically underestimates average %BF in older adults; the predictive accuracy of DXA for assessing body composition of older individuals ranges from ±9.1% to 10.3% BF.

- More research is needed to determine the extent of agreement between hydrodensitometry and ADP for measuring Db of older adults. Preliminary findings suggest that these two methods yield similar estimates of average Db for groups of older men and women.

- SKF and NIR methods should not be used in field and clinical settings to assess the body composition of older adults.

- To obtain a fairly close estimate of FFM_{4-C} for older Caucasian adults, use the Baumgartner (1991) BIA equation. To estimate FFM_{DXA} for older Caucasian adults, use the Geneva (Kyle, Genton, Karsegard et al. 2001) equation.

- As an alternative to BIA, you can use the generalized anthropometric equations of Tran and Weltman (1988, 1989) to assess body composition of older Caucasian adults in field and clinical settings.

- More research is needed to establish the validity and applicability of field method prediction equations for assessing body composition of older adults from various ethnic groups.

- Newly developed prediction equations for older adults should be based on multicomponent models (i.e., 4-C or 3-C water models) to account for interindividual variability in the mineral and water components of the FFB.

Key Terms

Learn the definition for each of the following key terms. Definitions of terms can be found in the Glossary, page 227.

osteopenia

sarcopenia

Review Questions

1. Describe the general pattern of changes in body composition due to aging.
2. Describe the changes in the water and mineral content of the FFB of older adults. How do these changes affect the overall FFB density of older adults?
3. Explain why Siri's 2-C model overestimates the average %BF of older men and women.
4. Identify two multicomponent molecular models that can be used to obtain reference measures of body composition for older adults.
5. What are the limitations of using HW for assessing body composition of older adults?

6. How accurate is DXA for estimating body composition of older individuals? Is DXA a suitable alternative reference method? Explain.

7. Identify two field methods and the respective prediction equations that you may use to estimate body composition of older men and women. Specify the age and ethnic groups for whom these equations are applicable.

8. Compare the predictive accuracy of field methods for estimating body composition of older individuals. Which field method is best?

BODY COMPOSITION AND ETHNICITY

There are strong links between ethnicity, obesity, and disease. The prevalence of overweight and obesity in the United States varies among ethnic groups. Certain ethnic groups have a relatively higher risk for obesity, predisposing them to cardiovascular disease, hypertension, non-insulin-dependent diabetes mellitus, certain cancers, and osteoarthritis. For this reason, population subgroups, particularly African American women, Hispanic women, and American Indian women and men, have been targeted for weight loss in the nationwide health promotion and disease prevention initiative *Healthy People 2010: Understanding and Improving Health* (U.S. Dept. of Health and Human Services 2000c). Accordingly there is a need for practical methods and equations that can be used to accurately assess the degree of adiposity of individuals from different ethnic origins.

Crude indices of total body fatness, such as BMI and abdominal obesity (waist-to-hip ratio or WHR), fail to identify at-risk individuals and do not always accurately predict all-cause mortality in certain ethnic groups (Stevens et al. 1992). For example, we found that only 25% of American Indian women in our sample were identified as obese with use of BMI

(BMI > 27.8). In contrast, 78% of these women were classified as obese (≥32% BF) using hydrodensitometry and a 3-C body composition model to estimate body fat (unpublished observation). Likewise, for a given BMI, Asians have a higher body fat than Caucasians, indicating that the prevalence of obesity (typically estimated by BMI cutoff values derived for Caucasians) is likely underestimated in the Asian population (Deurenberg et al. 2002; Deurenberg-Yap et al. 2000; Wang et al. 1994). Also, BMI is not sensitive to differences in fat patterning among ethnic groups. For example, ethnic differences in BMI among a multiethnic sample of 498 adolescent girls were not significant, but there was a significant difference in the proportion of subcutaneous adipose tissue deposited on the trunk (Asian > Mexican American > Caucasian > African American), with ethnicity accounting for 5.7% of this variability (Malina et al. 1995). It is becoming clear that the relationship between %BF and BMI differs among ethnic groups, suggesting that BMI-based criteria for "overweight" and "obesity" classifications need to be ethnic specific (Deurenberg & Deurenberg-Yap 2002; Deurenberg & Yap 1999). This also illustrates the importance of using more precise and accurate methods for assessing body composition and identifying individuals at risk due to obesity.

Ethnic differences in fat patterning and body proportions may also affect the validity of generalized field method prediction equations. Differences exist across ethnic groups in the ratio of trunk to extremity subcutaneous fat distribution and in limb lengths relative to total height (Deurenberg & Deurenberg-Yap 2001; Wagner & Heyward 2000). This obviously has implications for each of the field methods—SKFs, bioelectrical impedance analysis, near-infrared interactance (NIR), and anthropometry—as they rely on assumptions of consistent subcutaneous fat distribution and relative limb length. For example, after analyzing the bioimpedance vectors (i.e., resistance and reactance) from previously published studies of 10 different ethnic groups, Ward and colleagues (2000) found significant differences among the mean vectors of these groups, indicating that there are ethnic-specific differences for body impedance.

This chapter is subdivided into five ethnic classifications: African American, American Indian, Asian, Caucasian, and Hispanic. In addition to the differences in body composition among these classifications, differences are likely among ethnic groups within each classification. For example, several body composition variables differ significantly among blacks of African descent living in the United States, Nigeria, and Jamaica (Luke et al. 1997). Also, for the same BMI, there is a difference in %BF between Beijing Chinese and Singaporean Chinese (Deurenberg et al. 2000). However, it is beyond the scope of this chapter to attempt further subdivision of these ethnic classifications; moreover, data for additional subdivisions are limited. The chapter begins with a comparison of body composition models and reference methods for each ethnic group. It then presents the FFB composition and field method prediction equations for each ethnic classification. These prediction equations were selected using the evaluation criteria outlined in chapter 2.

Body Composition Models and Reference Methods

Because of the variability in the proportions and densities of the FFB components among various ethnic groups, the multicomponent model is highly recommended as the reference method. Deurenberg and Deurenberg-Yap (2001) went so far as to call it "obligatory" for body composition studies of ethnic groups. However, to date, such studies have often used 2-C models or the dual-energy X-ray absorptiometry (DXA) model as the criterion.

Two-Component Models

As presented throughout this chapter, it is likely that the density of the FFB (FFBd) varies from the assumed value for 2-C models (1.100 g/cc) and may not be consistent among ethnic groups. Thus, it is not surprising for a systematic bias to occur when 2-C models are applied to an ethnic group that varies from this assumption. For example, research suggests that the average FFBd of African Americans exceeds 1.100 g/cc. In fact, several 4-C model studies have indicated that the commonly used 2-C model equations of Siri (1961) and Brozek et al. (1963), which were derived from cadaver analyses of a few Caucasians (see chapter 1), systematically underestimated the %BF of their African American samples (Ortiz et al. 1992; Thompson & Moreau 2000; Wagner & Heyward 2001). Thompson and Moreau (2000) reported that using a 2-C model conversion formula developed specifically for African Americans (i.e., Schutte et al. [1984] equation) reduced this systematic bias (CE = –2.5% BF, 95% limits of agreement = ±2.8% BF for the Brozek et al. [1963] model and CE = 0.5% BF, 95% limits of agreement = ±2.6% BF for the Schutte [1984] ethnic-specific formula) for their African American sample. Additionally, the average FFBd has not been clearly established for many ethnic groups. For example, in 4-C model studies of African American women, the average FFBd reportedly ranges from 1.100 g/cc (Visser et al. 1997) to 1.109 g/cc (Thompson & Moreau 2000). Thus, choosing a 2-C model formula for such a population is questionable.

The Bod Pod air displacement plethysmography (ADP) system (see chapter 3) has also been used to estimate Db in studies involving several different ethnic groups. Generally, the results from these studies have been similar to those for Caucasian samples (see chapter 3); the average differences between the Bod Pod and other reference methods have typically been small, but statistically significant. For example, using the Siri 2-C model, %BF$_{ADP}$ underestimated %BF$_{4-C}$ by an average of 0.5% to 3.2% BF for various gender and ethnic subsamples of Singaporeans (Deurenberg-Yap et al. 2001). We found an acceptable prediction error (0.0072 g/cc) and a small but statistically significant average difference (0.0045 g/cc) between Db$_{ADP}$ and Db$_{HD}$ in African American men (Wagner et al. 2000). There was good agreement between hydrodensitometry (HD) (15.8% BF) and DXA (16.1% BF), but the average %BF from the Bod Pod (17.7% BF) was significantly greater than that from the other two methods in this study. Similarly, a small but significant difference in Db (0.004 g/cc) between the Bod Pod and HD was reported in a sample of collegiate football players that included both African Americans and Caucasians (Collins et al. 1999). The authors noted an average underestimation of 1.8% BF for the Bod Pod compared to a 3-C mineral model, with half of the participants accurately estimated within ±2% BF. Finally, the relationship between $\%BF_{ADP}$ and $\%BF_{DXA}$ has been strong (r = .89-.91) in Japanese samples (Koda et al. 2000; Miyatake et al. 1999, but the average difference in %BF between these two methods (−3.3% to 2.2% BF) varies with age and gender (Koda et al. 2000).

DXA Model

Compared to multicomponent models, DXA varies in its predictive accuracy among ethnic groups. There were no significant differences for average $\%BF_{DXA}$ and $\%BF_{4-C}$, and the prediction errors were acceptable for American Indian (CE = 0.4% BF, SEE = 2.5% BF, TE = 2.9% BF) and African American (CE = 0.3% BF, SEE = 2.3% BF, TE = 2.4% BF) men (Heyward et al. 1998; Wagner & Heyward 2001). Similarly, Hicks et al. (1993) reported no significant difference between $\%BF_{DXA}$ and $\%BF_{3-C}$ in American Indian women. However, on average, DXA underestimated $\%BF_{4-C}$ by 2.1% to 4.2% BF in male and female groups of Singaporean Chinese, Malays, and Indians (Deurenberg-Yap et al. 2001). Also, in a multiethnic study of girls (9-17 yr), DXA overestimated $\%BF_{4-C}$ by an average of 3.9% BF with an SEE of 3.3% BF (Wong et al. 2002). A similar bias existed for each of the ethnic groups (African American = 3.7%, Asian = 3.9%, Caucasian = 4.2%, Hispanic = 3.0%), and the authors noted that the differences between DXA and the 4-C model are not affected by ethnicity, age, or body fatness. After correction for the average bias (3.9% BF), the individual error was still large (95% limits of agreement = ±6.7% BF) for the entire sample.

The DXA method has also been compared to 2-C model methods. He et al. (1999) reported no significant difference in FFM estimated by DXA and hydrometry in obese (0.4 kg) and non-obese (0.5 kg) Chinese women. The limits of agreement were −3.1 to 4.1 kg and 4.4 to 5.2 kg for non-obese and obese participants, respectively. In contrast, Van Loan (1998) noted that DXA significantly underestimated average FFM from HD (−1.2 kg) and bioimpedance spectroscopy (−0.7 kg) for a combined sample of Chinese Americans and Caucasians.

Assessing Body Composition of African Americans

Although the prevalence of obesity in African American men is similar to that for Caucasian men, the incidence of obesity in African American women is considerably higher than in Caucasian and Mexican American women. Over 50% of African American women are considered obese (BMI ≥ 30), and about 77% are overweight (BMI > 25) (National Center for Health Statistics 2002). This is a great health concern because obesity is related to elevated cardiovascular risk factors in the African American population (Folsom et al. 1991). In 18- to 30-yr-old African Americans, obesity was associated with a 1.5-fold increase in hypercholesterolemia, a twofold increase in hypertension, and a fourfold increase in diabetes. This section first addresses ethnic variation in the FFB composition of African Americans and then examines the applicability of various body composition methods and prediction equations in this population. Recommended equations for this ethnic group are summarized in table 10.1.

FFB Composition of African Americans

We reviewed the biological differences in body composition between African Americans and Caucasians (Wagner & Heyward 2000). Research from this review consistently showed no significant difference between the two ethnic groups in the TBW portion of the FFB; however, African Americans have significantly greater bone mineral content (BMC), bone mineral density (BMD), and total body potassium (TBK) (indicative of protein content) than Caucasians. Although the magnitude of these ethnic differences varies, they remain significant across the life span in both men and women.

Although the mineral fraction of the FFM was greater in African American men (7.1%) and women (7.7%) than in Caucasian men (6.6%) and women (7.0%), Visser and colleagues (1997) reported that the FFBd was not significantly different among the four groups. However, all other multicomponent studies have reported a significantly greater FFBd for African Americans compared to Caucasians (Cote & Adams 1993; Evans, Prior et al. 2001; Ortiz et al. 1992).

For many years, the Schutte et al. (1984) formula, which assumes a FFBd of 1.113 g/cc for African American men, was used to convert Db to %BF for this population. However, this FFBd value was derived from a small sample (N = 15) using hydrometry rather than a multicomponent model to obtain reference measures of body composition. We cross-validated this formula, as well as other 2-C model equations, against a 4-C model. Although the errors were not large, the Schutte formula significantly overestimated the average %BF of 30 African American men (CE = 1.3% BF, TE = 2.0% BF). In contrast, the Siri (1961) (CE = 1.9% BF, TE = 2.6% BF) and Brozek et al. (1963)

(CE = 1.8% BF, TE = 2.3% BF) formulas, derived from Caucasian samples, significantly underestimated $\%BF_{4-C}$ of this sample (Wagner & Heyward 2001). Subsequently, we derived a new conversion formula to predict %BF from Db for African American men (see table 1.4, p. 9). Our formula assumes that the average FFBd of African American men is 1.106 g/cc and is based on the following values for the components of the FFB: 71.73%, 6.81%, and 21.46% for water, mineral, and protein, respectively. We recommend using our formula to convert Db to %BF for African American men because it was derived from a multicomponent model and a sample size that was double that of Schutte et al. (1984). However, the Wagner and Heyward (2001) formula still needs to be cross-validated on additional independent samples of African American men.

In studies of African American women, the relative mineral content in the FFB (7.5-7.8%) has consistently been greater than that of their Caucasian counterparts (6.7-7.3%) (Cote & Adams 1993; Ortiz et al. 1992; Visser et al. 1997). Likewise, when weight- and height-matched premenopausal African American and Caucasian women were compared using a model that combined DXA and elemental analyses from neutron activation (see chapter 3), TBM was 10.7% greater for the African American women (Aloia et al. 1997). The average BMD of African American women (1.18-1.25 g/cm²) was also significantly greater than that of Caucasian women (1.09-1.16 g/cm²) (Cote & Adams 1993; Ortiz et al. 1992). Additionally, African American women have 5.6% to 8.0% more TBK than Caucasian women (Aloia et al. 1997; Ortiz et al. 1992).

The average FFBd of African American women is generally considered to be greater than that of Caucasian women, but there is some variability in this value. Thompson and Moreau (2000) reported an average FFBd of 1.109 g/cc for their sample of 11 African American collegiate female athletes. Ortiz et al. (1992) also reported a value (1.106 g/cc) considerably higher than that assumed for the reference body. In contrast, Cote and Adams (1993) noted very close agreement between multicomponent and 2-C model estimates of %BF in young African American women, suggesting that the FFBd of these women is close to 1.100 g/cc. Also, an average FFBd of 1.100 g/cc was reported for a sample of 99 African American women (Visser et al. 1997) and a combined sample of African American men and women (Evans, Prior et al. 2001). Because of the high degree of variability in the FFBd values reported in these studies, it is difficult to pinpoint the average FFBd of African American women. These findings further illustrate the need to use multicomponent models whenever possible.

Like that of the adults, the BMC and LTM of African American children also appears to be greater than for children from other ethnic groups. In multiethnic studies of boys (Ellis 1997) and girls (Ellis et al. 1997) aged 3 to 18 yr, BMC and LTM were higher in African Americans than Caucasians, but the differences between Caucasians and Mexican Americans were not significant for either gender.

SKF Equations for African Americans

Some researchers have reported that the relationship between SKF and Db differs for African American and Caucasian men (Schutte et al. 1984; Vickery et al. 1988). Schutte et al. (1984) reported that the average Db, estimated from six different age-specific SKF equations for men, was significantly underestimated in African American men 18 to 32 yr of age. On the basis of this observation, Vickery et al. (1988) developed an ethnic-specific SKF equation for African American males, 18 to 32 yr of age, using the same seven SKF sites as in the generalized equation developed by Jackson and Pollock (1978). The SEE for this ethnic-specific equation was good (0.0070 g/cc); however, it was not cross-validated on a separate sample from this population. Recently, we cross-validated the Vickery et al. (1988) and Jackson and Pollock (1978) Σ3SKF and Σ7SKF equations on a sample of 30 African American men. All of the equations accurately estimated the average Db from HD, but the prediction errors were unacceptable (SEE = 0.0089-0.0101 g/cc). However, when the analysis was rerun on only non-obese men (≤20% BF, N = 22), the prediction errors were reduced to <0.0080 g/cc for all equations, and all three equations predicted Db equally well (unpublished observations).

Others have also reported that the generalized Σ3SKF and Σ7SKF equations of Jackson and Pollock (1978) accurately predict %BF of African American men with use of the ethnic-specific formula of Schutte et al. (1984) to convert Db to %BF. The prediction errors have been low for both the Σ3SKF (SEE = 1.9-2.6% BF) and Σ7SKF (SEE = 1.8-2.5% BF) equations when used to predict the %BF of young, nonathletic African American men (Stout, Eckerson, Housh, & Johnson 1994) and African American collegiate football players (Hortobagyi, Israel, Houmard, O'Brien et al. 1992; Houmard et al. 1991). The body fatness of the majority of participants in these studies was average (13.0-14.7% BF).

SKF equations have not been cross-validated against a multicomponent model in African American women. Researchers have used different reference methods and different 2-C conversion formulas to cross-validate SKF equations for African American women, making it difficult to evaluate the results. In a study that used TBW to obtain reference measures of FFM, the Jackson et al. (1980) generalized Σ7SKF equation systematically underestimated the average %BF of African American women, 19 to 44 yr of age, by 3% BF (SEE = 3.7% BF) (Zillikens & Conway 1990). Using HD and the 2-C conversion formula of Ortiz et al. (1992), which assumes a FFBd of 1.106 g/cc, the Jackson et al. (1980) Σ7SKF equation accurately estimated the average %BF$_{HD}$ (CE = 2.6% BF) in African American women 19 to 45 yr of age (Brandon & Bond 1999). However, in both this study and another using the same methodology (Brandon 1998), the prediction errors were large (SEE = 4.3-6.0% BF, TE = 9.3-14.1% BF) for this equation. Brandon (1998) also noted that for each of the SKF equations evaluated, the prediction errors were larger for African American women than for Caucasian women. Other researchers who used the 2-C conversion formula of Siri (1961), which assumes a FFBd of 1.100 g/cc, have reported that the Jackson et al. (1980) Σ7SKF equation accurately estimates average %BF$_{HD}$ (0.7% BF) and Db (0.0012 g/cc) of African American women (Irwin et al. 1998; Sparling et al. 1993).

A consistent finding is that the CE is small, but the prediction errors in these SKF studies of African Americans are large and unacceptable. However, close inspection of these studies reveals that often a large percentage of the participants were obese. The SKF method is not recommended for clients who are obese (see chapter 13). Thus, these large prediction errors are likely due, in part, to body fatness rather than ethnic variation. When samples have been limited to non-obese participants, the prediction errors have been acceptable. On the basis of these findings, we recommend using the Σ7SKF equations (Jackson & Pollock 1978; Jackson et al. 1980) to predict Db for your non-obese African American clients. Although research consistently shows that African Americans deposit subcutaneous fat in a different pattern than Caucasians, with African Americans having a higher ratio of trunk-to-extremity SKF thickness (Wagner & Heyward 2000), these generalized equations still accurately predict Db because they use the sum of SKF measurements from both the trunk and the extremities. However, the SKF method should be used only on clients who are not obese. Furthermore, ethnic-specific conversion formulas (see table 1.4, p. 9) should be used to estimate %BF from Db in African American men and women.

As discussed in chapter 8, the Slaughter et al. (1988) SKF equations can be used to estimate %BF of African American boys and girls (see table 10.1). These equations were based on a 4-C model of %BF for this population.

BIA Equations for African Americans

The majority of research indicates that the BIA method is not accurate for estimating the body composition of African Americans when one is using generalized equations. For African American men, significant bias and large prediction errors were reported for the RJL manufacturer's equation (CE = 7.0% BF, TE = 9.4% BF) (Stout, Eckerson, Housh, & Johnson 1994). Numerous BIA equations have significantly underestimated TBW by an average of 1.9 to 2.9 L (Zillikens & Conway 1991; unpublished observations) and average FFM by 1.5 to 5.1 kg with unacceptable prediction errors (TE = 3.1-6.0 kg) for African American men (Heyward et al. 1994).

Hortobagyi and colleagues reported that several BIA equations developed for Caucasian men had acceptable prediction errors (<3.5 kg) but systematically underestimated the average FFM of African American collegiate football players (Hortobagyi, Israel, Houmard, O'Brien et al. 1992). We also found acceptable prediction errors (SEE = 2.1 kg, TE = 2.7 kg) and a small, but statistically significant, bias (−1.8 kg) for the Segal et al. (1988) fatness-specific BIA equations as modified by Stolarczyk et al. (1997) for a sample of 37 African American men, 19 to 50 yr of age (Wagner et al. 1997). In this study, a 2-C model and the Schutte et al. (1984) ethnic-specific conversion formula were used to obtain reference measures. More recently, we used a 4-C model to evaluate the predictive accuracy of the modified Segal et al. fatness-specific equations in an independent sample of 30 African American men. Once again we found an acceptable SEE (2.2 kg) but a significant average difference in FFM (2.4 kg) (unpublished observations).

The BIA results for African American women are similar to those for the men. Generally, the BIA method has resulted in significant biases with unacceptable prediction errors for estimates of %BF (4.0-5.7% BF), TBW (1.2-2.0 L), and FFM (3.3-3.9 kg) in this population (Ainsworth et al. 1997; Brandon & Bond 1999; Sparling et al. 1993; Zillikens & Conway 1991). As with the men, Ainsworth et al. (1997) reported acceptable prediction errors (SEE = 2.8 kg, TE = 2.9 kg) and a small, statistically significant, underestimation (−0.6 kg) of FFM$_{HD}$ for the modified Segal et al. (1988) fatness-specific equations.

Table 10.1 Summary of Recommended Equations for African Americans

	Method/Equipment	Population	Equation	Reference
SKF	Σ7SKF[a] (Lange)	Non-obese males (18-61 yr)	1. Db (g/cc) = 1.112 − 0.00043499 (Σ7SKF) + 0.00000055 (Σ7SKF)^2 − 0.00028826 (age)	Jackson & Pollock 1978
	Σ7SKF[a] (Lange)	Non-obese females (18-55 yr)	2. Db (g/cc) = 1.097 − 0.00046971 (Σ7SKF) + 0.00000056 (Σ7SKF)^2 − 0.00012828 (age)	Jackson et al. 1980
	Σ2SKF[b] (Harpenden)	Boys (8-17 yr)	3. %BF$_{4-C}$ = 0.735 (Σ2SKF) + 1.0	Slaughter et al. 1988
	Σ2SKF[b] (Harpenden)	Girls (8-17 yr)	4. %BF$_{4-C}$ = 0.610 (Σ2SKF) + 5.1	Slaughter et al. 1988
BIA	RJL 101A	Males <17% BF	5. FFM (kg) = 0.00066360 (HT^2) − 0.02117 (R) + 0.62854 (BW) − 0.1238 (age) + 9.33285	Segal et al. 1988
		Males 17-25% BF	6. FFM (kg) = [0.00066360 (HT^2) − 0.02117 (R) + 0.62854 (BW) − 0.1238 (age) + 9.33285] + [0.0008858 (HT^2) − 0.02999 (R) + 0.42688 (BW) − 0.07002 (age) + 14.52435] / 2	Modified Segal equations[c]
		Males >25% BF	7. FFM (kg) = 0.0008858 (HT^2) − 0.02999 (R) + 0.42688 (BW) − 0.07002 (age) + 14.52435	Segal et al. 1988
		Females <25% BF	8. FFM (kg) = 0.00064602 (HT^2) − 0.01397 (R) + 0.42087 (BW) + 10.43485	Segal et al. 1988
		Females 25-35% BF	9. FFM (kg) = [0.00064602 (HT^2) − 0.01397 (R) + 0.42087 (BW) + 10.43485] + [0.00091186 (HT^2) − 0.01466 (R) + 0.2999 (BW) − 0.07012 (age) + 9.37938] / 2	Modified Segal equations[c]
		Females >35% BF	10. FFM (kg) = 0.00091186 (HT^2) − 0.01466 (R) + 0.2999 (BW) − 0.07012 (age) + 9.37938	Segal et al. 1988

[a]Σ7SKF (mm) = sum of seven skinfolds: chest + midaxillary + triceps + subscapular + abdomen + anterior suprailiac + thigh; [b]Σ2SKF (mm) = sum of two skinfolds: triceps + calf; [c]As modified by Stolarczyk et al. (1997).

Key: HT = height (cm); BW = body weight (kg); R = resistance (Ω).

These BIA equations underestimate FFM in African Americans for several reasons:

• Differences in body proportions may affect the relationship between the resistance index (height squared [HT^2]/resistance [R]) and FFM. African Americans, on the average, tend to have relatively shorter trunks and longer limbs than Caucasians (Wagner & Heyward 2000). Because total body resistance is largely determined by segmental resistances in the extremities, the total body resistance is greater in African Americans compared to Caucasians, resulting in an underestimate of FFM.

• The skeletal muscle mass and bone mass in the limbs is relatively greater in African Americans than in Caucasians. Therefore, it is incorrect to assume that the specific resistivity of the body is the same for both ethnic groups. The specific resistivity of the lower limbs is significantly greater in African American women compared to Caucasian women, reflecting differences in the tissue composition in the lower extremity (Stokes et al. 1993).

• The amount of bone composing the FFM is relatively greater in African Americans than in Caucasians, thereby influencing the cross-sectional area of the FFM and the body's geometry.

• African Americans have a relatively greater amount of FFM per unit of height (Vickery et al. 1988), suggesting that the relationship between HT^2/R and FFM will differ for African American and Caucasian populations.

These findings explain, in part, why BIA equations, developed primarily using Caucasian samples, significantly underestimate the FFM in African Americans. Several researchers have tried to develop ethnic-specific BIA equations with varying degrees of success. Schoeller and Luke (2000) created separate BIA equations for African Americans and Caucasians to estimate TBW. However, when the African American equation was cross-validated in Nigerians and Jamaicans of African origin, the mean errors were significant (1.3-1.9 kg). Ainsworth et al. (1997) developed an ethnic-specific BIA equation to estimate the FFM of African American women, but the prediction error for this equation (3.3 kg) was greater than that of the modified Segal et al. (1988) fatness-specific equations (2.8 kg). Finally, Jakicic et al. (1998) reported an ethnic difference when using the Segal fatness-specific equations to estimate the LBM of African American and Caucasian women who were obese. This equation overestimated the LBM_{DXA} of the Caucasian women by 0.9 kg and underestimated the LBM_{DXA} of the African American women by 1.4

kg. The authors subsequently derived a BIA equation that included the variables of weight, height, and resistance index, as well as a coded variable for ethnicity. This equation accurately predicted the average LBM of their cross-validation samples, but the prediction errors were not reported.

Generally, we do not recommend using the BIA method to estimate the body composition of your African American clients. However, if BIA is your only option, we recommend the fatness-specific BIA equations of Segal et al. (1988) as modified by Stolarczyk et al. (1997). Although these equations have small prediction errors, they tend to underestimate average FFM of African American men (1.8-2.4 kg) and African American women (0.6 kg).

NIR Equations for African Americans

At present, the NIR method and the Futrex-5000 manufacturer's equation should not be used to assess body composition in the African American population. We recently cross-validated the manufacturer's equation against a multicomponent model on a sample of 30 African American men 19 to 45 yr of age. The equation significantly underestimated $\%BF_{4-C}$ by 1.9% BF. For this group, the prediction errors were unacceptable (SEE = 4.1% BF, TE = 4.5% BF), and the prediction error for individuals exceeded ±3.5% BF in 50% of the sample (unpublished observations).

Similarly, the Futrex-5000 manufacturer's equation significantly underestimated $\%BF_{2-C}$ on average by 2.1% to 7.2% BF (SEE = 3.7-3.9% BF) for young, nonathletic African American men (Stout, Eckerson, Housh, & Johnson 1994) and collegiate football players (Hortobagyi, Israel, Houmard, O'Brien et al. 1992; Houmard et al. 1991). The errors were even greater (CE = –8.5% BF, SEE = 7.2% BF, TE = 11.0% BF) when this equation was applied to 16 African American women (Brandon & Bond 1999). Hortobagyi and colleagues developed a NIR prediction equation for African American collegiate football players that included body mass and an optical density (OD2) measurement from the Futrex analyzer (Hortobagyi, Israel, Houmard, O'Brien et al. 1992). Although they reported a low SEE (2.1% BF), this equation needs to be cross-validated before it can be recommended.

In short, we do not recommend using the NIR method to estimate body composition of your African American clients. Additional research is needed to develop ethnic-specific NIR equations for this population. It is highly probable that the relationship between biceps OD and %BF differs for African Americans and Caucasians because African Americans have relatively larger arm circumferences (Stevens et al. 1992), less subcutaneous fat at the biceps site (Mueller et al. 1982; Zillikens & Conway 1990), and darker skin tones (Wilson & Heyward 1993) than Caucasians.

Anthropometric Equations for African Americans

We are unaware of any anthropometric prediction equations developed specifically for African American adults or of any generalized anthropometric equations applicable to this group. With DXA as the reference method, prediction equations using age, weight, and height were developed to estimate the FM of African American boys (Ellis 1997) and girls (Ellis et al. 1997) aged 3 to 18 yr. However, the prediction errors were fairly large (SEE = 4.3 kg for boys, SEE = 2.5 kg for girls). Also, in African American and Caucasian children, different combinations of anthropometric measures (SKFs, circumferences, and skeletal diameters) were associated with total Db, depending on ethnicity and gender (Harsha et al. 1978). This finding, which is consistent with many other examples of fat patterning and limb length differences between these ethnic groups (Wagner & Heyward 2000), suggests that ethnic-specific anthropometric equations need to be developed before this method can be used to estimate body composition in the African American population.

Assessing Body Composition of American Indians

Non-insulin-dependent diabetes mellitus (NIDDM), or type 2 diabetes, is a major cause of death and morbidity in the American Indian population. The death rate due to this disease may be 4.3 times greater in American Indians than in the general population in the United States (Harris et al. 1995). American Indian adults are 2.8 times more likely to develop NIDDM than Caucasians of the same age (Centers for Disease Control 1998), and the incidence of this disease in the American Indian population is growing at a near epidemic rate. The prevalence of diabetes is especially high for American Indians living in the southwestern region of the United States. Despite many intervention efforts, the Pima Indians in Arizona continue to have the highest reported prevalence of NIDDM in the world (Krosnick 2000).

There is a strong association between obesity and NIDDM, and the prevalence of both obesity and NIDDM in American Indians has escalated over the past 30 yr. According to self-reported height and weight data from American Indians and Alaska

Natives, nearly one-fourth of the men and one-fifth of the women surveyed were categorized as obese (Centers for Disease Control 2000). Similarly, in their ethnically diverse sample of women from the U.S. Southwest, Thomas et al. (1997a) noted that one in four American Indian women has an at-risk WHR. Additionally, Lohman et al. (1999) reported a high prevalence of obesity in American Indian children from six different Indian nations, with an average %BF for boys and girls of 35.6% and 38.8%, respectively.

Because of the high prevalence of NIDDM and obesity in American Indians, this population has been targeted for weight loss. This section presents information about the FFB composition of American Indians. We also summarize studies concerning the applicability of field methods and prediction equations (see table 10.2) that can be used to assess the body composition of this ethnic group.

FFB Composition of American Indians

The water and mineral proportions of the FFB of 91 American Indian men (18-62 yr) from New Mexico were determined from isotope dilution (73.8%) and DXA (6.7%), respectively (Heyward et al. 1996). The remaining 19.5% of the FFB was assumed to be protein. Based on these values, the average FFBd for this sample of American Indian men with an average %BF of 26.4% was estimated to be 1.099 g/cc.

In contrast, the average BMD (1.18 g/cm^2) and BMC (2,596 g) of American Indian women were significantly greater than age-matched reference data (McHugh et al. 1993). The average relative total body mineral (TBM/FFM) of American Indian women was 8.1%, compared to 6.8% FFM assumed for the reference body. In fact, American Indian women have one of the highest relative TBM values reported in the literature to date (Stolarczyk et al. 1994). Based on this value, the FFBd for American Indian women was estimated to be 1.108 g/cc (table 1.4, p. 9). Therefore, the commonly used 2-C model equations of Siri (1961) or Brozek et al. (1963) will significantly underestimate the %BF of American Indian women.

SKF Equations for American Indians

Unfortunately, existing SKF prediction equations have not been cross-validated on populations of American Indian men. Therefore, we cannot recommend using the SKF method for estimating the body composition of men in this group at this time. Significant bias and prediction errors exceeding the acceptable limit (0.0080 g/cc) were reported when the Jackson et al. (1980) Σ3SKF and Σ7SKF were cross-validated in a group of 150 American Indian women from New Mexico (Hicks et al. 2000). Similarly, the prediction error for the Jackson et al. (1980) SKF equations was large (SEE = 0.019 g/cc) when applied to an ethnically mixed sample of American Indian, African American, Mexican American, and Caucasian women from the U.S. Southwest (Thomas et al. 1997b). The American Indian women in these studies had a greater android-type fat patterning, with more fat deposited on the trunk relative to the extremities, than women from the other ethnic groups. This may have contributed to the failure of the Jackson et al. SKF equations to accurately predict Db of American Indian women in these studies (Hicks et al. 2000; Thomas et al. 1997a, 1997b).

Therefore, a SKF equation predicting Db for American Indian women was developed from a sample of 99 women and cross-validated on 48 women (Hicks et al. 2000). The prediction errors for this equation were "very good" (SEE = 0.0068 g/cc, TE = 0.0070 g/cc), and there was no significant difference between the measured and SKF-predicted Db means. Data from the validation and cross-validation groups were pooled to develop the final prediction equation listed in table 10.2. We recommend using this equation for estimating the Db of American Indian women 18 to 60 yr of age. To convert Db into %BF, we suggest using the ethnic-specific conversion formula for American Indian women (table 1.4, p. 9).

Lohman, Caballero et al. (2000) reported that previously published field method prediction equations developed from samples of non-American Indian children all underestimated the average %BF in American Indian children by 5% to 14%. Lohman et al. subsequently developed prediction equations for grade school (7-11 yr) American Indian children using hydrometry as the reference method. An equation that includes both SKF and BIA measures was recommended; however, the authors noted that an equation that includes only two SKF measures (i.e., triceps and suprailiac SKF sites) was nearly as accurate. Because of the simplicity of using only one method, we recommend this SKF equation for American Indian children (see table 10.2).

BIA Equations for American Indians

BIA is one clinical method that is widely used on reservations and pueblos for assessing body composition of American Indians. Because some American Indians are severely obese, it may not be possible to accurately measure their SKF thicknesses. Therefore, the BIA method may be a more suitable field method for body composition assessment in this population.

Rising and colleagues (1991) developed a race-specific BIA prediction equation for Pima Indians (men and women) who varied greatly in body fatness (11-52% BF). When cross-validated, this equation was found to be superior to the RJL manufacturer's equation in estimating reference FFM (SEE = 3.2 vs. 6.9 kg, respectively). However, the Rising race-specific equation was developed using the Keys and Brozek (1953) 2-C model formula to estimate reference FFM. The Keys and Brozek equation was theoretically derived using hypothetical values for relative body fatness (14% BF for Db = 1.063 g/cc) of a "reference man." This equation was later revised (Brozek et al. 1963) based on empirical data for a reference body (15.3% BF for Db = 1.064 g/cc). Thus, the Keys and Brozek equation systematically underestimates fat content at any given Db compared to the revised Brozek equation.

The predictive accuracy of the Rising equation was examined in 71 American Indians (40 women, 31 men) in New Mexico (Hicks et al. 1992). Although the prediction error (SEE = 3.1 kg) was good, the average FFM estimated by this equation was significantly greater than the average FFM estimated by the Brozek et al. (1963), Siri (1961), and Lohman (1992) age-gender conversion formulas. The data indicated that the Rising equation underestimates the average %BF of American Indian men by approximately 2% to 3% BF.

Subsequently, the predictive accuracy of other BIA equations (i.e., Gray et al. 1989; Lohman 1992; Rising et al. 1991; Segal 1988; Van Loan & Mayclin [1987] equations) was assessed for American Indian men (Heyward et al. 1994). Generally, these BIA equations underestimated the FFM of American Indian men, but the Lohman (1992) age-specific BIA equations and the generalized equation of Van Loan and Mayclin (1987) accurately estimated average FFM with acceptable prediction errors (TE = 2.5-3.1 kg). However, in a multiethnic sample that was 41% American Indian, Stolarczyk et al. (1997) reported that the Segal et al. (1988) fatness-specific BIA equations, as well as an averaging of these equations for individuals who were neither clearly lean nor obese (i.e., modified Segal fatness-specific equations), produced more accurate estimations of FFM than either the Lohman (1992) or Van Loan and Mayclin (1987) equations. In this study, the modified Segal equation underestimated FFM by only 0.5 kg with a prediction error of 3.6 kg; the TEs for the equations of Lohman (5.1 kg) and Van Loan and Mayclin (7.2 kg) were much greater. Given the conflicting results of these studies, it is difficult to confidently recommend a BIA equation for American Indian

Table 10.2 Summary of Recommended Equations for American Indians

	Method/Equipment	Population	Equation	Reference
SKF	Σ3SKF[a] (Lange)	Females (18-60 yr)	1. Db (g/cc) = 1.06198316 − 0.00038496 (Σ3SKF) − 0.00020362 (age)	Hicks et al. 2000
	SKF (Lange)	Boys and girls (7-11 yr)	2. %BF = −0.08 (age) + 2.4 (sex[b]) + 0.21 (BW) + 0.38 (triceps SKF) + 0.20 (suprailiac SKF) + 17.66	Lohman, Caballero et al. 2000
BIA	RJL 101A	Females <25% BF	3. FFM (kg) = 0.00064602 (HT^2) − 0.01397 (R) + 0.42087 (BW) + 10.43485	Segal et al. 1988
		Females 25-35% BF	4. FFM (kg) = [0.00064602 (HT^2) − 0.01397 (R) + 0.42087 (BW) + 10.43485] + [0.00091186 (HT^2) − 0.01466 (R) + 0.2999 (BW) − 0.07012 (age) + 9.37938]/2	Modified Segal equations[e]
		Females >35% BF	5. FFM (kg) = 0.00091186 (HT^2) − 0.01466 (R) + 0.2999 (BW) − 0.07012 (age) + 9.37938	Segal et al. 1988
	Valhalla 1990B	Females (18-60 yr)	6. FFM (kg) = 0.001254 (HT^2) − 0.04904 (R) + 0.1555 (BW) + 0.1417 (Xc) − 0.0833 (age) + 20.05	Stolarczyk et al. 1994
NIR	Futrex-5000	Females (18-60 yr)	7. Db (g/cc) = 1.0707606 − 0.0009865 (hip C) − 0.0369861 (Σ2ΔOD2) + 0.0004167 (HT) + 0.0000866 (FIT index) − 0.0001894 (age)	Hicks et al. 2000

[a]Σ3SKF (mm) = sum of three skinfolds: triceps + suprailiac + axilla; [b]Girls = 1, boys = 0; [e]As modified by Stolarczyk et al. (1997).

Key: HT = height (cm); BW = body weight (kg); R = resistance (Ω); Xc = reactance (Ω); C = circumference (cm); OD = optical density; (Σ2ΔOD2) = [(biceps OD2 − standard OD2) + (chest OD2 − standard OD2)].

men. Furthermore, these equations were evaluated against 2-C models rather than a multicomponent model, and the single data set (Heyward et al. 1996) they rely on shows that the FFBd of American Indian men is close to the 2-C model assumption of 1.100 g/cc. Clearly, additional cross-validation analysis using a multicomponent model to obtain reference measures is needed.

In another study involving 151 American Indian women (Stolarczyk et al. 1994), the Rising equation was cross-validated against a 3-C model that adjusted Db for TBM. The Rising equation overestimated average FFM in this sample by almost 4 kg. Additionally, this equation failed to correctly identify 30% of the women in the sample who were obese (≥32% BF). Therefore, Stolarczyk et al. (1994) generated a new ethnic-specific BIA equation to estimate FFM of American Indian women (18-60 yr) with a wide range of body fatness (14-57% BF). Cross-validation of this equation on a subsample from this population indicated good predictive accuracy (SEE = 2.6 kg), and there was no significant difference between 3-C model reference values and BIA-predicted FFM. Additionally, the Stolarczyk et al. (1997) modification of the Segal et al. (1988) fatness-specific BIA equations accurately estimated the FFM (CE = 0.2 kg, SEE = 2.2 kg, TE = 2.3 kg) of an ethnically diverse sample that included 153 American Indian women. We recommend using either the Stolarcyzk et al. (1994) ethnic-specific equation or the Segal et al. (1988) fatness-specific equations, as modified by Stolarczyk (1997), to estimate FFM of American Indian women ages 18 to 60 yr (see table 10.2).

NIR Equations for American Indians

As with the SKF method, the NIR method should not be used to assess body composition of American Indian men. We cross-validated the Futrex-5000 NIR manufacturer's equation on a small sample of American Indian men from the Southwest. This equation had a large prediction error (SEE = 8.3% BF) and systematically underestimated average %BF of this sample by 2.4% BF (unpublished observations).

In addition to assessing the predictive accuracy of the SKF method for American Indian women (18-60 yr), Hicks et al. (2000) cross-validated a multisite NIR equation (Heyward, Jenkins et al. 1992) and the Futrex-5000 manufacturer's NIR equation for this sample. Prediction errors for both the multisite equation and the Futrex-5000 equation exceeded 0.0080 g/cc or 3.5% BF. Additionally, the multisite equation significantly overestimated the average Db, and the Futrex-5000 equation significantly underestimated %BF. Therefore, a new NIR equation was

developed and cross-validated. This new equation yielded an acceptable prediction error (SEE = 0.0076 g/cc) and small, but significant, differences between NIR-predicted and measured Db.

Although we prefer the ethnic-specific SKF equation or the fatness-specific BIA equations over this NIR equation to assess body composition of American Indian women, you could use the ethnic-specific NIR equation if NIR was the only available method in your clinical setting. However, we do not recommend using the Futrex-5000 manufacturer's equation or Heyward's multisite NIR equation for this population.

Anthropometric Equations for American Indians

No anthropometric equations have been developed or cross-validated for American Indians. Therefore it is necessary to use alternative methods like BIA or SKF.

Assessing Body Composition of Asians

The prevalence of obesity in the Asian and Pacific Islander population (6%) is much less than that for other ethnic groups in the United States (Schoenborn et al. 2002). However, as Asians adopt the Western lifestyle and diet, the incidence of obesity in this population is likely to increase. For example, Japanese men in Hawaii and California have higher BMI and SKF thicknesses than Japanese Natives (Curb & Marcus 1991). Furthermore, the prevalence of obesity may be underestimated in Asians. Based on 4-C model estimates of %BF, Deurenberg-Yap et al. (2000) suggested that the obesity cutoff for Chinese from Singapore should be lower (27 kg/m^2) than 30 kg/m^2 (current cutoff value for obesity as defined by the World Health Organization). Such a change would more than double the prevalence of obesity for Chinese women (6.5-15.4%) and Chinese men (5.2-17.3%) from Singapore.

The Asian population consists of many different nationalities and ethnic subgroups, but the majority of body composition research has been done on Chinese and Japanese participants. This section presents the FFB composition of Asians, as well as body composition methods and prediction equations (see table 10.3) applicable to this population.

FFB Composition of Asians

Much of the body composition research on the Asian population has been done by the same research group. Recently, these researchers assessed the FFB composition of Chinese, Malays, and Indians from Singapore using a 4-C model (Deurenberg-Yap et al. 2001). The Chinese (1.0987 g/cc) and Malay (1.1011 g/cc) men had an average FFBd that was similar to the reference value of 1.100 g/cc. The TBW/FFM ratio was close to 73% (72.1-73.8%) for all subgroups.

However, the mineral fraction of the FFM (7.8-7.9%) and the FFBd (1.1038-1.1082 g/cc) for Singaporean women from all three ethnic subgroups exceeded the assumed value (1.100 g/cc) for the 2-C model. There were no significant differences in the average FFBd and the relative composition of the FFB for women from these different subgroups. The average FFBd for all Singaporean women was 1.1068 g/cc.

In another study, this research team used a 4-C model to compare the body composition of Singaporean Chinese and Dutch Caucasians (Werkman et al. 2000). Once again, the Chinese women had an average FFBd (1.1074 g/cc) that exceeded the assumed reference value and was greater than that of the Chinese men (1.1027 g/cc), as well as that of Caucasian women (1.1012 g/cc) and men (1.1004 g/cc). For the females, there were ethnic differences in all components of the FFB, with the Chinese women having a greater percentage of mineral (7.9% vs. 7.6%) and protein (19.2% vs. 18.3%) compared to the Dutch women. The Singaporean men also had a greater mineral fraction (7.3% vs. 6.6%) but a lower protein percentage (19.6% vs. 20.5%) than the Dutch men.

Finally, Koda et al. (2000) reported that the average relative mineral content of the FFM (BMC/FFM) of Japanese men and women 40 to 70 yr of age was 4.3% and 4.4%, respectively. The relative hydration of the FFM (TBW/FFM) was 70.0% and 68.3% for men and women, respectively. Although the authors did not report average Db and FFBd values, these data suggest that the average FFBd of Japanese women (40-59 yr) may be higher than 1.100 g/cc. Based on these studies, it appears that the average FFBd of Asian women (1.107 g/cc) is greater than the assumed value for 2-C models; but the FFBd of Asian men is probably close to or only slightly greater than 1.100 g/cc.

SKF Equations for Asians

The Durnin and Womersley (1974) Σ4SKF equations have been cross-validated in many samples of Chinese. Using a 4-C model to obtain reference body composition measures, Deurenberg and Deurenberg-Yap (2002) reported that the Durnin and Womersley equation significantly underestimated the %BF of Malay and Indian women from

Singapore by 2.3% BF and 3.1% BF, respectively; however, the average bias was small (–0.4 to 0.9% BF) for both Chinese men and women, as well as Malay and Indian men. Bias among ethnic groups was primarily due to differences in body fatness and age. For the total sample of Singaporeans, the relationship between $\%BF_{SKF}$ from the Durnin and Womersley equation and $\%BF_{4-C}$ was $r_{xy} = 0.88$, and the SEE was 4.2% BF.

Wang et al. (1994) reported similar prediction errors (SEE = 4.0-4.3% BF) and a significant underestimation of average %BF (1-2.9% BF) for Asian women and men using the Durnin and Womersley (Σ4SKF) and Jackson et al. (Σ3SKF) equations. They noted that Asians distribute a greater proportion of their subcutaneous fat on the upper body than Caucasians, and this may have contributed to the error.

Generally, the Durnin and Womersley Σ4SKF equation closely predicts (<1% BF) average %BF from 2-C and 3-C models in Chinese men and women aged 18 to 68 yr (Deurenberg et al. 2000; Wang & Deurenberg 1996). Deurenberg et al. (2000) commented that this SKF equation gave the best group and individual results of all field methods tested, but the SEE and limits of agreement were not reported. They also found that with age the increase in body fat was less in Chinese than in Caucasians. On the basis of data from all of these studies, the Durnin and Womersley Σ4SKF equation appears to provide an acceptable estimate of average %BF for Chinese clients, but the prediction error can be large (>3.5% BF).

Several researchers have developed ethnic-specific SKF equations for the Asian population; however, none of these were derived from a multicomponent model. After finding that the Jackson and Pollock (1978) Σ3SKF equation underestimated (CE = –4.2% BF, SEE = 3.1% BF) and the Durnin and Womersley equation overestimated (CE = 2.0% BF, SEE = 2.9% BF) average %BF in 47 Chinese men (24-43 yr), Eston et al. (1995) developed their own equation for this population. They noted that the medial calf had the highest correlation with $\%BF_{HD}$ in Chinese men, and subsequently developed a prediction equation using only this site to estimate %BF. Although this equation had good predictive accuracy (SEE = 2.1% BF), it was derived using only 32 participants and cross-validated on only 15 men.

Using DXA as the reference method, Kwok et al. (2001) developed a prediction equation for Chinese aged 69 to 82 yr. The best estimate of %BF (SEE = 4.1% BF) came from an equation that used gender, BMI, and a combination of triceps and biceps SKFs as the predictors. Cross-validation on a separate sample showed that the bias was significant for women (–2.1% BF) but not for men (–0.1% BF).

Wang et al. (1994) developed SKF prediction equations for Asian American men (N = 110) and women (N = 132) ages 18 to 98 yr. The sample was 93% Chinese, 4% Japanese, 2% Korean, and 1% Filipino; 97% of this sample was born in Asia. These equations used BMI, age, and four SKF measures to estimate %BF, measured by dual-photon absorptiometry. Although the prediction errors for these equations were fairly good (SEE = 3.5% BF for women, 3.7% BF for men), they still need to be cross-validated on additional samples of Asian men and women before we can recommend using them.

Nagamine and Suzuki (1964) developed gender- and ethnic-specific SKF equations for young Japanese Native adults (18-27 yr) using HD and a 2-C model to obtain reference measures. Cross-validation of these equations yielded accurate body fat estimates (SEE = 3.7% BF) for Japanese Native men (18-56 yr) but underestimated the average body fatness of Japanese Native women (18-54 yr) by 3% BF (SEE = 4.3% BF) (Nakadomo et al. 1990). The predictive accuracy of these equations was slightly better (SEE = 3.4% BF) in another sample of Japanese Native adults (Sawai et al. 1990), but had only a fair degree of accuracy when applied to a sample of obese Japanese women (Tanaka et al. 1992). Most recently, this equation was cross-validated in a sample of 50 Japanese men, 18 to 27 yr of age, using the 2-C model of Brozek et al. (1963) and HD as the reference method (Demura, Yamaji, Goshi, Kobayashi et al. 2002). The SKF equation underestimated average %BF$_{2-C}$ by 1.9%, and there was a tendency to overestimate %BF of participants with low body fat and to underestimate the %BF of those with high body fat. Because of these findings, the authors developed a new SKF equation for this sample and cross-validated it on an independent sample of 30 young Japanese men. The mean difference between the SKF and HD estimates of %BF in the cross-validation group was −0.5% BF, but the correlation was only moderate (r = .69). In addition, the researchers reported that the predictive accuracy of BIA equations was better than that of this SKF equation. Thus, we recommend using the BIA method, rather than the SKF method, for Japanese clients.

BIA Equations for Asians

The BIA equations of Deurenberg and colleagues (Deurenberg, van der Kooy et al. 1991; Deurenberg et al. 1995), which were derived from Caucasian samples, were cross-validated against HD and the Siri 2-C model for Chinese men and women (Deurenberg et al. 2000; Wang & Deurenberg 1996). Wang and Deurenberg (1996) reported small CEs for Chinese men (0.6 kg) and women (−1.2 kg),

but Deurenberg et al. (2000) reported average underestimations in %BF of 3.0% BF to 3.6% BF in all groups of males and females from Beijing and Singapore. They hypothesized that the underestimation may be due to relatively shorter legs in Chinese compared to Caucasians, and suggested that ethnic-specific BIA equations are needed.

Nakadomo et al. (1990) cross-validated other generalized BIA equations (Lukaski et al. 1986; Segal et al. 1988) on Japanese Natives. Both equations systematically overestimated the average FFM of Japanese men and women by 3 to 4 kg. Thus, Nakadomo et al. developed gender-specific BIA equations that have acceptable prediction errors (1.9 and 2.2 kg FFM, respectively, for men and women). However, this equation failed to accurately estimate the body composition of obese Japanese women (SEE = 0.0089 g/cc or ~4% BF) (Tanaka et al. 1992). Consequently, Tanaka and associates developed and cross-validated a fatness-specific BIA equation for obese Japanese Native women.

Variations on the traditional whole-body, hand-to-foot BIA method (i.e., hand-to-hand and foot-to-foot BIA methods; see chapter 6) have also been tested on Chinese and Japanese participants. The correlation between the Omron BF306 hand-to-hand BIA unit and a 4-C model was 0.87 (SEE = 4.5% BF) in a sample of 298 Singaporeans (Deurenberg & Deurenberg-Yap 2002). The authors noted that the bias of this device correlated with level of body fatness, age, and relative arm span.

Demura and colleagues evaluated hand-to-foot (Selco, SIF-891), foot-to-foot (Tanita, TBF-102), and hand-to-hand (Omron, HBF-300) BIA analyzers against HD using Brozek's 2-C model (Demura, Yamaji, Goshi, Kobayashi et al. 2002). They found that the manufacturers' equations overestimated %BF, on average, by 2.2% to 3.3% BF; therefore, they developed new BIA prediction equations for their sample of 50 Japanese men aged 18 to 27 yr. Because of a higher correlation with HD, the hand-to-foot (r = .96) equation was recommended over the foot-to-foot equation (r = .71) and SKF equation (r = .72). The whole-body BIA equation was cross-validated on an independent sample of 30 Japanese men (18-23 yr) and yielded a 95% confidence interval of −0.03% to 1.58% BF.

The BIA method has also been evaluated for Asian children. In a sample of 94 Chinese children, aged 11 to 17 yr, there was good agreement (r = .96, SEE = 2.2 kg) between the Houtkooper et al. (1989) BIA equation and the Boileau et al. (1985) SKF equation; however, no reference method was used in this study (Eston et al. 1993). Kim and colleagues (1994) developed BIA equations to estimate FFM from a sample of 84 Japanese boys

Table 10.3 Summary of Recommended Equations for Asians

	Method/Equipment	Population	Equation	Reference
SKF	SKF (calf) (Holtain)	Chinese males (24-43 yr)	1. %BF = 5.87 + 1.37 (calf SKF)	Eston et al. 1995
BIA	Selco SIF-881	Japanese Native females (18-54 yr)	2. Db (g/cc) = 1.1628 − 0.1067 BW × Z / HT²	Nakadomo et al. 1990
	Selco SIF-881	Japanese Native obese females (18-68 yr)	3. Db (g/cc) = 1.1307 − 0.0719 (BW × Z / HT²) − 0.0003 (age)	Tanaka et al. 1992
	Selco SIF-891	Japanese Native males (18-27 yr)	4. Db (g/cc) = 1.4 (age) + 0.66 (HT) − 1.51 (BW) − 0.26 (Z) + 1.13857	Demura, Yamaji, Goshi, Kobayashi et al. 2002
	Selco SIF-891	Japanese Native males (9-14 yr)	5. FFM (kg) = 0.56 (HT² / Z) + 0.20 (BW) + 1.66	Kim et al. 1994
	NR	Japanese Native females (9-15 yr)	6. FFM$_{2-C}$ (kg) = 0.42 (HT² / Z) + 0.60 (BW) − 0.75 (arm C) + 7.72	Watanabe et al. 1993

Key: HT = height (cm); BW = body weight (kg); SKF = skinfold; Z = impedance (Ω); C = circumference (cm); NR = not reported.

(9-14 yr) and cross-validated it on 57 boys of the same age. HD and the Brozek et al. (1963) 2-C model were used to obtain reference measures. An equation that included abdominal SKF, impedance index (HT^2/Z), and body mass produced the lowest prediction errors in both the validation and cross-validation groups. Pooled data from these two groups (N = 141) yielded an R^2 = 0.98 and SEE = 1.5 kg. Additionally, the 95% limits of agreement ranged from 0.2 to 3.6 kg. Another equation that included only impedance index and body mass was nearly as accurate (SEE = 1.6 kg). Until equations based on multicomponent models are developed, we recommend the BIA equations of Kim et al. (1994) to estimate the FFM_{2-C} of Japanese boys (see table 10.3). As mentioned in chapter 8, one can estimate the FFM_{2-C} of Japanese Native girls (9-15 yr) using a BIA equation that includes upper arm circumference as a predictor (see table 10.3).

NIR Equations for Asians

As in other ethnic groups, there is a lack of research substantiating the validity of the NIR method for assessing body composition in Asian Americans and Japanese Natives. The Futrex-5000A was used in a study that included premenarcheal (10-11 yr old) and postmenarcheal (16-17 yr old) Japanese and Caucasian girls, but no reference method was used (Sampei et al. 2001). The average differences between NIR and other field methods (BIA and SKF) ranged from –1.4% BF to –6.3% BF between the age and ethnic groups. We obtained one abstract (from Futrex) that had been translated from Japanese to English and reprinted from the annals of Tokyo University (Sawai et al. 1990). These researchers found no significant difference between %BF estimated from the Futrex-5000 NIR equation and HD (SEE = 3.3% BF) for Japanese men and women. Because we cannot verify the accuracy of this translation or provide a complete citation for this work, it may be better to use alternative methods to assess body composition of Japanese Native adults.

Anthropometric Equations for Asians

Although Nagamine and Suzuki (1964) measured circumferences and skeletal diameters of Japanese Natives and compared these values to those reported for men and women in the United States, they did not develop anthropometric equations for this population. Deurenberg et al. (2000) cross-validated the gender-specific anthropometric formulas of Lean et al. (1996), which use waist circumference and age as predictors of %BF, on 353 Chinese. These equations significantly underestimated aver-

age %BF in Beijing (–3.5% BF) and Singaporean (–5.0% BF) males, but were reasonably accurate (–1.1% to 0.6% BF) for females. Unfortunately, the group and individual prediction errors were not reported. Therefore it is advisable to use alternative methods, like SKF and BIA, instead.

Assessing Body Composition of Caucasians

Although the prevalence of obesity is slightly greater in the African American and Mexican American populations, it is still a serious and increasing health problem for Caucasians. Over a quarter of the Caucasian population in the United States is obese, and approximately two-thirds are overweight (National Center for Health Statistics 2002). The odds of Caucasians having at least one obesity-associated risk factor (low high-density lipoprotein-cholesterol, high low-density lipoprotein-cholesterol, high blood pressure, or high glucose) are increased by 2.37 times for men and 3.16 times for women at a BMI of 30 kg/m^2 (Zhu et al. 2002).

A vast amount of research has focused on quantifying the FFB composition of Caucasian men and women and on developing methods and equations for estimating body fatness in this population. This section summarizes information on the FFB composition of Caucasians and the methods and prediction equations (table 10.4) most applicable to this ethnic group.

FFB Composition of Caucasians

As discussed in chapter 8, the FFB and overall density of the FFB increase from childhood to adulthood for both females and males. Research consistently shows that the FFBd of younger and middle-aged men, estimated from multicomponent models, is remarkably similar to the assumed value (1.100 g/cc) of Siri's (1961) 2-C model. Therefore, many prediction equations based on 2-C model estimates of body composition work quite well in this population subgroup.

According to Lohman (1986), the average FFBd of Caucasian females at any age is somewhat less than that of males because of the relatively greater hydration of their FFB (Lohman 1986). Ortiz et al. (1992) calculated the average FFBd in younger and middle-aged Caucasian women to be 1.097 g/cc. Using this value, the Siri 2-C model systematically overestimates average body fatness in Caucasian women. However, other researchers have reported the average FFBd of Caucasian women to be 1.1 g/cc, a value similar to that for Caucasian me making the Siri 2-C model acceptable (Visser et al. 1997; Werkman et al. 2000). Because of the differences in these studies and interindividual variability in FFB composition, we recommend using multicomponent models to develop and cross-validate prediction equations.

SKF Equations for Caucasians

Although there are many age-specific SKF equations for Caucasian men (Lohman 1992; Sloan 1967), an abundance of research demonstrates the applicability of Jackson and Pollock's (1978) generalized SKF equations for this population subgroup. Cross-validation of these equations indicates prediction errors ranging from 2.6% to 3.5% BF for the $\Sigma 3$SKF equation (McLean & Skinner 1992; Paijmans et al. 1992) and the $\Sigma 7$SKF equation (Hortobagyi, Israel, Houmard, McCammon, & O'Brien 1992; Israel et al. 1989; Jackson et al. 1988). Typically, there are small differences (up to 1.4% BF) between measured and SKF-predicted body fat in these samples.

In contrast, one study reported that the $\Sigma 3$SKF equation significantly underestimated average %BF of Caucasian men 22 to 75 yr of age (Clark, Kuta, & Sullivan 1993). However, some of the subjects exceeded the age range of the equation (>61 yr) and were obese (up to 30% BF). Wilmore et al. (1994) reported that the $\Sigma 7$SKF equation significantly underestimated the average %BF of overweight and obese men by 4% BF, with a prediction error (SEE) of 4.1% BF. Jackson and Pollock (1985) warned that these equations may not be accurate for estimating %BF of men whose $\Sigma 3$SKF (chest, abdomen, and thigh SKFs) exceeds 118 mm.

The Jackson et al. (1980) generalized SKF equations also show good predictive accuracy (SEE = 2.9-3.5% BF) and small differences (0.3-1.3% BF) when they are used to estimate the body fatness of Caucasian women (Eaton et al. 1993; Heyward, Cook et al. 1992; Jackson et al. 1988; McLean & Skinner 1992). As mentioned for men, these equations generally work best when they are applied to non-obese women. Wilmore et al. (1994) noted that average %BF of overweight and obese women was significantly underestimated by 2.8% BF using the Jackson $\Sigma 7$SKF equation. We found that the Jackson $\Sigma 3$SKF equation significantly underestimates the average body fat of obese women by 3.7% BF (Heyward, Cook et al. 1992). Likewise, Paijmans et al. (1992) reported that the average body fatness of obese Caucasian women and men was underestimated by 5.5% BF. In a meta-analysis of body composition studies limited to adult Caucasians, the equations of Jackson and Pollock (1978) and Jackson et al. (1980) underestimated

%BF$_{HD}$ on average by 1.2% BF; however, some of the studies in this analysis included overweight or obese participants (Fogelholm & van Marken Lichtenbelt 1997).

On the basis of these studies, you may use either the $\Sigma 3$SKF or $\Sigma 7$SKF equation to estimate Db of non-obese Caucasian men and women. However, for practical purposes, we recommend using the $\Sigma 3$SKF equation. To convert Db to %BF, use the conversion formulas presented in table 1.4, page 9.

BIA Equations for Caucasians

There are many age-specific and generalized equations that were derived from adult Caucasian samples. Before using the estimates of FFM or %BF from your BIA analyzer, you should find out which equations the manufacturer has programmed into the accompanying software version. The overwhelming majority of BIA equations are derived from 2-C models, and we are aware of only two BIA equations based on multicomponent body composition models (Lohman 1992; Van Loan et al. 1990). Lohman's age-specific equation for 50- to 70-yr-old women adjusts for changes in bone mineral with aging; Van Loan's equation accounts for interindividual differences in TBW. However, the Van Loan et al. (1990) equation was not cross-validated.

Several BIA equations (i.e., EZ Comp 1500 manufacturer's equation; Segal et al. [1988] fatness-specific equations; Gray et al. [1989] generalized equation; Lukaski et al. [1986] equation), as well as a BMI equation (Deurenberg, Weststrate et al. 1991) and a SKF equation (Durnin & Womersley 1974), were cross-validated against FFM determined from HD in a sample of 150 Caucasian men and women aged 17 to 71 yr (Han et al. 1996). Of all prediction equations, the Segal et al. (1988) fatness-specific BIA equations had the highest precision (as measured by R^2 of 83% and 75% for men and women, respectively) and least error (as measured by slopes of .88 and .81 for men and women, respectively). Additionally, the individual errors, measured by 95% limits of agreement, were smaller for this equation than for others and were within ±6 kg regardless of gender or quantity of FFM. However, the authors acknowledged one shortcoming of Segal's fat-specific equations; the client needs to be preclassified as above or below 20% BF for men and 30% BF for women to allow selection of the appropriate equation. To do this, Segal recommends measuring the sum of four SKFs, but this negates the advantage of using BIA as a simple method requiring less technician skill than the SKF method.

To circumvent this shortcoming of the Segal fatness-specific equations, Stolarczyk et al. (1997)

Table 10.4 Summary of Recommended Equations for Caucasians

	Method/Equipment	Population	Equation	Reference
SKF	Σ3SKF[a] (Lange)	Non-obese males (18-61 yr)	1. Db (g/cc) = 1.109380 − 0.0008267 (Σ3SKF) + 0.0000016 (Σ3SKF)^2 − 0.0002574 (age)	Jackson & Pollock 1978
	Σ3SKF[b] (Lange)	Non-obese females (18-55 yr)	2. Db (g/cc) = 1.0994921 − 0.0009929 (Σ3SKF) + 0.0000023 (Σ3SKF)^2 − 0.0001392 (age)	Jackson et al. 1980
BIA	RJL 101A	Males <17% BF	3. FFM (kg) = 0.00066360 (HT^2) − 0.02117 (R) + 0.62854 (BW) − 0.1238 (age) + 9.33285	Segal et al. 1988
		Males 17-25% BF	4. FFM (kg) = [0.00066360 (HT^2) − 0.02117 (R) + 0.62854 (BW) − 0.1238 (age) + 9.33285] + [0.0008858 (HT^2) − 0.02999 (R) + 0.42688 (BW) − 0.07002 (age) + 14.52435] / 2	Modified Segal equations[c]
		Males >25% BF	5. FFM (kg) = 0.0008858 (HT^2) − 0.02999 (R) + 0.42688 (BW) − 0.07002 (age) + 14.52435	Segal et al. 1988
		Females <25% BF	6. FFM (kg) = 0.00064602 (HT^2) − 0.01397 (R) + 0.42087 (BW) + 10.43485	Segal et al. 1988
		Females 25-35% BF	7. FFM (kg) = [0.00064602 (HT^2) − 0.01397 (R) + 0.42087 (BW) + 10.43485] + [0.00091186 (HT^2) − 0.01466 (R) + 0.2999 (BW) − 0.07012 (age) + 9.37938] / 2	Modified Segal equations[c]
		Females >35% BF	8. FFM (kg) = 0.00091186 (HT^2) − 0.01466 (R) + 0.2999 (BW) − 0.07012 (age) + 9.37938	Segal et al. 1988
NIR	Futrex-5000	Females (20-72 yr)	9. Db (g/cc) = 1.02823066 − 0.080035 (ΔOD2) − 0.000459 (age) − 0.000754 (BW) + 0.000493 (HT)	Heyward, Jenkins et al. 1992
ANTHROPOMETRY		Males (18-40 yr)	10. FFM (kg) = 39.652 + 1.0932 (BW) + 0.8370 (bi-iliac D) + 0.3297 (AB$_1$ C) − 1.0008 (AB$_2$ C) − 0.6478 (knee C)	Wilmore & Behnke 1969
		Females (15-79 yr)	11. Db (g/cc) = 1.168297 − 0.002824 (AB C) + 0.0000122098 (AB C)^2 − 0.000733128 (hip C) + 0.000510477 (HT) − 0.000216161 (age)	Tran & Weltman 1989

[a]Σ3SKF (mm) = sum of three skinfolds: chest + abdomen + thigh; [b]Σ3SKF (mm) = sum of three skinfolds: triceps + suprailiac + thigh; [c]As modified by Stolarczyk et al. (1997).

Key: HT = height (cm); BW = body weight (kg); R = resistance (Ω); OD = optical density; ΔOD2 = OD2 standard − OD2 biceps; C = circumference (cm); D = diameter (cm); AB C = average abdominal circumference: [(AB$_1$ + AB$_2$)/2], where AB$_1$ = abdominal circumference anteriorly midway between the xiphoid process of the sternum and the umbilicus and laterally between the lower end of the rib cage and iliac crests, and AB$_2$ = abdominal circumference at the umbilicus level.

modified these equations by using a "BIA averaging method" for clients who are not obviously lean or obese (i.e., modified Segal fatness-specific equations). To use this averaging method, one visually classifies clients as either non-obese (<17% BF for men and <25% BF for women), not obviously lean or obese (17-25% BF for men and 25-35% BF for women), or obese (>25% BF for men and >35% BF for women). Trained technicians can correctly categorize clients by visual assessment about 75% of the time (Lockner et al. 1999). For individuals categorized as not obviously lean or obese, FFM is calculated using both the non-obese and obese equations. These values are then averaged and used as estimates of the FFM of clients who are not obviously lean or obese.

This averaging method produced smaller prediction errors (TE = 2.3 kg for women and 3.6 kg for men) than other commonly used BIA equations (i.e., Deurenberg, van der Kooy et al. 1991; Gray et al. 1989; Lohman 1992; Lukaski & Bolonchuk 1987; Segal et al. 1988; Van Loan & Mayclin [1987] equations) in a multiethnic sample that was about 40% Caucasian (Stolarczyk et al. 1997). Lockner et al. (1999) reported excellent prediction errors for men (TE = 2.7 kg) and fairly good prediction errors for women (TE = 3.1 kg) when using this approach on an ethnically mixed sample in which 27% of the men and 46% of the women were Caucasian.

Cross-validation studies have reported large prediction errors (SEE = 3.6-5.0% BF) for other BIA equations (i.e., Deurenberg, van der Kooy et al. 1990; Lukaski et al. 1986; Segal et al. [1985] equations) when they were applied to Caucasian men and women (Eckerson et al. 1992; Graves et al. 1989; Jackson et al. 1988; Jenkins et al. 1994; Pierson et al. 1991; Van Loan & Mayclin 1987). On the basis of these findings, we recommend using the Segal fatness-specific BIA equations, as modified by Stolarczyk (1997), for estimating the FFM of Caucasian men and women (see table 10.4).

NIR Equations for Caucasians

As discussed in chapter 7, there are many questions concerning the use of the NIR method to assess body composition. Much NIR research has focused on the validity of the Futrex-5000 manufacturer's equation for estimating body fat of Caucasian men and women. Regardless of gender, the prediction error of the manufacturer's equation in the Caucasian population is unacceptable (SEE > 3.5% BF). On the basis of the information presented in chapter 7, we do not recommend using the %BF estimates from your Futrex-5000 analyzer.

Using a predominantly Caucasian sample (85%) of women ages 20 to 72 yr, a NIR equation was developed with fairly good predictive accuracy (SEE = 0.0082 g/cc or 3.6% BF), but additional cross-validation studies are needed to evaluate the extent of its applicability to other samples of Caucasian women (Heyward, Jenkins et al. 1992). In the meantime, we suggest using this equation (table 10.4) to assess the Db of your Caucasian female clients only when other methods, such as SKF and BIA, are unavailable to you. To convert Db to %BF, use the conversion formula for Caucasian women 20 to 80 yr of age (table 1.4, p. 9).

Anthropometric Equations for Caucasians

Wilmore and Behnke (1969, 1970) developed age-specific anthropometric equations to estimate Db and LBM of young women and men using combinations of skeletal diameter and circumference measures. Cross-validation of the equations indicated that the male equation overestimated LBM by 1.8 kg (SEE = 2.5 kg). The accuracy of these equations was generally poorer (SEE = 3.0 kg) for women than for men (Katch & McArdle 1973).

As an alternative, we recommend a generalized anthropometric equation to estimate the Db of Caucasian women 15 to 79 yr of age (Tran & Weltman 1989). This equation has good predictive accuracy (SEE = 0.0082 g/cc or 3.6% BF) in this population. On the other hand, it is not advisable to use the generalized anthropometric equation developed for Caucasian men (21-78 yr) because of its large prediction error (SEE = 4.4% BF) (Tran & Weltman 1988).

FM can be predicted from height, weight, and age in Caucasian boys (Ellis 1997) and girls (Ellis et al. 1997). The prediction error is fairly large for boys (SEE = 3.6 kg), but small enough (SEE = 1.1 kg) for girls to have some utility.

Assessing Body Composition of Hispanics

On the basis of BMI and the sum of two SKFs, Pawson et al. (1991) reported that the prevalence of overweight for Hispanic men and women was, respectively, 36% and 42% for Mexican Americans in the Southwest, 31% and 41% for Puerto Ricans in the New York City area, and 34% and 38% for Cubans in Florida. With use of more recent data from the National Center for Health Statistics (2002), about 72% of Mexican American men and 70% of Mexican American women have BMIs greater than 25 kg/m²; and 30% and 38%, respectively, have BMIs indicative of obesity (≥30 kg/m²). Additionally, the prevalence of obesity in Mexican

American boys ages 6 to 19 yr is greater than that for other ethnic groups (National Center for Health Statistics 2002).

In addition to having high BMI values, Hispanics may be at further risk for cardiovascular disease because they tend to deposit relatively more fat on the trunk than the extremities. Greater total and truncal fat was observed in middle-aged Hispanic women compared to a similar group of Caucasian women matched for age, body mass, socioeconomic status, and estimated physical activity (Casas et al. 2001). Additionally, older (60-86 yr) Mexican American women have a higher ratio of trunk to limb fat (0.93) compared to age-matched Caucasian women (0.77) as measured by DXA (Taaffe et al. 2000). Because the Hispanic population is the fastest growing minority group in the United States, it is reasonable to speculate that the prevalence of obesity-linked diseases will also increase, placing a relatively greater burden on the public health care system (Pawson et al. 1991).

Although there is a great need for information concerning appropriate methods for body fat assessment in Hispanics, there are few published research studies on this topic. In this section, we present what is known about the FFB composition of Hispanics. Body composition methods and equations applicable to the Hispanic population are summarized in table 10.5.

FFB Composition of Hispanics

If current population trends continue, the Hispanic population may soon become the largest minority group in the United States. Despite this fact, few studies have used multicomponent models to assess the FFB composition of this population. The water, mineral, and protein fractions of the FFB and the FFBd of Hispanic men are still unknown. However, the BMC, BMD, and TBW of 30 healthy premenopausal Hispanic women (20-40 yr) in New Mexico were assessed using DXA and deuterium dilution methods (Heyward et al. 1995). In this sample the average body weight and %BF, as measured by HD and a 4-C model, were 60.0 kg and 30.6% BF. The average total body BMD, BMC, and TBM and TBW were, respectively, 1.161 g/cm², 2.41 kg, 3.08 kg, and 30.1 L. These values are close to those reported for Caucasian women (Hansen et al. 1993; Ortiz et al. 1992; Van Loan & Mayclin 1992) with similar body weight (60-64 kg) and %BF (29-32%). Likewise, similar BMD was found for middle-aged Hispanic (1.16 g/cm²) and Caucasian (1.18 g/cm²) women who were matched for age, body mass, socioeconomic status, and estimated weekly physical activity (Casas et al. 2001). However, as Forbes (2001) pointed out, Casas and colleagues did not

adjust for the 4-cm difference in height between ethnic groups, potentially limiting the results of their study. For older (60-86 yr) women, Mexican Americans had higher hip and whole-body BMD values (1.000 g/cm²) than non-Hispanic Caucasians (0.941 g/cm²) after adjustment for height differences (Taaffe et al. 2000).

The relative mineral content of the FFB of Hispanic women in New Mexico (7.4% FFB) was similar to that reported for Brazilian women (7.5% FFB) of the same age (Bottaro et al. 2002). Although this value exceeds the assumed reference value of 6.8%, it is only slightly greater than the average value reported by Hansen et al. (1993) for premenopausal Caucasian women (7.3% FFB). Comparison of Siri's 2-C model and a 4-C model that adjusts Db for body mineral and body water yielded significantly different estimates of %BF for the Hispanic women in the New Mexico study (26.9% and 30.6%, respectively). These data suggest that the FFBd of Hispanic women is greater than the assumed value for the reference body (1.100 g/cc). We estimate the FFBd of Hispanic women to be 1.105 g/cc (table 1.4, p. 9) using a FFB-mineral content of 7.4% and FFB-water content of 72.8%.

SKF Equations for Hispanics

Generalized SKF equations (Jackson & Pollock 1978; Jackson et al. 1980) have proven to be good predictors of Db in other ethnic groups (Hortobagyi, Israel, Houmard, O'Brien et al. 1992; Paijmans et al. 1992; Sparling et al. 1993) and are currently being used to assess the body composition of Hispanic individuals. The Jackson Σ3SKF and Σ7SKF equations, used in conjunction with Siri's (1961) 2-C model formula to convert Db to %BF, significantly underestimated %BF$_{DXA}$ (−2.8% BF) for a diverse sample of Brazilian women (20-40 yr) whose %BF values ranged from 14.6% to 39.6% BF (Bottaro et al. 2002). However, when the sample was limited to women with ≤30% BF and an ethnic-specific conversion formula (which assumes an average FFBd of 1.105 g/cc) was used, the average difference in %BF$_{DXA}$ for both the Σ3SKF and Σ7SKF equations was <0.3% BF, and the prediction errors (2.3-2.5% BF) were acceptable. Likewise, the Σ7SKF equation accurately estimated the average Db of Hispanic women (20-40 yr) from New Mexico with an acceptable prediction error (TE = 0.0078 g/cc) (unpublished observations). On the basis of these studies, we recommend using the Jackson et al. (1980) Σ7SKF equation to estimate the Db of young, healthy, non-obese Hispanic women; however, this equation needs to be cross-validated against a multicomponent model for the Hispanic population. Unfortunately, at this time,

Table 10.5 Summary of Recommended Equations for Hispanics

	Method/Equipment	Population	Equation	Reference
SKF	Σ7SKF[a] (Lange)	Non-obese females (18-55 yr)	1. Db (g/cc) = 1.097 − 0.00046971 (Σ7SKF) + 0.00000056 (Σ7SKF)^2 − 0.00012828 (age)	Jackson et al. 1980
BIA	RJL 101A	Males <17% BF	2. FFM (kg) = 0.00066360 (HT^2) − 0.02117 (R) + 0.62854 (BW) − 0.1238 (age) + 9.33285	Segal et al. 1988
		Males 17-25% BF	3. FFM (kg) = [0.00066360 (HT^2) − 0.02117 (R) + 0.62854 (BW) − 0.1238 (age) + 9.33285] + [0.0008858 (HT^2) − 0.02999 (R) + 0.42688 (BW) − 0.07002 (age) + 14.52435]/2	Modified Segal equations[b]
		Males >25% BF	4. FFM (kg) = 0.0008858 (HT^2) − 0.02999 (R) + 0.42688 (BW) − 0.07002 (age) + 14.52435	Segal et al. 1988
		Females <25% BF	5. FFM (kg) = 0.00064602 (HT^2) − 0.01397 (R) + 0.42087 (BW) + 10.43485	Segal et al. 1988
		Females 25-35% BF	6. FFM (kg) = [0.00064602 (HT^2) − 0.01397 (R) + 0.42087 (BW) + 10.43485] + [0.00091186 (HT^2) − 0.01466 (R) + 0.2999 (BW) − 0.07012 (age) + 9.37938]/2	Modified Segal equations[b]
		Females >35% BF	7. FFM (kg) = 0.00091186 (HT^2) − 0.01466 (R) + 0.2999 (BW) − 0.07012 (age) + 9.37938	Segal et al. 1988

[a]Σ7SKF (mm) = sum of seven skinfolds: chest + midaxillary + triceps + subscapular + abdomen + anterior suprailiac + thigh; [b]As modified by Stolarczyk et al. (1997).
Key: HT = height (cm); BW = body weight (kg); R = resistance (Ω).

there are no research data to support the use of the SKF method for assessing the body composition of Hispanic men.

BIA Equations for Hispanics

Previously published BIA equations (Gray et al. 1989; Lohman 1992; Rising et al. 1991; Segal et al. 1988; Van Loan & Mayclin 1987) were cross-validated on 42 Hispanic men and 30 Hispanic women from New Mexico. Total Db was assessed by HD and converted to %BF using Siri's 2-C model equation. The total prediction errors of these equations ranged from 2.8 to 4.9 kg for men (Heyward et al. 1994) and 1.6 to 4.6 kg for women (Stolarczyk et al. 1995). All of these equations, with the exception of the Rising Pima Indian equation, significantly underestimated the average FFM of Hispanic men. Cross-validation of the Rising equation produced an acceptable prediction error (SEE = 3.4 kg) and closely estimated the average FFM of Hispanic men (63.4 vs. 63.3 kg). However, as mentioned earlier, this prediction equation was developed using the Keys and Brozek (1953) 2-C model conversion formula, which systematically underestimates %BF in each individual.

In the Stolarczyk et al. (1997) multiethnic BIA study, 23% of the men in the sample were Hispanic (N = 53). The modified Segal et al. (1988) fatness-specific BIA equations provided the best estimate of FFM for this multiethnic sample (CE = 0.5 kg, SEE and TE = 3.6 kg). In another multiethnic BIA study in which 81% of the men were Hispanic, the modified Segal fatness-specific equations accurately estimated the average FFM_{2-C} with small prediction errors (SEE and TE = 2.7 kg) (Lockner et al. 1999). Until additional studies for Hispanic men are undertaken, we recommend using the Segal et al. (1988) fatness-specific BIA equations, as modified by Stolarczyk et al. (1997), to estimate the FFM of Hispanic men.

For Hispanic women, the Lohman, Gray, and Segal equations accurately predicted 2-C model estimates of FFM within ±1 kg (Stolarczyk et al. 1995). The prediction errors for these equations were similar, ranging from 1.9 to 2.1 kg. In another multiethnic BIA study that included 58 Hispanic women (16% of the female sample), the modified Segal et al. (1988) fatness-specific equations resulted in a slightly more accurate prediction of FFM (TE = 2.3 kg) than either the Lohman or Gray prediction equations (TE = 2.8-3.0 kg) (Stolarczyk et al. 1997). Thus, we recommend using the modified Segal equations to estimate the FFM of Hispanic women.

NIR Equations for Hispanics

We do not recommend using the Futrex-5000 NIR manufacturer's equation for estimating body fat of Hispanic women and men. Cross-validation of this equation yielded an unacceptable prediction error (SEE = 6.4% BF) and significantly underestimated the average body fat of our samples by 2.4% BF (unpublished observations).

NIR equations were developed and cross-validated for a multiethnic sample of women (86% Caucasian, 9% Hispanic, 4% American Indian, and 1% African American) (Heyward, Jenkins et al. 1992). This equation accurately estimated the average Db of Hispanic women with a marginally acceptable prediction error (TE = 0.0081 g/cc). Given that this prediction equation was derived using a 2-C model and a relatively small number of Hispanic women, we would like to see this equation cross-validated on independent samples of Hispanic women before endorsing its use. It is preferable to use other field methods to estimate the body composition of Hispanic women.

Anthropometric Equations for Hispanics

As noted for other ethnic groups, there are no data supporting the use of the anthropometric method (circumference and bony diameter measures) to assess the body composition of your Hispanic clients. Equations using only age, height, and weight as predictors of FM were derived for Hispanic girls and boys aged 3 to 18 yr (Ellis 1997; Ellis et al. 1997). However, in light of the total body mass of these children, the prediction errors were fairly large (2.4 kg for girls and 3.7 kg for boys). Therefore, we do not recommend using these equations to assess the body composition of Hispanic children.

Key Points

- Certain ethnic groups, specifically African American and Hispanic women and American Indian men and women, have an especially high incidence of obesity and are at increased risk for developing obesity-related diseases.

- African American women have a higher prevalence of obesity than Caucasian or Hispanic women.

- A high incidence of type 2 diabetes coincides with the increasing prevalence of obesity in the American Indian population.

- The relationship between %BF and BMI differs among ethnic groups; thus disease risk stratification based on BMI may not be accurate for your ethnic-minority clients.

- Ethnic differences in fat patterning and body proportions place into question the applicability of generalized field method prediction equations derived primarily from Caucasian samples.

- Because of differences in the proportions of the FFB composition among ethnic groups and interindividual differences within ethnic groups, multicomponent model body composition assessment is recommended as a reference measure. Unfortunately, most studies evaluating the applicability of methods and equations for specific ethnic groups have not used multicomponent models.

- The FFBd of certain ethnic groups is not clearly defined. Because of elevated BMC and BMD values, it is likely that the average FFBd of African American men and women, as well as American Indian, Asian, and Hispanic women, is greater than 1.100 g/cc.

- Even though African Americans tend to deposit relatively more fat on the trunk than the extremities, the generalized ΣSKF equations of Jackson et al. (1980) and Jackson and Pollock (1978) provide an acceptable estimate of Db for non-obese African American women and men, respectively. These equations can also be used for non-obese Hispanic and Caucasian clients.

- The Segal et al. (1988) fatness-specific BIA equations, as modified by Stolarczyk et al. (1997), have the best predictive accuracy for estimating FFM of African Americans, American Indians, Caucasians, and Hispanics.

- For Asian populations, ethnic-specific SKF and BIA equations have been developed.

- Ethnic-specific SKF and NIR equations have been developed for American Indian women.

- To date the FFB composition of Hispanic males has not been assessed; therefore, 4-C model-based prediction equations are lacking for this population subgroup.

Key Terms

All of the terms used in this chapter have been identified and defined in previous chapters.

Review Questions

1. Identify ethnic groups that have a high incidence of obesity and that have been targeted for weight loss.

2. Explain how ethnic differences in fat patterning and body proportions potentially affect SKF and BIA estimates of body composition.

3. Why is the 4-C model recommended as the reference measure when one is determining the body composition of non-Caucasians?

4. How does the FFB composition of African Americans differ from that of Caucasians?

5. Are generalized SKF and BIA equations applicable to African Americans? If so, identify these equations.

6. What obesity-related disease is at near epidemic proportions in the American Indian population?

7. Which body composition prediction equations are most appropriate for American Indians?

8. Describe how the FFB composition, fat patterning, and body proportions of Asians differ from those of Caucasians.

9. Are generalized prediction equations applicable to Asians? Which methods and equations are recommended?

10. The body composition of the Caucasian population has been extensively studied. Identify suitable field methods and prediction equations for this population subgroup.

11. The Hispanic population is one of the largest minority groups in the United States. To what extent has the FFB composition of this ethnic group been studied, and how prevalent is obesity in this group compared to Caucasians?

12. Can generalized field method prediction equations be applied to Hispanics? Identify the methods and equations recommended for use.

CHAPTER 11

BODY COMPOSITION AND ATHLETES

Key Questions

- How does the fat-free body composition of physically active individuals and athletes differ from the reference body?
- How do physical activity and exercise alter the overall density and composition of the fat-free body?
- What body composition model is recommended, and which reference methods have typically been used, to obtain body composition reference measures for athletes?
- What field methods and prediction equations may be used to estimate the body composition of athletes?
- What are some of the health risks associated with low body weight and low body fat levels in athletes, and what steps can be taken to limit these risks?
- What methods and steps can be used to predict minimal body weight for athletes?

In many sports, body composition is important for optimal physical performance. Generally, a relatively low body fat is desirable to optimize physical performance in sports requiring jumping and running. A large muscle mass enhances performance in strength and power activities. Because of these performance-related implications, coaches, parents, exercise scientists, sports medicine specialists, and of course the athletes themselves have an interest in body composition. For years, exercise scientists and sports medicine professionals have examined the physiological profiles of elite athletes. Typically, athletes and physically active individuals are leaner than sedentary individuals, regardless of gender. However, female athletes have relatively greater body fat than male athletes in a given sport, and the average body fatness depends on the type of sport and the athlete's position (Wilmore 1983).

In addition to establishing physiological profiles and identifying the body composition characteristics common among elite athletes within a specific sport, body composition data for athletes are useful in a variety of ways. You can use such data for

- tracking changes in an athlete's body composition to monitor the effectiveness of a training or dietary regimen;
- estimating an optimal body weight or competitive weight class for certain sports, such as wrestling, boxing, and bodybuilding; and
- screening and monitoring the health status of athletes to detect and prevent disorders associated with excessively low body fat levels.

This chapter describes the FFB composition of physically active individuals and compares body composition models and reference methods for this population. Field methods and prediction equations applicable to athletes are recommended. The chapter concludes with a discussion of low body fat levels and the associated health risks, as well as a strategy for predicting minimal competitive body weight for athletes.

Fat-Free Body Composition of Athletes

Physical activity and exercise training result in moderate weight loss, moderate-to-large losses in body fat, and small-to-moderate gains in FFM (Wilmore 1983). The degree of alteration in body composition depends on the mode of exercise, as well as the frequency, intensity, and duration of training. In general, athletes have greater bone mineral content (BMC), bone mineral density (BMD), and FFM and a lower %BF than nonathletes (Evans, Prior et al. 2001). Numerous studies show that females participating in high-impact activities have BMDs significantly higher than those of nonathletic women (Alfredson et al. 1997; Dook et al. 1997; Duppe et al. 1996; Pichard, Kyle, Gremion et al. 1997; Taaffe et al. 1997). Generally, these are modest increases (2-6%), but research also showed that the average BMD of the United States Olympic women's field hockey team and elite female heptathletes was 13% to 14% greater than age- and race-matched BMD reference values (Houtkooper et al. 2001; Sparling et al. 1998). Additionally, lumbar BMD significantly increased (0.017 g/cm^2) following 27 weeks of gymnastics training (Nichols et al. 1994). Likewise, weight-trained men exhibited greater BMD and FFM than controls (Modlesky, Cureton et al. 1996).

Given the increases in BMC, BMD, and muscle mass, a likely assumption is that the density of the FFB (FFBd) in physically active individuals and athletes will be greater than that of their sedentary counterparts. However, Evans and colleagues suggested that BMC and BMD should not be used to infer differences in FFBd. BMC and BMD were not positively related to FFBd in their mixed sample of 132 athletes and 84 nonathletes (Evans, Prior et al. 2001). Results from several multicomponent model studies indicated that the FFBd of athletes appears to be similar to, or in some cases even less than, the FFBd of nonathletes (Arngrimsson et al. 2000; Evans, Prior et al. 2001; Modlesky, Cureton et al. 1996; Penn et al. 1994; Prior et al. 2001). In contrast, Thompson and Moreau (2000) reported a greater FFBd (1.109 g/cc) and bone mineral-to-FFM ratio (7.3%) in African American women collegiate athletes compared to the assumed reference values for men (1.100 g/cc and 6.8%, respectively). However, some of this increase in the FFBd could be attributed to ethnicity (see chapter 10). Arngrimsson et al. (2000) reported nonsignificant differences in the FFBd of collegiate male runners (1.097 g/cc) and male controls (1.100 g/cc) as well as female runners (1.100 g/cc) and female

controls (1.104 g/cc). Similarly, the average %BF of 10 recreational runners (>35 miles/week) was significantly lower than that of sedentary controls of similar age, height, weight, BMI, health status, and ethnicity (12.9% vs. 21.1% BF), but the FFBd remained unchanged (Penn et al. 1994). The water, protein, potassium, and mineral fractions of the FFM were similar between the endurance athletes and the controls. The ratio of skeletal muscle mass and bone to FFM was increased in the lower body but decreased in the upper body for the runners, providing further evidence that body composition alterations are dependent on exercise mode.

Variability in FFBd is more strongly related to the variance in the water fraction of the FFM than in the mineral portion of the FFM (Arngrimsson et al. 2000; Evans, Prior et al. 2001; Modlesky, Cureton et al. 1996; Prior et al. 2001; van der Ploeg et al. 2001). Because the density of water is less than that of the other components of the FFM (see table 1.3, p. 8), an increase in the relative hydration of the FFM results in a decrease in overall FFBd. Thus, in weight-trained athletes with a large amount of muscular hypertrophy, and therefore an increased water fraction, the FFBd may actually be *less* than that of athletes in other sports or sedentary people. Modlesky, Cureton et al. (1996) reported that the percentage of water was significantly greater (74.8% vs. 72.6%) and the percentages of mineral (5.3% vs. 5.9%) and protein (19.9% vs. 21.5%) significantly less in the FFM of weight trainers than in controls, resulting in a lower FFBd for the athletes (1.089 vs. 1.099 g/cc). This same pattern of proportionately greater water and less bone mineral relative to the FFM resulted in a lower FFBd for female bodybuilders preparing for competition compared to female athletic controls (van der Ploeg et al. 2001). However, in a more heterogeneous sample of both athletes and nonathletes whose FFBd ranged from 1.075 to 1.127 g/cc, Prior et al. (2001) reported that FFBd is not related to muscularity or musculoskeletal development.

In summary, athletes generally have a greater BMC, BMD, and FFM than nonathletes. However, the FFBd of athletes may be either higher or lower than the assumed reference value of 1.100 g/cc. The cause of this variance in FFBd is complex, as results from recent 4-C model body composition studies indicated that neither muscularity nor bone mineral is significantly related to FFBd (Evans, Prior et al. 2001; Prior et al. 2001). Deviations from the reference value may be training dependent, with athletes who engage in high-intensity, explosive training (e.g., gymnastics) having a higher mineral-to-FFM ratio and greater FFBd, and those training for muscular hypertrophy (e.g., bodybuilders) having an increased water-to-FFM ratio and lower FFBd.

Body Composition Models and Reference Methods

Given the changes that occur in the FFB composition with physical training and the variability in the FFBd, multicomponent model equations should be used to develop prediction equations and obtain accurate estimations of body fatness in physically active men and women. Unfortunately, the body composition prediction equations developed for athletes have been based on either the 2-C model using the hydrodensitometry (HD) method or the dual-energy X-ray absorptiometry (DXA) method.

Two-Component Models

Penn et al. (1994) reported that there was no significant difference among 2-C models (hydrodensitometry, total body potassium, and TBW) and a 4-C model in estimating the %BF of 10 white male recreational runners. Given that the average FFBd was similar to the assumed value (1.100 g/cc), the classic 2-C model approach was accurate for estimating %BF in this population. Similarly, Arngrimsson et al. (2000) found that the 2-C model using the HD method accurately predicted the average %BF$_{4-C}$ of male and female collegiate middle- and long-distance runners, but individual variation was substantial (–5.3% to 3.0% body mass). Likewise, the average FFBd of female bodybuilders was close to 1.100 g/cc, resulting in only a small difference between 2-C (HD) and 4-C model estimates of %BF for these athletes (van der Ploeg et al. 2001). However, the authors noted that the variability in TBW could be large; there was greater interindividual variability between 2-C and 3-C models than between 3-C and 4-C models that measured TBW.

In contrast, the 2-C model equation of Siri (1961) overestimated the mean %BF of weight trainers by 4.1% BF (Modlesky, Cureton et al. 1996). Similarly, the 2-C model overestimated the %BF of male athletes from a variety of sports by 2.7% BF, but there was not a significant difference between %BF from densitometry and the 4-C model in nonathletic males or in female athletes and nonathletes (Prior et al. 2001). Individual variations in amenorrheic runners with low BMC and female bodybuilders with high BMC varied by as much as ±3% BF with 2-C model equations (Bunt et al. 1990). These studies suggest that the 2-C model is not appropriate in

athletic samples that have a FFBd different from 1.100 g/cc. Furthermore, differences exist between 2-C methods. The average %BF estimated from the Bod Pod (10.9% BF) was significantly less than that estimated from HD (13.3% BF) and a 3-C model (12.7% BF) in collegiate football players (Collins et al. 1999).

Selecting the appropriate conversion formula is an additional concern with use of the 2-C model to assess the body composition of athletes. For example, the 2-C model conversion formula of Brozek et al. (1963) significantly underestimated the average %BF of African American female athletes by 2.6% BF, whereas the race-specific formula of Schutte et al. (1984) slightly overestimated %BF (0.5% BF) (Thompson & Moreau 2000). Also, several research studies of teenage athletes showed better agreement between field methods and the HD method when the 2-C conversion formula of Brozek et al. (1963) was used rather than the age-specific 2-C model conversion formulas of Lohman (1986) (Eckerson et al. 1997; Smith et al. 1997). Lohman (1986) acknowledged that these age-specific Db-conversion constants may not be applicable to athletes. Because of the increase in muscle mass and BMC likely to occur with physical training, we recommend using the adult conversion formulas (Brozek et al. 1963; Siri 1961) rather than Lohman's age-specific constants when converting the Db to %BF for high school athletes.

DXA Model

On average, body fat estimates from DXA are less than 2-C, 3-C, and 4-C model estimates of %BF in female athletes and bodybuilders and male and female collegiate runners (Arngrimsson et al. 2000; van der Ploeg et al. 2001). However, some have reported good agreement between DXA and 4-C model estimates, with mean differences ≤1% BF in both male distance runners (Penn et al. 1994) and weightlifters (Modlesky, Cureton et al. 1996). Modlesky, Cureton et al. (1996) reported that individual differences between these models ranged from –3.0% to 2.6% BF. Additionally, Sparling et al. (1998) reported that the average difference between DXA and the Siri (1961) 2-C model (HD) was only 1.5% BF for elite female field hockey players.

Fornetti and coworkers (1999) derived sport-specific bioimpedance, near-infrared interactance, and height-weight equations from DXA reference measures. Many other researchers have used DXA as the reference method to cross-validate field method equations for athletes (De Lorenzo et al. 1998, 2000; Houtkooper et al. 2001; Pichard, Kyle, Gremion et al. 1997; Stewart & Hannan 2000; Yannakoulia et al. 2000). Since DXA also measures BMC and BMD and since variations in FFM density of athletes are likely due, at least in part, to increased bone mineral, we believe that DXA is preferable to HD (2-C model) as a reference method for assessing the body composition of athletes.

Field Methods and Prediction Equations

Although some athletes may have access to university laboratories, the body composition of athletes is typically assessed with field methods. Sport-specific equations, as well as the generalized equations that are often applied to athletes, have been developed and cross-validated using the HD (2-C model) or DXA methods. We are unaware of any athletic-specific equations derived from multicomponent model reference measures.

SKF Equations for Athletes

Research indicates that population-specific and generalized SKF equations developed for women and men accurately estimate the Db of athletes in many different sports (Hortobagyi, Israel, Houmard, O'Brien et al. 1992; Houmard et al. 1991; Oppliger et al. 1992; Sinning et al. 1985; Sinning & Wilson 1984; Thorland et al. 1991; Withers, Whittingham et al. 1987). The generalized sum of seven SKFs (Σ7SKF) equation of Jackson and Pollock (1978) has been shown to accurately estimate the average body fatness of physically active men (Israel et al. 1989), African American and Caucasian collegiate football players (Hortobagyi, Israel, Houmard, O'Brien et al. 1992; Houmard et al. 1991), and males participating in 12 different collegiate sports (Sinning et al. 1985). The prediction error of this equation ranged from 2.2% to 2.9% BF. In comparison, the SKF equations of Katch and McArdle (1973) and Sloan (1967) accurately estimated average body fatness of elite Australian athletes participating in 18 different sports (Withers, Craig et al. 1987) and collegiate football players (Oppliger et al. 1992), but with slightly larger prediction errors (SEE = 2.9-3.5% BF). For male adolescent athletes, 14 to 19 yr of age, Thorland and colleagues (1984) recommended using the SKF equations of Forsyth and Sinning (1973), Lohman (1981), and Pollock et al. (1980). With respect to these equations, the Forsyth and Sinning equation had the least variability between predicted and measured Db in their sample. The SKF equations of Slaughter et al. (1988) or Deurenberg, Pieters et al. (1990) also provided good group estimates of %BF for active adolescent males, but individual

differences (95% limits of agreement = –7.7% to 9.3% BF for Slaughter et al. equation and –9.1% to 9.6% BF for Deurenberg et al. equation) were large (De Lorenzo et al. 1998).

For female athletes, we recommend using the generalized Σ4SKF equation of Jackson et al. (1980). Lohman (1992) also recommended this equation for this population. This equation accurately estimated the average body fatness of elite female heptathletes (SEE = 2.4% BF) (Houtkooper et al. 2001) as well as female athletes from 10 different collegiate sports (SEE = 3.2% BF) (Sinning & Wilson 1984). Also, similar prediction errors (SEE = 3.0% BF, TE = 3.4% BF) were reported when the Σ4SKF equation was used to assess %BF of high school gymnasts (Housh et al. 1996). Likewise, Thorland et al. (1984) noted that this equation accurately predicted the average Db (SEE = 0.0072 g/cc) of female adolescent athletes, ages 11 to 19 yr. Compared to that of the Jackson et al. Σ4SKF equation, the group and individual predictive accuracy of other SKF equations for estimating the body fat of athletic women is not as good (Pacy et al. 1995; Williams & Bale 1998; Withers, Whittingham et al. 1987).

BIA Equations for Athletes

The applicability of the bioelectrical impedance (BIA) method for predicting body composition is highly dependent on testing under controlled conditions, but the physiology of athletes is often altered or in an "uncontrolled state." For example, carbohydrate loading results in increased water retention; and exercise can affect skin temperature, electrolyte concentration, and hydration status. All of these factors alter impedance measurements (see chapter 6), and Segal (1996) stressed the importance of controlling these variables with use of the BIA method to assess the body composition of athletes.

Hypohydration and superhydration may result in significant errors in the estimation of FFM and %BF from BIA (Saunders et al. 1998). Lukaski et al. (1990) reported a 30% reduction in the error of predicting FFM of athletes when performing BIA under controlled conditions (i.e., 2 hr postconsumption of a light meal and no preceding exercise). We recommend following the BIA pretesting guidelines presented in chapter 6 (p. 94). Given that it may be difficult or impractical for the athlete in training to meet these guidelines, the BIA method may not be the most suitable field method for testing athletes. Furthermore, several researchers reported that the SKF method is a better predictor of %BF for athletes than the BIA method (Houtkooper et al. 2001; Stewart & Hannan 2000).

Many population-specific and generalized BIA equations developed for the average population have been tested on athletic samples. Overall, cross-validation studies indicated that these equations cannot be used to accurately estimate the FFM of athletic men and women. Several such equations systematically underestimated the FFM of African American and Caucasian collegiate football players (Hortobagyi, Israel, Houmard, O'Brien et al. 1992) and lean Caucasian men (Eckerson et al. 1992), with prediction errors (SEEs) ranging from 2.9 to 6.3 kg. Likewise, BIA manufacturers' equations (BMR, RJL, and Valhalla) significantly underestimated the FFM of African American and Caucasian football players (Oppliger et al. 1992), male and female college athletes (Williams & Bale 1998), and male and female bodybuilders (Colville et al. 1989) by 3.8 to 4.8 kg, with SEEs ranging from 3.3 to 5.8 kg.

In contrast, Lukaski et al. (1990) reported that a BIA equation developed for women and men 18 to 74 yr of age (Lukaski & Bolonchuk 1987) accurately predicted the average FFM of female and male collegiate athletes (SEE = 2.0 kg). However, Houtkooper et al. (2001) reported that this equation overestimated the average %BF of elite female heptathletes by 4.4% BF (SEE = 3.0% BF, TE = 6.0% BF). Also, Stewart and Hannan (2000) reported that the Lukaski and Bolonchuk (1987) BIA equation explained only 41% of the variance in %BF$_{DXA}$ of male athletes compared to 79% for the Jackson and Pollock (1978) Σ3SKF equation.

There have been few efforts to develop sport-specific BIA equations for male athletes (see table 11.1). Hortobagyi and colleagues developed equations for African American and Caucasian football players, but their cross-validation samples consisted of only 15 African Americans and 13 Caucasians (Hortobagyi, Israel, Houmard, O'Brien et al. 1992). Eckerson et al. (1992) reported that a BIA equation developed for high school wrestlers (Oppliger, Nielsen, Hoegh et al. 1991) could be used to predict the FFM of lean men (SEE = 1.70 kg) ages 19 to 40 yr. However, they also mentioned that body weight alone is as accurate as this BIA equation for estimating FFM of lean men. Furthermore, in another study of male athletes, the Oppliger, Nielson, Hoegh et al. (1991) equation overestimated the FFM$_{DXA}$ by an average of 2.3 kg, with large individual variation (95% limits of agreement = −9.8 to 5.1 kg) (De Lorenzo et al. 2000).

Several researchers examined the applicability of the BIA method for assessing body composition of female athletes (see table 11.1). A BIA equation that estimates FFM of female college athletes was developed using a double cross-validation technique and DXA as the reference method (Fornetti et al. 1999). This equation had an SEE of only 1.1 kg for the sample of 132 athletic women. Similar methods were used to develop a BIA equation specifically for female dancers (Yannakoulia et al. 2000). Although the prediction error for this equation was low (SEE = 1.5 kg), it was developed using a relatively small sample (N = 42) of professional dance students and therefore needs to be cross-validated on independent groups of dancers.

Previously published BIA equations, as well as the RJL manufacturer's equation, have been cross-validated for female high school gymnasts (Eckerson et al. 1997) and elite female runners (Pichard, Kyle, Gremion et al. 1997). There were acceptable prediction errors (SEE and TE < 2.8 kg of FFM) for several of the equations evaluated in these studies. The equation of Van Loan et al. (1990) met all of the cross-validation criteria for the teenaged gymnasts. Interestingly, an equation developed for anorexia nervosa patients (Hannan et al. 1993) produced low prediction errors (SEE = 1.2 kg, TE = 1.6 kg) for a sample of very lean (14.8% BF) elite female runners. The authors also noted that the best BIA equations for estimating body composition of the runners had poor predictive accuracy for age- and height-matched sedentary controls.

Near-Infrared Interactance Equations for Athletes

Cross-validation of the Futrex-5000 manufacturer's near-infrared interactance (NIR) equation indicated that this equation systematically underestimates the body fatness of physically active men (Israel et al. 1989), as well as African American and Caucasian collegiate football players (Hortobagyi, Israel, Houmard, O'Brien et al. 1992; Houmard et al. 1991), with prediction errors ranging from 3.6% to 4.2% BF. In contrast, for female high school gymnasts, there was good agreement between this NIR equation and %BF$_{HD}$ (r = 0.78, CE = −0.3% BF) using the Brozek et al. (1963) 2-C model conversion formula, and prediction errors were acceptable (SEE = 3.1% BF, TE = 3.1% BF) (Smith 1997). The researchers noted, however, that other Futrex devices (i.e., models 5000A and 1000) produced large, unacceptable errors. Additional cross-validation studies on female athletes in various sports are needed before the Futrex-5000 manufacturer's equation can be recommended for this population.

Several researchers have used the optical density (OD) values from the Futrex-5000 to develop prediction equations for estimating body composition in athletic populations (Fornetti et al. 1999;

Table 11.1 Comparison of Bioelectrical Impedance Analysis Equations for Athletes

Name of equation	Reference model and method	Validation sample	Equation and validation statistics for original study	Cross-validation analysis		
				Sample	Method	Statistics
Fornetti et al. 1999	2-C; DXA (Hologic QDR-1000W; SV 6.10)	132 women athletes, 18-27 yr, various sports	FFM = 0.282 (HT) + 0.415 (BW) − 0.037 (R) + 0.096 (Xc) − 9.734; R^2 = 0.96; SEE = 1.1 kg	66 female athletes	Double XV	d = 0.3 kg; SEE = 1.1 kg
				66 female athletes		d = 0.2 kg; SEE = 1.1 kg
Hortobagyi, Israel, Houmard, O'Brien et al. 1992	2-C; HD (Brozek; Schutte)	40 AA and 22 CA college football players	AA: %BF = −46.6 + 1.576 (BMI) + 0.071 (R) − 1.753 (ϕ); R^2 = 0.80; SEE = 2.6% BF	AA: 15	Internal XV	AA: d = −3.6 kg; SEE = 2.7 kg
			CA: %BF = −12.6 + 1.601 (BMI) − 2.389 (ϕ); R^2 = 0.92; SEE = 2.1% BF	CA: 13		CA: d = −2.9 kg; SEE = 3.2 kg
Oppliger, Nielson, Hoegh et al. 1991	2-C; HD	110 high school wrestlers	FFM = 1.949 + 0.701 (BW) + 0.186 (HT²/R); R^2 = 0.96; SEE = 1.9 kg	**	**	**
Yannakoulia et al. 2000	2-C; DXA (Lunar DPX; SV 1.3z)	42 female dance students, 18-26 yr	FFM = 0.247 (BW) + 0.214 (HT²/R) − 0.191 (HT) − 14.96; R^2 = 0.83; SEE = 1.5 kg	21 dancers	Double XV	d = 2.8 kg; LA = −2.8 to 2.7 kg
				21 dancers		d = 3.0 kg; LA = −3.1 to 2.8 kg

** Equation not cross-validated in original study.

Key: DXA = dual-energy X-ray absorptiometry; HD = hydrodensitometry; FFM = fat-free mass (kg); %BF = percent body fat; BW = body weight (kg); HT = height (cm); R = resistance (Ω); Xc = reactance (Ω); ϕ = phase angle; XV = cross-validation; SV = software version; R^2 = multiple correlation coefficient squared; SEE = standard error of estimate; LA = 95% limits of agreement; d = constant error or mean difference between BIA-predicted and reference values; AA = African American; CA = Caucasian.

Hortobagyi, Israel, Houmard, O'Brien et al. 1992; Oppliger et al. 2000) (see table 11.2). Hortobagyi developed separate NIR equations for African American and Caucasian football players (Hortobagyi, Israel, Houmard, O'Brien et al. 1992). For African Americans, body weight and the OD at the biceps site accounted for 87% of the variance in %BF, with an acceptable prediction error (SEE = 2.7% BF). For Caucasian football players, OD measures failed to improve the predictive accuracy of the equation beyond that explained by body weight and height only (SEE = 2.3% BF). These equations need to be cross-validated on additional samples of football players before their use can be recommended.

Using DXA as the reference method, Fornetti et al. (1999) developed an equation that includes height, weight, and an OD measurement at the biceps to estimate the FFM of female athletes (18-27 yr). The SEE (1.2 kg) for this equation was rated as "ideal" (see table 2.1, p. 20). To estimate the %BF of high school wrestlers, Oppliger et al. (2000) developed four NIR equations, ranging in complexity from one OD measure at the biceps site to three OD measures at the biceps, triceps, and thigh. These OD measures were used with age and BMI to estimate the %BF. The multisite equation that included BMI and age explained the greatest amount of variance in %BF, but all four NIR equations had similar predictive accuracy when cross-validated (SEE = 3.1% BF and 2.2 kg of minimum weight). The authors also noted that these NIR equations were slightly better than the SKF method (SEE = 3.3% BF and 2.5 kg minimum weight) for estimating body composition and minimal body weight of wrestlers.

The Futrex-5000A/WL analyzer has a prediction equation developed specifically for high school wrestlers and lean high school male athletes. To date, there are no published research studies validating this NIR equation.

Anthropometric Equations for Athletes

Several researchers reported that body weight and stature alone provide a reasonable estimate of body composition in lean and athletic populations (Eckerson et al. 1992; Fornetti et al. 1999; Hergenroeder et al. 1993; Hortobagyi, Israel, Houmard, O'Brien et al. 1992). For example, Hergenroeder et al. (1993) noted that the average FFM of female ballet dancers is accurately estimated by body weight alone (SEE = 1.5 kg), and the addition of bioimpedance measures does not improve the predictive accuracy of this equation. Likewise, body weight alone accounted for 96% of the variance in the FFM of lean men, and the prediction

errors were similar for equations using weight only (SEE = 1.68 kg) and bioimpedance measures (SEE = 1.70 kg) (Eckerson et al. 1992). Although Fornetti et al. (1999) reported that the predictive accuracy of BIA (SEE = 1.1 kg) and NIR (SEE = 1.2 kg) equations was better than use of height and weight (1.6 kg) alone for estimating FFM of female collegiate athletes, these values correspond to an improvement in predictive accuracy of only about 1% BF. These findings suggest that FFM accounts for much of the total body weight in lean individuals, regardless of gender.

Table 11.3 presents anthropometric equations developed for athletic samples. Some research shows that the predictive accuracy of anthropometric equations is not as good as that of other field methods for assessing the body composition of athletic women and men. Yannakoulia et al. (2000) reported that the Hergenroeder et al. (1993) equation, which predicts FFM of dancers from body weight alone, has greater limits of agreement than other field methods.

Also, Thorland et al. (1991) noted that the Db of high school wrestlers is accurately estimated using the Katch and McArdle (1973) anthropometric equation; but this research group also found that this equation does not adequately estimate the Db of male and female athletes in other sports (Thorland et al. 1984). In addition, Mayhew et al. (1983) developed an anthropometric equation to predict FFM of female collegiate athletes participating in seven different sports. Using a combination of body weight, neck circumference, and thigh circumference as predictors, this equation accurately estimated average FFM (SEE = 2.6 kg), but the size of the cross-validation sample (N = 32) was relatively small. Also, large prediction errors (SEE = 0.1000 g/cc) have been reported with estimation of Db from combinations of skeletal breadth measures (Forsyth & Sinning 1973).

Summary of Field Method Prediction Equations for Athletes

Table 11.4 contains a summary of the prediction equations that we recommend to estimate body composition of athletes. None of these equations are based on multicomponent model reference measures; instead, they were developed and cross-validated using 2-C model reference methods. We recommend using the Jackson and Pollock (1978) Σ7SKF and the Jackson et al. (1980) Σ4SKF equations to estimate the body composition of male and female athletes, respectively. In the hands of a skilled technician, the SKF method is probably preferable to BIA for assessing the %BF of athletes (Houtkooper et al. 2001; Stewart & Hannan 2000).

Table 11.2 Comparison of Near-Infrared Interactance Equations for Athletes

Name of equation	Reference model and method	Validation sample	Equation and validation statistics for original study	Cross-validation analysis		
				Sample	Method	Statistics
Fornetti et al. 1999	2-C; DXA (Hologic QDR-1000W; SV 6.10)	132 women athletes, 18-27 yr, various sports	FFM = 0.111 (HT) + 0.641 (WT) − 12.397 (biceps OD2) − 8.423 $R^2 = 0.95$; SEE = 1.2 kg	66 female athletes 66 female athletes	Double XV	$d = 0.1$ kg; SEE = 1.5 kg $d = 0.2$ kg; SEE = 1.1 kg
Hortobagyi, Israel, Houmard, O'Brien et al. 1992	2-C; HD (Brozek; Schutte)	55 AA college football players	%BF = 23.2 + 0.184 (BW) − 22.68 (biceps OD2) $R^2 = 0.87$; SEE = 2.7% BF	**	**	**
Oppliger et al. 2000	2-C; HD (Brozek)	150 high school wrestlers	%BF = 14.061 − 40.786 (OD)[a] $R^2 = 0.55$; SEE = 3.3% BF	53 high school wrestlers	Internal XV	$d = 0.3$% BF; TE = 3.1% BF

[a]Only the simplest of the four NIR equations from Oppliger et al. (2000) is listed. The other equations have slightly higher R^2 (0.68) and lower SEE (2.9% BF), but similar cross-validation statistics with no clinically significant difference among the equations. The OD for the Oppliger et al. equations is the average of OD1 and OD2 at the biceps site.

**Equation not cross-validated in original study.

Key: DXA = dual-energy X-ray absorptiometry; HD = hydrodensitometry; FFM = fat-free mass (kg); %BF = percent body fat; BW = body weight (kg); OD = optical density; XV = cross-validation; SV = software version; R^2 = multiple correlation coefficient squared; SEE = standard error of estimate; TE = total error; d = constant error or mean difference between NIR-predicted and reference values; AA = African American.

Table 11.3 Comparison of Anthropometric Equations for Athletes

Name of equation	Reference model and method	Validation sample	Equation and validation statistics for original study	Cross-validation analysis		
				Sample	Method	Statistics
Fornetti et al. 1999	2-C; DXA (Hologic QDR-1000W; SV 6.10)	132 women athletes, 18-27 yr, various sports	FFM = 0.143 (HT) + 0.565 (BW) − 10.03 R^2 = 0.92; SEE = 1.6 kg	**	**	**
Hergenroeder et al. 1993	2-C; TOBEC	112 female dancers, 11-25 yr	FFM = 0.73 (BW) + 3.0 R^2 = 0.88; SEE = 1.5 kg	23 female dancers	Internal XV	d = 0.2 kg; SEE = 0.9 kg; LA = ±1.8 kg
Hortobagyi, Israel, Houmard, O'Brien et al. 1992	2-C; HD (Brozek)	35 CA college football players	%BF = 55.2 + 0.481 (BW) − 0.468 (HT) R^2 = 0.91; SEE = 2.3% BF	**	**	**
Mayhew et al. 1983	2-C; HD (Brozek); estimated RV	75 college female athletes, 18-23 yr	FFM = 0.757 (BW) + 0.981 (neck C) − 0.516 (thigh C) + 0.79 R^2 = 0.85; SEE = 2.6 kg	32 female athletes, 17-21 yr	Internal XV	d = 1.1 kg

**Equation not cross-validated in original study.

Key: DXA = dual-energy X-ray absorptiometry; HD = hydrodensitometry; TOBEC = total body electrical conductivity; FFM = fat-free mass (kg); %BF = percent body fat; BW = body weight (kg); HT = height (cm); C = circumference (cm); RV = residual lung volume; XV = cross-validation; SV = software version; R^2 = multiple correlation coefficient squared; SEE = standard error of estimate; LA = 95% limits of agreement; d = constant error or mean difference between anthropometric-predicted and reference values; CA = Caucasian.

Table 11.4 Summary of Recommended Equations for Athletes

	Method/Equipment	Population	Equation	Reference
SKF	∑7SKF[a] (Lange)	Male athletes (>18 yr)	1. $Db\ (g/cc) = 1.112 - 0.00043499\ (\Sigma7SKF) + 0.00000055\ (\Sigma7SKF)^2 - 0.00028826\ (age)$	Jackson & Pollock 1978
	∑4SKF[b] (Lange)	Female athletes, college age; elite hept-athletes; female adolescent athletes, 11 to 19 yr; female high school gymnasts	2. $Db\ (g/cc) = 1.096095 - 0.0006952\ (\Sigma4SKF) + 0.0000011\ (\Sigma4SKF)^2 - 0.0000714\ (age)$	Jackson et al. 1980
	∑3SKF[c]	High school & collegiate wrestlers; male adolescent athletes (14-19 yr)	3. $Db\ (g/cc) = 1.0973 - 0.000815\ (\Sigma3SKF) + 0.00000084\ (\Sigma3SKF)^2$	Lohman[d] 1981
BIA	RJL 101A	Female athletes (18-27 yr)	4. $FFM\ (kg) = 0.282\ (HT) + 0.415\ (BW) - 0.037\ (R) + 0.096\ (Xc) - 9.734$	Fornetti et al. 1999
	BIO-Z	Elite female distance runners (16-37 yr)	5. $\%BF = 7.32 - 0.572\ (HT^2/R) + 0.664\ (BW)$	Hannan et al. 1993
	RJL 106	Female high school gymnasts, (13-17 yr)	6. $FFM\ (kg) = 0.52\ (HT^2/R) + 0.23\ (BW) + 7.49$	Van Loan et al. 1990
NIR	Futrex-5000	Female athletes (18-27 yr)	7. $FFM\ (kg) = 0.111\ (HT) + 0.641\ (BW) - 12.397\ (biceps\ OD2) - 8.423$	Fornetti et al. 1999
	Futrex-5000	High school wrestlers	8. $\%BF = 14.061 - 40.786\ (avg.\ OD)$	Oppliger et al. 2000

[a]∑7SKF (mm) = sum of seven skinfolds: chest + midaxillary + triceps + subscapular + abdomen + anterior suprailiac + thigh; [b]∑4SKF (mm) = sum of four skinfolds: triceps + anterior suprailiac + abdomen + thigh; [c]∑3SKF (mm) = sum of three skinfolds: triceps + subscapular + abdomen; [d]Modified by Thorland et al. (1991) for high school wrestlers.

Key: HT = height (cm); BW = body weight (kg); R = resistance (Ω); Xc = reactance (Ω); OD = optical density (avg. OD is average of OD1 and OD2 at biceps site).

BIA is more valuable as a tool for obtaining group data on athletes than for detecting small changes in %BF within individual athletes (Segal 1996). However, the Fornetti at al. (1999) BIA equation, as well as their NIR equation, may be used to estimate the FFM of collegiate female athletes. Additionally, the SKF (Lohman 1981) and NIR (Oppliger et al. 2000) equations developed for wrestlers (see following section on predicting competitive and minimal body weight) have good predictive accuracy and are recommended for this population. Although the anthropometric equations listed in table 11.3 have acceptable prediction errors, they either were not cross-validated or were derived from inferior reference methods (e.g., total body electrical conductivity, estimated residual lung volume during HD). Use these equations only when other field methods are not available.

Low Body Fat and Health Risks

Because FM does not contribute to force production and because excess body fat makes it more difficult to move one's body mass, a low body fat is generally considered advantageous in most sports. However, very low body fat can result in serious health complications, and some athletes turn to unhealthy behaviors in an effort to achieve a more desirable body composition. The issue of disordered eating and controlling body weight by athletes is addressed in a text by Brownell et al. (1992).

Concerns for the Female Athlete

Essential body fat for females is estimated to be about 12% (American College of Sports Medicine 1996; Behnke & Wilmore 1974). Lohman (1992) suggested that the lower limits for most female athletes should range between 12% and 16% BF, depending on the sport and the individual. In an effort to achieve low body fat levels, some female athletes, especially adolescents and those participating in sports that emphasize low body weight, are at risk for developing a syndrome known as the **female athlete triad.** The interrelated components of the triad are disordered eating (e.g., anorexia, bulimia), **amenorrhea** (absence of three or more consecutive menstrual cycles after menarche), and **osteoporosis** (loss of BMC). The female athlete triad is of such concern that the American College of Sports Medicine (1997) has issued a position statement on this syndrome.

The body composition of young, developing female elite athletes is of special concern. The physique and physiology of some of these athletes approaches that of individuals with anorexia ner-

vosa. For example, there was no significant difference in the Db or lean body mass relative to total mass among elite female runners and gymnasts ages 12 to 16 yr and recovering anorexics ages 9 to 16 yr (Bale et al. 1996). Furthermore, menarche was delayed in these athletes, and one-third of the athletes who had started menstruating developed amenorrhea. The late menarche and delayed pubertal development typical in young elite athletes may be linked to low levels of **leptin** (energy-regulating protein secreted by adipocytes) caused, in part, by low fat reserves (Weimann 2002). Weimann (2002) reported low leptin levels and delayed maturation in elite female gymnasts.

Female athletes are at greater risk of developing eating disorders than nonathletes (Beals & Manore 1994; Sundgot-Borgen 1994). Likewise, the prevalence of amenorrhea is higher in the athletic population (3.4-66%) than in the general female population (2-5%) (Nattiv et al. 1994). The large discrepancy in the prevalence of amenorrhea among female athletes may be due to differences in the sample population (e.g., type of sport, level of competition) and differences in the definition of amenorrhea among studies. However, **athletic amenorrhea** is not always the result of an eating disorder or low body fat. The etiology of menstrual irregularities in female athletes is complex and multifactorial, including additional factors such as low body weight, delayed menarche, nutrition, intensity and volume of training, and psychological stress (Fruth & Worrell 1995; Sanborn 1986; Yeager et al. 1993). Low body weight may be a more significant factor than low body fat (Fruth & Worrell 1995). One thing is clear, however; there is a strong relationship between athletic amenorrhea and a decrease in BMD. This relationship, as well as the resulting increase in stress fractures due to the lower BMD in amenorrheic athletes, was a consistent finding reported in research studies reviewed by Fruth and Worrell (1995).

Strategies for Developing Healthy Body Fat Levels in Athletes

Pressure from others, such as coaches, and the athlete's own preoccupation with reaching a prescribed target weight or %BF can be contributing factors to the development of disordered eating and the cascade of the female athlete triad (Sundgot-Borgen 1994; Yeager et al. 1993). The opinion of some authorities is that body composition testing further promotes the athlete's anxiety; thus, the Canadian Academy of Sport Medicine (2001) has issued a position statement calling for the abandonment of routine body composition testing of female athletes. We believe that this is an extreme measure. Body composition assessment can be a beneficial tool to screen athletes at risk for disorders associated with low body fat, to safely and effectively monitor athletes as they progress through diet and training programs, and to help athletes set and reach realistic body weight and body composition goals. However, the clinician must be sensitive to the issue of disordered eating when evaluating and interpreting body composition data for female athletes (Oppliger & Cassady 1994). All body composition assessments should be done by well-trained, highly skilled technicians who have experience interpreting body composition results and counseling athletes. In addition, we suggest the following strategies to help you develop healthy body fat levels for your athletes:

• Educate athletes and coaches about minimal body fat values (5% for males and 12-14% for females) and the health risks associated with dropping below those values. Additionally, you should provide athletes and coaches with information about "normal or healthy" body fat values (table 1.2, p. 6), as well as ranges of %BF for specific sports (table 11.5). Emphasize that "optimal" body fat differs among individual athletes.

• Recognize that small changes in body fat percentage can be due to measurement error rather than a true change in the athlete's body composition.

• To minimize measurement error, use the same method and instrument to monitor changes in the athlete's body composition over time.

• Encourage athletes and coaches to seek professional advice from exercise physiologists and sport nutritionists when developing exercise prescriptions and planning dietary recommendations for weight loss, weight gain, and body composition changes.

• Educate athletes, coaches, parents, and athletic trainers about eating disorders so that they will be able to recognize the signs of an eating disorder. To screen athletes for eating disorders, use the Female Athlete Screening Tool (FAST), a questionnaire validated for women athletes (McNulty et al. 2001).

Prediction of Competitive and Minimal Body Weight

The American College of Sports Medicine (1996) recommends **minimal body fat** values of 7% for males 16 yr and younger, 5% for males over 16 yr, and 12% to 14% for female athletes. These values are probably a close estimate of the lower limit of body fat needed for normal physiological and metabolic functions. Typical ranges of %BF for athletes in a given sport are listed in table 11.5. One should use caution when applying %BF values from tables to athletes. In many cases, the values listed in a table may have been derived from studies of elite athletes who have undergone years of high-level training in a particular sport. It is also important to recognize that the minimal body fat values and those listed in table 11.5 should be viewed only as *guidelines;* optimal body weight and body composition to maximize performance vary among individuals. Thus, it is wiser to set individual goals for athletes that fall within a range of body fat values than to expect all athletes in a given sport to achieve the same level of body fatness. Otherwise, some athletes may feel pressured to engage in unsafe weight loss practices or may develop eating disorders in an attempt to meet unrealistic weight and fat loss goals.

Once the body composition of the athlete is measured, you can use this information to determine a competitive or **minimal body weight (MW)** for the athlete. Competitive body weight is based on the athlete's present FFM and desired %BF level. Use table 11.5 as a guideline for the athlete's %BF range in a given sport. Figure 11.1 provides a sample calculation of competitive body weight for a female gymnast. Assuming that her FFM does not change, this athlete must lose 7 lb (fat and water weight) to achieve her competitive body weight and body fat goals. Remember that this is a guideline and that it is necessary to monitor the athlete's health to detect warning signs of excessive loss of body weight such as menstrual irregularities.

The concept of predicting MW has probably been studied and applied more in wrestling than in any other sport. In a position statement addressing weight loss in wrestlers, the American College of Sports Medicine (1996) recommends that the body composition of wrestlers be assessed prior to the start of the season as a means of determining a MW. Assuming a minimal fat level of 5% BF, MW is calculated as FFM × 1.05. In judging the accuracy of equations that predict MW, the prediction error should not exceed 2.0 kg given that competitive weight class intervals for high school wrestlers are 2.5 to 3.0 kg (Thorland et al. 1987).

In a landmark cross-validation study, Thorland et al. (1991) evaluated the predictive accuracy of numerous equations on 860 high school wrestlers. One of the equations with the lowest prediction error was their modification of Lohman's (1981) Σ3SKF equation. Other researchers also recommended using this equation to estimate MW in high school (Clark, Kuta, Sullivan et al. 1993) and

Table 11.5　Average Body Fat of Male and Female Athletes

Sport	Females %BF	Males %BF
Ballet dancing	13-20	8-14
Baseball		12-15
Basketball	20-27	7-11
Bodybuilding	9-13	6-9
Cycling	15	8-10
Football		
Backs		9-12
Linebackers		13-14
Linemen		16-19
Quarterbacks/Kickers		14
Gymnastics	10-17	5-10
Ice hockey		8-15
Racquetball	14	8-9
Rock climbing	10-15	5-10
Rowing	14-18	8-15
Skiing		
Alpine	21	7-14
Cross country	16-22	7-12
Jumping		14
Soccer		10
Softball	22	
Speed skating	15-24	11
Swimming	14-24	9-12
Tennis	20	15-16
Track and field		
Discus throwers	25	16
Jumpers	8-14	7-8
Long distance runners	10-19	6-13
Middle distance runners	10-14	7-12
Shot putters	20-28	16-20
Sprinters	11-19	8-16
Decathletes		8-9
Pentathletes	11	
Triathalon	7-17	5-11
Volleyball	16-25	11-12
Weightlifting		
Power lifters		9-16
Olympic lifters		10-12
Wrestling		5-12

Data from Fleck (1983, pp. 399-400) and Wilmore (1983, pp. 23-24).

Athlete: female gymnast

Precompetition data	Competition data
Body weight = 120 lb	Body weight = 113 lb
%BF = 21%	%BF = 16%
FFM = 94.8 lb	FFM = 94.8 lb

Steps:

1. Calculate the present FFM of the athlete:
   120 lb − (120 × .21) = 94.8 lb or 120 × .79
   = 94.8 lb

2. Set a reasonable competitive %BF goal
   for the athlete based on the individual and
   a value that falls within the range reported
   for female gymnasts (see table 11.5):
   16%

3. Calculate the athlete's competitive body
   weight by dividing her present FFM by the
   competitive percent FFM (100% − 16% BF
   = 84% FFM): 94.8 lb/0.84 = 113 lb

4. Calculate the athlete's body weight loss by
   subtracting the competitive body weight
   from the precompetitive body weight:
   120 − 113 = 7 lb

FIGURE 11.1   Sample calculation of competitive body weight using the body composition method.

collegiate (Carey 2000) wrestlers. The Wisconsin wrestling minimum weight project, a model program for weight control among high school wrestlers, adopted this equation to predict MW (Oppliger et al. 1995). Detailed instructions for the measurement and calculation of MW using this equation are available (see Wagner 1996).

The SKF equations of Katch and McArdle (1973) and Tipton and Oppliger (1984) also provide good estimates of MW in wrestlers (Carey 2000; Oppliger, Nielsen, & Vance 1991; Thorland et al. 1991). However, significant biases and large prediction errors (TE = 0.0106-0.0229 g/cc) were reported when generalized SKF equations developed for adults, children, or nonathletic adolescents were applied to youth wrestlers (Stout et al. 1995). We recommend the Thorland et al. (1991) modification of the Lohman (1981) Σ3SKF equation for estimating the MW of wrestlers (see table 11.4). Although this equation was developed from a 2-C model estimate of FFM, multicomponent models that adjust Db for hydration and mineral status did not improve estimates of MW in adolescent males compared to the 2-C model (Horswill et al. 1990).

As mentioned earlier in this chapter, Oppliger et al. (2000) developed and cross-validated NIR equations to estimate MW of high school wrestlers (see table 11.4). The prediction error for these NIR equations (TE = 2.2 kg for MW) was slightly better than that of the Lohman (1981) Σ3SKF equation (TE = 2.5 kg for MW). Thus, NIR may be an alternative to the SKF method for estimating MW of high school wrestlers.

Although good agreement (R = 0.67-0.83, SEE = 2.1-3.5% BF) between the leg-to-leg BIA (see chapter 6) and SKF methods was reported for collegiate wrestlers tested at different competitions (Utter et al. 2001), no acceptable reference method was used in this study. Given the likelihood of dehydration among wrestlers trying to make weight and the effect that this could have on BIA estimates, we do not recommend the BIA method for determining body fat in this population. Nevertheless, by monitoring large changes in resistance and reactance measures, the BIA method may be useful in providing clinically relevant information about the hydration status of athletes (Segal 1996).

Key Points

- Physical activity, especially high-impact activities and resistance training, results in increased BMC, BMD, and muscle mass.

- In many well-conditioned athletes the FFBd is likely to be greater than that for the reference body; however, large amounts of muscular hypertrophy may actually lower FFBd due to a disproportionately greater increase in the water component compared to the mineral component of the FFB.

- The SKF method is the preferred field method for assessing the body composition of athletes; the generalized SKF equations of Jackson et al. (Σ7SKF and Σ4SKF) are appropriate for adult and adolescent male and female athletes.

- The individual predictive accuracy of the SKF method is better than that of the BIA method for estimating body composition of athletes.
- NIR equations have been developed for female collegiate athletes and high school wrestlers.
- In lean athletes, FFM accounts for much of the total body weight; therefore, equations that include only weight and height are sometimes as accurate as ones that include additional variables from field methods.
- Female athletes, especially those in sports in which leanness is emphasized, are at increased risk of developing the female athlete triad, a syndrome of disordered eating, menstrual dysfunction, and bone demineralization.
- Body composition data can be used to predict competitive weight and MW for athletes. Sport-specific SKF and NIR equations have been developed to estimate minimal competitive body weights for wrestlers.

Key Terms

Learn the definition for each of the following key terms. Definitions of terms can be found in the Glossary, page 227.

amenorrhea

athletic amenorrhea

female athlete triad

leptin

minimal body fat

minimal body weight (MW)

osteoporosis

Review Questions

1. Identify ways in which body composition data can be used by athletes, coaches, and sports medicine professionals.
2. Describe the effects of physical training on FFB composition.
3. Explain scenarios in which the FFBd of athletes may be less than, equal to, or greater than the reference FFBd of 1.100 g/cc.
4. Identify the limitations of using hydrodensitometry with a 2-C model to obtain reference measures of body composition for athletes.
5. Which reference methods are used most often to develop and cross-validate body composition prediction equations for athletes? When 4-C model reference measures cannot be obtained, what method or model is preferred for assessing the body composition of athletes and why?
6. Can generalized SKF and BIA equations be applied to athletes? If so, identify the equations.
7. For which athletic groups can NIR and anthropometric equations be used?
8. What are the minimal body fat levels for males and females? How do low body weight or low body fat levels contribute to the health risks of female athletes?
9. Identify strategies for helping athletes achieve a healthy body composition.
10. Describe the procedure for predicting the minimal competitive body weight for a high school wrestler.

Body Composition Methods and Equations for Clinical Populations

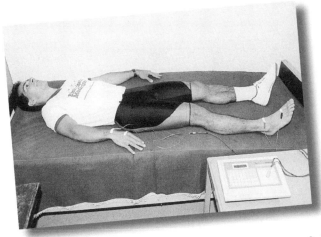

This section addresses reference and field methods to assess the body composition of clinical populations. In each chapter, you will find information about the fat-free composition, suitable reference methods and models, as well as recommended field methods and prediction equations for assessing the body composition of individuals with the following diseases and disorders:

- Cardiopulmonary dieases including coronary artery disease and heart failure, heart and lung transplants, chronic obstructive pulmonary and restrictive pulmonary diseases, and cystic fibrosis (chapter 12)
- Metabolic diseases such as obesity, diabetes mellitis, and thyroid diseases (chapter 13)
- Wasting diseases and disorders including anorexia nervosa, HIV and AIDS, cancer, kidney failure and dialysis, cirrhosis and other liver diseases, spinal cord injury, and selected neuromuscular diseases (chapter 14)

The final chapter addresses reference and field methods, as well as prediction equations, that you can use to assess and monitor changes in your client's body composition (chapter 15).

CHAPTER 12

BODY COMPOSITION AND CARDIOPULMONARY DISEASES

Key Questions

- How is body composition affected by cardiopulmonary diseases such as chronic heart failure, chronic obstructive pulmonary disease, and cystic fibrosis?
- What are the effects of heart and lung transplantation on body composition?
- What methods are commonly used to obtain reference measures of body composition for patients with cardiopulmonary diseases?
- Are there viable field methods that can be used by clinicians to estimate the body composition of patients with various types of cardiopulmonary disease?

In clinical settings, the measurement of body composition and regional fat distribution is important for patients with cardiopulmonary diseases. Individuals with cardiopulmonary diseases like chronic heart failure, chronic obstructive pulmonary disease, and cystic fibrosis often lose body weight and lean body tissues because of muscle wasting and the depletion of bone mineral. Assessing and monitoring body composition of these patients enable clinicians to provide adequate nutritional support to counter these disease effects.

At the opposite extreme, individuals with coronary artery disease (CAD) tend to be overweight or obese. Also, heart and lung transplant patients taking immunosuppressant drugs after surgery typically gain weight due to an increase in FM. Obesity is an independent risk factor for developing CAD. Although obesity is strongly associated with other CAD risk factors such as hypertension, glucose intolerance, and hyperlipidemia, the contribution of obesity to CAD appears to be independent of the influence of obesity on these risk factors (Grundy et al. 1999). The American Association of Cardiovascular and Pulmonary Rehabilitation (AACVPR) has identified controlling body fat through regular exercise and proper nutrition as one goal and beneficial outcome of cardiopulmonary rehabilitation programs (AACVPR Outcomes Committee 1995). To assess this outcome, noninvasive and inexpensive methods for measuring body composition of cardiopulmonary patients in clinical settings are necessary.

In this chapter, we address the following types of cardiopulmonary disease: CAD and heart failure, heart and lung transplants, chronic obstructive pulmonary disease and restrictive pulmonary disease, and cystic fibrosis. For each disease, we present a brief definition and describe how the disease and its treatment affect body composition. We also provide suitable reference methods, field methods, and prediction equations for estimating body composition of patients with these diseases.

Coronary Artery Disease and Heart Failure

This section provides a brief explanation of CAD and heart failure. On the basis of research findings, we identified suitable field methods for estimating the body composition of patients with these diseases.

Definitions of CAD and Heart Failure

Coronary artery disease (CAD) is caused primarily by the deposition of atherosclerotic lesions on the inner lining of large- to medium-sized arteries supplying the heart muscle (**myocardium**). These lesions interfere with the flow of blood to the cardiac muscle tissue. **Angina pectoris** (chest pain), **myocardial infarction** (heart attack), and sudden death due to **arrhythmias** (abnormal heart rhythms) are major complications of CAD (Beers & Berkow 1999). Many CAD patients are overweight or obese (Douphrate et al. 1994; Katch et al. 1986; Young et al. 1998) and tend to distribute body fat in the upper body (Young et al. 1998).

Although there is no clear-cut definition, Beers and Berkow (1999, p. 1682) describe **heart failure** as a "symptomatic myocardial dysfunction resulting in a characteristic pattern of hemodynamic, renal, and neurohormonal responses." Patients with acute and chronic heart failure have altered fluid status due to increased TBW and extracellular fluids (Massari et al. 2001; Steele et al. 1998). With congestive heart failure, plasma volume increases and fluid accumulates in the lungs, abdominal organs, and peripheral tissues (Beers & Berkow 1999). In addition, some patients with chronic heart failure develop **cachexia,** a wasting syndrome that results in the loss of body weight and lean body tissues (Anker et al. 1999; Massari et al. 2001). Medical treatment for heart failure patients usually consists of various combinations of diuretics, beta-blockers, angiotensin-converting enzyme inhibitors, digitalis, aspirin, calcium antagonists, and anti-arrhythmia drugs.

Influence of Heart Failure on Body Composition

Steele et al. (1998) compared the body composition and energy expenditure of patients with stable chronic heart failure (i.e., on drug treatment regimen) to values for age-matched, sedentary healthy individuals with no cardiac or pulmonary disease. None of these patients had clinical evidence of fluid overload (peripheral edema), wasting, or cachexia. The authors reported no significant differences in body weight, BMI, TBW, extracellular fluid volume, and %BF between these two groups. However, cachectic heart failure patients have diminished bone mineral content, Db, LBM, and FM compared to noncachectic patients and control subjects (Anker et al. 1999).

Assessing Body Composition of CAD and Heart Failure Patients

In the past, 2-C models have been used to obtain reference measures of body composition for indi-

viduals with CAD or heart failure. In some studies, reference measures of FFM were estimated from TBW measured by isotope dilution techniques using the formula FFM = TBW/0.73. Other studies used hydrodensitometry (with or without head submersion) and the Siri (1961) 2-C model conversion formula to estimate %BF from Db. Compared to the TBW method, hydrodensitometry (HW) significantly underestimates (≅3%BF) the average $\%BF_{2-C}$ of patients with chronic heart failure because many of these patients have difficulty complying with underwater weighing testing procedures (Steele et al. 1998). Patients may not be able to bend forward to assume the proper body position or expel air completely from the lungs to obtain an accurate measure of residual lung volume (RV). In such cases, a satisfactory alternative would be to underwater weigh your patients at total lung capacity (i.e., vital capacity + RV) while their heads remain above water level. The average difference between Db from HW at RV and the predicted Db from weighing at total lung capacity with the head above water is less than 0.0014 g/cc or 0.7% BF (Donnelly et al. 1988). Alternatively, future studies should explore the possibility of using air displacement plethysmography (see chapter 3) to measure Db of this clinical population.

The SKF (Durnin and Womersley Σ4SKF equation) and bioelectrical impedance analysis (BIA) (Holtain manufacturer's equation) methods tend to underestimate average $\%BF_{TBW}$ (2-8% BF) in men with stable chronic heart failure (Steele et al. 1998). Other researchers reported similar findings for cardiac patients and concluded that BIA manufacturer's equations (i.e., Bio-Analogics ELG, RJL, and Tanita-150 analyzers), SKF equations (Durnin & Womersley [1974] Σ4SKF and Jackson & Pollock [1978] Σ3SKF equations), and NIR Futrex-5000 equations do not yield valid estimates of body composition for cardiac populations (Douphrate et al. 1994; Katch et al. 1986; Young et al. 1998). The only field method with acceptable predictive accuracy (SEE = 3.8% BF; TE = 3.9% BF) for estimating the average $\%BF_{2-C}$ for a small group (N = 24) of Caucasian males with stable cardiac disease was the Tran and Weltman (1988) anthropometric circumference equation (Young et al. 1998). Keep in mind that this equation provides only a 2-C model estimate of %BF.

Additional research is needed to develop and cross-validate field method prediction equations applicable to cardiac populations and individuals with CAD and heart failure. In these studies, reference measures of body composition should be obtained using dual-energy X-ray absorptiometry (DXA) or multicomponent molecular models that adjust Db for interindividual variability in TBW and TBM. For patients who cannot tolerate underwater weighing, hydrostatic weighing without head submersion or air displacement plethysmography may be viable alternatives for measuring Db. Until this work is completed, clinicians should take periodic measurements of body weight, height, and SKF thicknesses to monitor changes in these measures for their cardiac patients. To assess regional adiposity and fat distribution of these patients, use one of the anthropometric measures (e.g., waist circumference or sagittal abdominal diameter) recommended in chapter 5 (see pp. 78-79).

Heart and Lung Transplants

This section describes body composition of heart and lung transplant patients before (pretransplant) and after (posttransplant) surgery. On the basis of research findings, we identified suitable field methods for estimating the body composition of these patients.

Definition of Heart or Lung Transplant

For end-stage heart failure or end-stage respiratory diseases, a heart or lung transplant may be required. **Heart and lung transplantations** are surgical procedures that immediately restore cardiac and pulmonary function. Immunosuppressant medications are typically prescribed after transplantation. Patients receiving these drugs typically show body weight gains (as much as 20-25 kg) after surgery (Ragsdale 1987). To ensure that gain in FM is not excessive, clinicians need to assess body composition of heart and lung transplant patients throughout their treatment. As a clinician, you can use information about the body fat and FFM of pre- and posttransplant patients to design effective nutrition programs for patients who are malnourished or obese.

Influence of Transplants on Body Composition

Stellato et al. (2001) monitored changes in body weight, TBW, and the ratio of intracellular water (ICW) and extracellular water (ECW) in a small sample (N = 8) of heart transplant patients before and after (3, 7, 12, 15, and 180 days posttransplant) surgery. Single-frequency BIA was used to estimate TBW, ICW, and ECW. Six months following surgery, heart transplant patients showed an average weight gain of approximately 5.0 kg. TBW increased after surgery and remained slightly higher than baseline values throughout the posttransplantation follow-up period. The ICW/ECW ratio decreased from 1.4 to 0.81 immediately following surgery but progressively increased during the follow-up period. The

initial drop in ICW/ECW was primarily due to an increase in ECW following surgery.

A retrospective clinical evaluation study compared the body composition of heart, lung, and liver transplant patients before and after surgery (Kyle, Genton, Mentha et al. 2001). Although the difference was not statistically significant, the average body weight of posttransplant patients was higher than that of pretransplant patients (1.8 kg for men, 5.7 kg for women). FFM did not differ significantly between pre- and posttransplant patients. However, the average absolute and relative body fat (FM and %BF) of posttransplant patients was significantly greater than that of pretransplant patients, suggesting that the higher body weight in posttransplant patients was due to increased fat rather than FFM.

Assessing Body Composition of Heart and Lung Transplant Patients

Few studies have dealt with the development and cross-validation of field methods and equations for estimating the body composition of heart and lung transplant patients. In most of these studies, patients with visible edema or ascites (i.e., an abnormal accumulation of fluid in the abdominal cavity) were excluded.

Cross-validation of the Jackson and Pollock (1978) Σ3SKF and Σ7SKF equations against Db measured by hydrodensitometry indicated that these equations overestimated Db_{HW} by 0.010 g/cc and underestimated $\%BF_{2-C}$ by 4.5% BF in male heart transplant patients (Keteyian et al. 1992). Adjusting the y-intercepts of these equations improved the prediction errors; the SEE for revised equations was $\cong$ 4% BF compared to 6% to 8% BF for the original equations. These revised SKF equations, however, need to be cross-validated on additional samples from this clinical population.

In a study of 357 chronically ill women and men (Pichard et al. 1999), the predictive accuracy of two BIA equations developed specifically for patients with respiratory insufficiency (Kyle et al. 1998) and HIV (Kotler et al. 1996) was tested on heart (29 men) or lung (35 men and 39 women) transplant patients. DXA was used to obtain reference measures of FFM. Neither equation had acceptable predictive accuracy (TE = 6.5-6.9 kg) for male heart transplant patients. However, for lung transplant patients, the Kyle et al. equation reasonably estimated FFM_{DXA} in men (CE < 1 kg; TE = 3.4 kg; limits of agreement = $\pm$6.8 kg) and women (CE = $-$1.3 kg; TE = 2.8 kg; limits of agreement = $\pm$5.0 kg).

As an alternative, you may use a single, generalized BIA equation (i.e., Geneva equation), developed and cross-validated for healthy adults (22-94 yr), to accurately estimate FFM_{DXA} in male and female heart, lung, and liver pre- and post-transplant patients (Kyle, Genton, Karsegard et al. 2001). The group and individual predictive accuracy of this equation is excellent (CE < 1.0 kg; TE = 1.7 kg; limits of agreement = ±4.5 kg). In addition to HT^2/R, body weight, and sex, reactance (Xc) was identified as an important predictor variable in this BIA equation. Xc was sensitive to changes in FFM due to aging and disease; as FFM decreases with age, Xc increases, reflecting age- or disease-related changes in electrical conductivity and capacitance of cell membranes (Kyle, Genton, Karsegard et al. 2001; Stellato et al. 2001).

On the basis of these results, we suggest that you use the Geneva generalized BIA equation (Kyle, Genton, Karsegard et al. 2001) to assess the body composition of pre- and posttransplantation heart, lung, and liver transplant patients (see table 12.1). Keep in mind that this equation is based on DXA reference measures and impedance measures (i.e., resistance and reactance) obtained from a Xitron 4000B analyzer in a single-frequency (50 kHz and 0.8 µA) mode. Other brands and models of whole-body BIA analyzers may yield different results.

Chronic Obstructive Pulmonary Disease and Restrictive Pulmonary Disease

This section describes body composition of patients with chronic obstructive and restrictive pulmonary diseases. Studies dealing with the predictive accuracy of various field methods for these clinical populations are addressed.

Definitions of Pulmonary Diseases

Pulmonary diseases and disorders such as chronic bronchitis and emphysema cause respiratory insufficiency. **Emphysema** is a condition characterized by a permanent, abnormal enlargement of air spaces in the lungs and destruction of terminal bronchiole walls. **Chronic bronchitis,** characterized by a chronic cough, is a persistent inflammation of the mucous membranes of the bronchi usually caused by a bacterial infection, lung carcinoma, or chronic heart failure (Beers & Berkow 1999). **Chronic obstructive pulmonary disease (COPD)** is a condition in which airflow to the lungs is obstructed. Most patients with COPD have both emphysema and chronic bronchitis, as well as obstructed airflow to the lungs. Patients who have emphysema or chronic bronchitis but

no airway obstruction do not have COPD (Beers & Berkow 1999).

Influence of Pulmonary Disease on Body Composition

Given that pulmonary diseases increase the energy cost of breathing and often limit daily physical activity, individuals with chronic obstructive or restrictive pulmonary disease commonly have an imbalance between food intake and energy expenditure. A negative energy balance in undernourished patients leads to the loss of body weight, especially FFM, whereas a positive energy balance in overnourished and sedentary patients increases body weight, particularly FM. Kyle et al. (1998) reported that half of the patients with either COPD or severe respiratory insufficiency were underweight (BMI ≤ 18.5 kg/m²); only a few of the patients were overweight (BMI = 27-32 kg/m²). Depletion of FFM and muscle wasting commonly occur in this clinical population. Compared to healthy, age-matched controls, COPD patients have a significantly lower FFM caused by muscle wasting and bone mineral loss (Engelen et al. 1998; Pichard et al. 1999). Engelen and colleagues (1998) reported that 82% of the COPD patients in their study had moderate to severe bone mineral loss. Therefore, it is extremely important to assess body composition of this clinical population, especially in cases in which the loss of FFM due to protein malnutrition or physical inactivity is masked (i.e., there is little change in body weight) by gains in FM.

Assessing Body Composition of COPD and Restrictive Pulmonary Disease Patients

With the exception of one study (Schols et al. 1991), all of the studies assessing the validity of field methods and prediction equations for pulmonary disease populations used DXA as the reference method to evaluate the predictive accuracy of BIA equations for estimating FFM. Engelen et al. (1998) compared FFM estimates obtained from TBW_{D20} (FFM = TBW/0.73) and DXA. The average FFM_{TBW} of COPD patients with no visible edema or fluid imbalance was 1.4 to 3.6 kg lower than FFM_{DXA}, suggesting that the 0.73 hydration constant may not be applicable in the COPD population. In fact, the average relative hydration of the FFM in COPD patients was only 68% compared to 71% for healthy controls (Engelen et al. 1998).

Schols et al. (1991) used single-frequency BIA to develop an equation that estimated TBW_{D20} of patients (N = 32) with severe, stable COPD (SEE

Table 12.1 Summary of Recommended Methods and Equations for Cardiopulmonary Populations

	Method/Equipment	Clinical population	Equation	Reference
ANTHROPOMETRY	Tape measure	Males with cardiac disease (46-82 yr)	1. $\%BF_{2-C} = -47.371817 + 0.57914807$ (abdominal C)[a] $+ 0.25189114$ (hip C) $+ 0.21366088$ (iliac C) $- 0.35595404$ (BW)	Tran & Weltman 1988
BIA	Xitron 4000B	Heart and lung transplant patients, males and females (average age 49 yr)	2. FFM_{DXA} (kg) $= 0.58$ (HT^2/R) $+ 0.231$ (BW) $+ 0.130$ (Xc) $+ 4.429$ (sex)[b] $- 4.104$	Kyle et al. 2001 Geneva equation
	BIO-Z[c]	COPD and restrictive pulmonary disease, males and females (45-86 yr)	3. FFM_{DXA} (kg) $= 0.283$ (HT) $+ 0.207$ (BW) $- 0.024$ (R) $+ 4.036$ (sex)[b] $- 6.06$	Kyle et al. 1998
	BIO-Z[c]	Adult and adolescent males with CF (average age 27 yr)	4. FFM_{DXA} (kg) $= 0.283$ (HT) $+ 0.207$ (BW) $- 0.024$ (R) $+ 4.036$ (sex)[b] $- 6.06$	Kyle et al. 1998
	BIO-Z[c]	Adult and adolescent females with CF (average age 34 yr)	5. FFM_{DXA} (kg) $= 0.88$ ($HT^{1.97}/Z^{0.49} \times 1.0/22.22$) $+ 0.081$ (BW) $+ 0.07$	Kotler et al. 1996
SKF	Harpenden calipers	Children with CF (10-18 yr)	6. Boys: Db_{HW} (g/cc) $= 1.1533 - 0.0643$ (log $\Sigma 4SKF$)[d] 7. Girls: Db_{HW} (g/cc) $= 1.1369 - 0.0598$ (log $\Sigma 4SKF$)[d]	Durnin & Rahaman 1967
			8. Boys: $\%BF_{4-C} = 1.21$ ($\Sigma 2SKF$)[e] $- 0.008$ ($\Sigma 2SKF$)^2 + I[f] 9. Girls: $\%BF_{4-C} = 1.33$ ($\Sigma 2SKF$)[e] $- 0.013$ ($\Sigma 2SKF$)$^2 - 2.5$	Slaughter et al. 1988

Key: HT = height (cm); BW = body weight (kg); R = resistance (Ω); Z = impedance (Ω); Xc = reactance (Ω); Db = body density (g/cc); %BF = relative body fat; COPD = chronic obstructive pulmonary disease; CF = cystic fibrosis.

[a]Abdominal C (cm) is the average of two abdominal circumferences measured (1) anteriorly midway between the xiphoid process of the sternum and the umbilicus and laterally between the lower end of the rib cage and iliac crests and (2) at the umbilical level; [b]Sex = 0 for women, 1 for men; [c]BIO-Z analyzer is equivalent to the Xitron 4000B, RJL 101, and RJL 109 analyzers; [d]RJL 109 analyzers; [e]$\Sigma 2SKF$ = triceps + subscapular; [d]$\Sigma 4SKF$ = triceps + biceps + suprailiac + subscapular; [f]I = intercept substitutions based on maturation and ethnicity for boys:

Age	African American	Caucasian
Prepubescent	–3.2	–1.7
Pubescent	–5.2	–3.4
Postpubescent	–6.8	–5.5

= 1.8 L). FFM was estimated from TBW using the assumed constant hydration factor of 73%. In an independent sample of patients with COPD and restrictive pulmonary disease (Pichard, Kyle, Janssens et al. 1997), this equation systematically overestimated average FFM_{DXA} by 4 to 5 kg, and the limits of agreement were not clinically acceptable (±6.2 kg). Also, Schols and coworkers reported that the Durnin and Womersley Σ4SKF equations significantly overestimated FFM_{D20} by 4.4 kg.

In addition to cross-validating Schols' TBW equation, Pichard, Kyle, Janssens et al. (1997) tested the validity of 25 previously published BIA equations developed primarily for healthy populations. Based on their analysis, none of these equations was satisfactory for estimating the FFM of patients with COPD or restrictive pulmonary disease, suggesting that disease-specific BIA equations need to be developed. Using the same sample of patients, this research team generated such an equation (Kyle et al. 1998). The group and individual prediction errors for this equation were excellent (SEE and TE = 1.7 kg; limits of agreement = ±3.3 kg). Cross-validation of this equation on another sample of patients with chronic pulmonary disease (Pichard et al. 1999) yielded slightly higher but clinically acceptable prediction errors (SEE = 2.1 to 2.4 kg; TE = 2.2-2.6 kg; limits of agreement = ±4-5 kg).

On the basis of these studies, we recommend using the Kyle et al. (1998) disease-specific BIA equation to assess body composition of your patients with COPD or restrictive pulmonary disease (see table 12.1). In the original validation and cross-validation studies, a BIO-Z bioimpedance analyzer was used to measure whole-body resistance. This brand of analyzer yields resistance values similar (within ±5 Ω) to those of the RJL 101, RJL 109, and Xitron 4000B (Kyle et al. 1998; Pichard, Kyle, Janssens et al. 1997; Pichard et al. 1999). If possible, use either a BIO-Z or an equivalent analyzer to avoid measurement error due to differences between BIA analyzers (see chapter 6).

Cystic Fibrosis

This section describes body composition of patients with cystic fibrosis. On the basis of research findings, we recommend field methods that can be used to assess the body composition of children, adolescents, and adults with this disease.

Definition of Cystic Fibrosis

Cystic fibrosis (CF) is a life-shortening genetic disease of the exocrine glands (e.g., pancreas and sweat glands), primarily affecting the gastrointesti-

nal and respiratory systems. This disease is usually characterized by COPD, exocrine pancreatic insufficiency, and abnormally high levels of electrolytes (i.e., sodium and chloride) in the sweat (Beers & Berkow 1999). CF is most common in the Caucasian population, and 70% of CF patients are children and adolescents.

Influence of CF on Body Composition

Children and adolescents with CF typically are malnourished and have retarded growth, delayed onset of puberty, reduced body weight, and a limited capacity for physical activity and exercise (Gulmans et al. 1997; McNaughton et al. 2000; Preece et al. 1986). Compared to age- and gender-matched healthy controls, children with CF are shorter, weigh less, and have a higher Db (lower %BF); the difference in Db is considerably greater for boys with CF than for girls with CF (Johnston et al. 1988). Also, compared to values for healthy children, the average body fat and total body bone mineral of children and adolescents with CF may be 18% to 20% less; the deficits in bone mineral and bone density appear to be related to disease severity and nutritional status (Henderson & Madsen 1999). In addition, diminished bone density is associated with elevated fracture rates in adolescents with CF (Henderson & Specter 1994). Clinicians must assess and monitor the body composition of patients to provide adequate nutritional support for maintaining normal growth in children with CF.

Assessing Body Composition of CF Patients

In studies of children and adults with CF, either isotope dilution methods for measuring TBW or DXA is typically used to obtain reference body composition measures for cross-validation of field method equations. Some adults with CF have difficulty expelling all of the air from the lungs (i.e., performing RV maneuver) during underwater weighing; this results in an overestimation of %BF (Newby et al. 1990). Thus, hydrostatic weighing is not a suitable method for assessing body composition of CF patients.

Two studies, evaluating the predictive accuracy of the Slaughter et al. (1988) Σ2SKF equations and the Durnin and Rahaman (1967) Σ4SKF equations (see chapter 8) for boys and girls with CF, reported only small differences in average estimates of FFM (0.06-0.7 kg) for the SKF method compared to dual-photon absorptiometry (Lands et al. 1993) and TBW_{D20} (de Meer et al. 1999). The group and individual predictive accuracy of the Slaughter SKF equations (triceps + subscapular sites) was SEE = 1.7 kg, and limits of agreement were –3.9 to 2.6 kg (Lands et al. 1993). For the Durnin and Rahaman four-site equation (triceps + biceps + subscapular + suprailiac SKFs), the limits of agreement were –1.9 to 3.2 kg for children and adolescents with mild to moderate CF (de Meer et al. 1999).

For the BIA method, studies show that there is a fairly good relationship (r_{xy} = 0.87) between the resistance index (HT^2/R) and reference measures of FFM in normal children and those with CF (Borowitz & Conboy 1994). However, the regression line depicting this relationship differs for CF patients compared to healthy children, suggesting that disease-specific BIA equations need to be used to accurately estimate the FFM of children with CF (Azcue et al. 1993; Borowitz & Conboy 1994). The prediction errors (SEE) for two different BIA manufacturers' equations ranged from 1.3 kg (Valhalla 1990A) to 3.2 kg (RJL 101). In a study of chronically ill patients that included 10 males (M = 26.8 ± 4.6 yr) and 9 females (M = 33.6 ± 9.8 yr) with CF, Pichard et al. (1999) compared FFM estimates from the Kotler et al. (1996) and Kyle et al. (1998) BIA equations to FFM_{DXA}. For females, the group and individual predictive accuracy of the Kotler et al. equation (CE = 2.7 kg; SEE = 1.7 kg; TE = 3.2 kg; limits of agreement = ±4.0 kg) was better than that of the Kyle et al. equation (CE = –4.4 kg; SEE = 2.2 kg; TE = 5.0 kg; limits of agreement = ±5.2 kg). However, for males with CF, the Kyle et al. equation had better group predictive accuracy (CE = –2.3 kg; SEE = 1.8 kg; TE = 3.6 kg) than the Kotler et al. equation (CE = 2.9 kg; SEE = 2.7 kg; TE = 4.0 kg). The two equations had similar individual predictive accuracy (limits of agreement = ±5.8 kg).

Additional research is needed to firmly establish the applicability of previously published SKF and BIA equations and to develop disease-specific field method equations for estimating the body composition of children, adolescents, and adults with CF. Until this work is completed, we suggest using the SKF method (either the Slaughter et al. or Durnin and Rahaman equations) to assess the body composition of children and adolescents with CF. For adults with CF, we recommend using the BIA equations of Kotler et al. for women and Kyle et al. for men (see table 12.1).

Table 12.1 summarizes recommended methods and equations for assessing the body composition of individuals with various types of cardiopulmonary diseases or disorders. For these equations, try to use the same type or an equivalent brand of equipment to minimize this potential source of measurement error.

Key Points

* Research is needed to develop and cross-validate clinical methods and prediction equations applicable to individuals with CAD and heart failure.

* The anthropometric equation of Tran and Weltman (1988) may be used to estimate the body composition of men with heart disease.

* Following heart or lung transplant surgery, patients tend to gain body weight due to an increase in FM rather than FFM.

* The BIA method can be used to assess the body composition of heart and lung transplant patients.

* Individuals with COPD have both emphysema and chronic bronchitis, as well as obstructed airflow to the lungs.

* Patients with COPD or restrictive pulmonary disease tend to be underweight; depletion of FFM and muscle wasting are common in these clinical populations.

* The BIA method can be used in clinical settings to assess the body composition of patients with COPD and restrictive pulmonary disease.

* Children and adolescents with CF tend to be malnourished and may have retarded growth and delayed onset of puberty, as well as reduced body weight, body fat, and bone mineral density.

* To assess the body composition of children and adolescents with CF, use the SKF method; for adults with CF, use the BIA method.

Key Terms

Learn the definition for each of the following key terms. Definitions of terms can be found in the Glossary, page 227.

angina pectoris

arrhythmias

cachexia

chronic bronchitis

chronic obstructive pulmonary disease (COPD)

coronary artery disease (CAD)

cystic fibrosis (CF)

emphysema

heart and lung transplantations

heart failure

myocardial infarction

myocardium

Review Questions

1. Describe the effects of chronic heart failure on body composition.

2. In clinical settings, which field method can be used to assess the body composition of males with stable cardiac disease?

3. Describe the effects of heart or lung transplantation on body composition. What method and equation are recommended for assessing body composition of patients receiving heart or lung transplants?

4. Which bioimpedance variable reflects age- or disease-related changes in electrical conductivity and capacitance of cell membranes in clinical populations?

5. What is the primary difference between COPD and emphysema?

6. Describe the effects of COPD and restrictive pulmonary disease on body composition. Identify the body composition method and equation recommended for these clinical populations.

7. Describe the effects of CF on the growth, maturation, and body composition of children. What methods and equations are recommended for assessing body composition of children, adolescents, and adults with CF?

CHAPTER 13

BODY COMPOSITION
AND METABOLIC DISEASES

Key Questions

- How do metabolic diseases such as obesity, diabetes, and thyroid disease affect body composition?
- In clinical settings, what field methods can be used to estimate body composition of persons who are obese?
- How does the body composition of individuals with type 1 and type 2 diabetes differ?
- How does the body composition of individuals with hyperthyroidism and hypothyroidism differ?
- What field methods can be used to assess body composition of patients with diabetes or thyroid disease?

Metabolic diseases such as obesity, diabetes, and hypo- or hyperthyroidism directly affect the metabolism of fat, carbohydrate, protein, and minerals, resulting in weight gain or loss and alterations in body composition. Obesity is on the rise worldwide, reaching epidemic proportions in some countries. The overall prevalence of obesity in adults for the year 2000 was 8.2% of the global population (World Health Organization 2001); since 1960, the prevalence of obesity has increased across all age, gender, and ethnic groups in the United States (Flegal et al. 1998). Alarmingly, the age-adjusted prevalence of obesity for U.S. adults is 30.5%. Approximately 33% of adult women and 28% of adult men are obese. Obesity is more prevalent among African-American women (50%) compared to Hispanic women (40%) and Caucasian women (30%) (Flegal et al. 2002). Persons who are obese, especially those with upper-body or abdominal obesity, have a relatively high risk for developing cardiovascular and pulmonary diseases, as well as diabetes.

In the United States, approximately 6.5% of adults, 18 yr of age or older, have been diagnosed with diabetes (Centers for Disease Control 2003). Research demonstrates a positive association between increased levels of body fatness (obesity) and type 2 diabetes (Pi-Sunyer 1999). Abdominal obesity has been identified as a significant risk factor for the development of insulin resistance, glucose intolerance, and type 2 diabetes (Despres 1993). Approximately 85% of patients with type 2 diabetes display increased levels of body fatness (Tsui et al. 1998).

Since the thyroid gland controls substrate metabolism and metabolic rate, over- or under-production of thyroid hormones disrupts energy balance. An overactive thyroid causes weight loss, whereas an underactive thyroid produces weight gain. Therefore, assessing body composition of individuals with metabolic diseases is important in clinical settings to determine their obesity-related disease risk. Accurate assessment of body composition can also help in planning weight management programs and in monitoring changes in body composition resulting from these interventions.

In this chapter we address obesity, diabetes, and thyroid diseases. For each disease, we present a definition and briefly describe how the disease affects body composition. Suitable reference methods, field methods, and prediction equations for estimating body composition of patients with each disease are provided.

Obesity

This section provides definitions of obesity and describes body composition of obese populations. Studies dealing with the predictive accuracy of various reference and field methods for this clinical population are summarized.

Definition and Causes of Obesity

Obesity is an excessive amount of body fat relative to body weight (see table 1.2, p. 6). In epidemiological studies, obesity is defined as a BMI (see table 5.3, p. 76) of 30 kg/m² or more (U.S. Dept. of Health and Human Services 2000c). Both the total amount of body fat and the way in which fat is regionally distributed in the body are principal causes for the high rate of chronic disease in individuals who are obese (Blair & Brodney 1999). Persons with **abdominal obesity,** or **upper-body obesity,** tend to deposit excess subcutaneous and visceral fat in the abdominal region and have a relatively higher risk of developing chronic diseases compared to individuals who deposit excess fat in the lower body (i.e., lower-body obesity).

Both genetic and environmental factors contribute to obesity. Although environmental factors, such as a decline in daily energy expenditure and an increase in energy intake, are major causes of obesity (Hill & Melanson 1999), approximately 25% to 40% of the variability in body fatness can be attributed to genetic factors (Bouchard 1994). Multiple genes, as well as the interaction of these genes with the environment, define one's predisposition for obesity (Pi-Sunyer 2000). Researchers have noted that mutations in genes associated with the metabolism of fat and amino acids may lead to obesity (Pi-Sunyer 2000).

Influence of Obesity on Body Composition

Obesity not only involves an increase in body fat, but may also affect the composition (i.e., water, mineral, and protein content) and density of the FFB. In addition to triglycerides, adipose tissue contains water, which may increase the relative hydration of the FFB in some persons who are obese (Deurenberg, Leenen et al. 1989). Earlier studies reported higher hydration levels in men and women who were obese (74.2-77% FFB) compared to leaner men and women (72-74% FFB) (Albu et al. 1989; Segal et al. 1987). In contrast, recent research

suggests that the average hydration of the FFB in women with obesity is approximately 71% to 73% FFB (Evans et al. 1999; Fogelholm et al. 1997; Fuller et al. 1994), but there is a high degree of interindividual variability (68-75% FFB). In addition, obesity may influence the mineral and protein content of the FFB. Some studies suggest that the average relative mineral (7.2-8.1%) and protein (20-21%) in women with obesity are greater than the assumed values (6.8% and 19.4%, respectively) for the Siri 2-C model (Fogelholm et al. 1997; Stolarczyk et al. 1994).

Overall, average FFB density values for groups of obese women range between 1.097 and 1.104 g/cc (Evans et al. 1999; Fogelholm et al. 1997; Fuller et al. 1994). Although these average group values do not differ significantly from the assumed value (1.10 g/cc) of 2-C models, there is substantial interindividual variability (1.093-1.117 g/cc) in the density of the FFB in women who are obese (Fuller et al. 1994).

Assessing Body Composition of Individuals Who Are Obese

Information about body composition assessment of obese populations is primarily based on studies of women. There is a lack of research quantifying the FFB composition of men with obesity. Most studies have used 2-C models with hydrodensitometry or isotope dilution methods to obtain reference measures of body composition. Fuller et al. (1994) reported good agreement between hydrodensitometry (2-C model), TBW (2-C model), and 3-C model (Siri Db-water) estimates of %BF in a small sample of obese Caucasian women. For this group, the average %BF differed from %BF$_{3-C}$ by less than 1% BF; for individuals, the 95% limits of agreement were good (±3.1% BF) for both methods. Using dual-energy X-ray absorptiometry (DXA) to obtain reference measures of %BF, He and colleagues (1999) reported that the average difference between %BF$_{DXA}$ and %BF$_{TBW}$ in obese Chinese women was less than 1% BF; but the limits of agreement between these two methods for individuals was fairly large (−7.6% to 6.4% BF).

The large amount of interindividual variability in the composition and density of the FFB in individuals with obesity points to the importance of using 3-C or 4-C models that account for biological variability in FFB components when one is assessing the body composition of people who are obese. Compared to 4-C model estimates, the Siri 3-C water model is more accurate than the Siri 2-C or Lohman 3-C mineral models for evaluating individual and group changes in body composition resulting from diet or diet/exercise weight loss programs (Evans et al. 1999).

To use these multicomponent models, one must measure Db. For this purpose, most studies use hydrodensitometry to assess Db. However, clients with large amounts of body fat are more buoyant than leaner individuals, and therefore they have more difficulty remaining motionless while under the water or may be unable to fully submerge. In this case, clients can undergo underwater weighing without having to submerge their heads (see chapter 3). Researchers reported fairly good agreement between hydrostatic weighing with and without head submersion in morbidly obese (>40% BF) women and men (Evans et al. 1989; Heath et al. 1998). As an alternative to hydrodensitometry, researchers recently have examined the viability of using air displacement plethysmography (ADP) to measure Db of persons with obesity. Vescovi et al. (2001) reported that the average group difference between these methods for obese men and women is less than 1% BF; however, for individuals, the limits of agreement between methods may be large (−7.5% to 9.4% BF). Additional research is needed to assess the validity of ADP for measuring Db of obese populations.

Most of the field method prediction equations developed for obese populations are based on 2-C model reference measures. When using these equations, keep in mind that errors for individual clients may be fairly large.

SKF Method

A number of investigators have tested the accuracy of population-specific (Durnin & Womersley 1974) and generalized (Jackson & Pollock 1978; Jackson et al. 1980) SKF equations in overweight and obese samples. Overall, it appears that the SKF method and these equations have limited applicability to persons who are obese.

Although some individuals with obesity were included in the original samples used to develop Jackson's generalized SKF equations, Jackson and Pollock (1985) warned that these equations may produce larger than expected prediction errors when they are applied to individuals whose sum of seven SKFs (Σ7SKF) exceeds 266 mm for women and 272 mm for men. In fact, the Σ3SKF, Σ4SKF, and Σ7SKF equations significantly underestimated the average %BF of obese men and women by 1.5% to 4% BF (Gray et al. 1990; Heyward, Cook et al. 1992; Teran et al. 1991; Wilmore et al. 1994). Also, the prediction errors for these equations were high, with SEEs ranging from 3.4% to 5.1% BF.

In general, similar results were observed for Durnin and Womersley's (1974) population-specific

SKF equations. These equations underestimated %BF in obese samples by as much as 12% BF. The group predictive accuracy (SEE) of these equations ranged from 3.4% to 7% BF, and the limits of agreement (±10% BF) for individuals were large (Fuller et al. 1994; Gray et al. 1990; McNeill et al. 1991; Teran et al. 1991).

Overall, experts agree that the SKF method should not be used to estimate the body composition of persons who are obese. With increasing levels of body fatness, the proportion of subcutaneous to total body fat changes, thereby affecting the relationship between the ΣSKF and total Db. Furthermore, the applicability of the SKF method in people who are obese is limited for the following reasons:

- Site selection and palpation of bony landmarks are more difficult in individuals with obesity.
- The SKF thickness may be larger than the jaw aperture of most calipers, and it may not be possible to lift the SKF from the underlying tissue in some clients who are obese.
- There is greater variation in the depth at which the caliper tips can be placed on the SKF, and the caliper tips may slide on larger SKFs.
- Variability in adipose tissue composition may affect SKF compressibility in clients who are obese.
- There is greater variability between testers when measuring larger SKF thicknesses.

In short, these factors limit the accuracy and precision of SKF measurements in persons with obesity. Therefore, we do not recommend using the SKF method to assess the body composition of these clients.

Bioelectrical Impedance Analysis

Unlike the SKF method, bioelectrical impedance (BIA) appears to be a promising method for assessing body composition of persons who are obese, provided that BIA equations developed specifically for obese populations are selected for this purpose. The first fatness-specific BIA equations were developed by Segal et al. (1988) for women (<30% and ≥30% BF) and men (<20% and ≥20% BF). Cross-validation of the equations for obese men and women yielded smaller prediction errors (SEE = 2.8 and 2.0 kg, respectively) compared to generalized BIA equations (SEE = 3.6 and 2.4 kg, respectively), indicating that fatness-specific equations are more accurate in this population. In addition, other researchers reported that Segal's fatness-specific equations accurately predict 2-C model estimates of average FFM in obese American Indian women (SEE = 2.7 kg), obese Caucasian women (SEE = 2.4 kg), slightly obese men (SEE = 3.2 kg), and obese women and men with less than 42% BF (SEE = 3.5 kg) (Gray et al. 1989; Jenkins et al. 1994; Ross et al. 1989; Stolarczyk et al. 1994). Compared to 3-C (Db-water) estimates of %BF, the individual predictive accuracy of this equation for women who were obese was ±6.5% BF (Fuller et al. 1994).

Gray et al. (1989, 1990) noted that Segal's fatness-specific equations tended to systematically overestimate FFM in more obese men and women, especially for women whose %BF exceeded 48% BF. As a result, Gray et al. (1989) developed generalized equations for men (9-45% BF) and women (19-59% BF) and a fatness-specific equation for severely obese women (>48% BF). Although these equations were not cross-validated in the original study, researchers reported that Gray's generalized equation for women accurately predicted 2-C model estimates of FFM in Caucasian women (11-58% BF) and American Indian women (14-54% BF) (Jenkins et al. 1994; Stolarczyk et al. 1994). In contrast, Gray's fatness-specific (>48% BF) equation significantly overestimated the average $\%BF_{3-C}$ of obese women by 6.5% BF; the 95% limits of agreement were ±7.0% BF (Fuller et al. 1994).

Generally, research shows poor group and individual predictive accuracy with use of equations provided by manufacturers of various types of bioimpedance analyzers (i.e., BMR-2000, EZ Comp 1500, Maltron BT-905, RJL 103, IMPB01, and Valhalla 1990B) to estimate body composition of individuals with obesity. These equations tend to significantly underestimate the average %BF of obese men and women by 2.6% to 9.0% BF, and the predictive accuracy for individuals ranges between ±8% and 25% BF (Fuller et al. 1994; Heath et al. 1998; McNeill et al. 1991; Panotopoulos et al. 2001). However, Fuller and colleagues (1994) reported that the Bodystat-500 manufacturer's equation accurately estimated average $\%BF_{3-C}$ within 1% BF for a small sample (N = 15) of obese Caucasian women. Also, in comparison to that of other manufacturers' equations, the individual predictive accuracy was good (±6.7% BF). Additional research is needed to substantiate these findings using larger samples of persons who are obese.

On the basis of these studies, we recommend using Segal's (1988) fatness-specific (≥20% for men and ≥30% BF for women) equations to obtain 2-C model estimates of body composition for clients who ʳᵉ obese (see table 13.1). To use Segal's equations, ᵗo visually determine that your client is ᵒ, we do not recommend using the manu-

facturers' equations programmed in the software accompanying BIA analyzers given that these equations typically overestimate average FFM in men and women who are obese.

Near-Infrared Interactance Method

There is much evidence indicating that the near-infrared interactance (NIR) method, particularly the Futrex-5000 and Futrex-6000 manufacturer's equations, should not be used to assess body composition of persons who are obese. In obese samples, these equations have large, unacceptable prediction errors (SEE = 3.8-4.3% BF) and systematically underestimate %BF by as much as 10% to 16% BF (Davis et al. 1989; Elia et al. 1990; Heyward, Cook et al. 1992; Wilmore et al. 1994). Also, the individual predictive accuracy of these equations is poor, with 95% limits of agreement ranging between ±8.0% and 11.0% BF (Fuller et al. 1994; Panotopoulous et al. 2001). It appears that fat layering, which is characteristic of individuals with obesity who have large amounts of subcutaneous fat, may affect the depth to which the near-infrared light penetrates the tissues at the measurement site. Quatrochi et al. (1992) reported that the relationship between optical density (OD) and SKF measures at the biceps site was stronger in leaner women compared to women who were obese. This observation may explain why the degree of underestimation of %BF is markedly increased in more obese individuals (Davis et al. 1989; Elia et al. 1990).

Anthropometric Method

In addition to BIA, anthropometric methods and equations appear to be suitable alternatives for obtaining 2-C model estimates of body composition for persons who are obese. This method is less expensive than BIA and does not require the client to adhere to strict pretesting guidelines. Unlike SKFs, circumferences can be easily measured regardless of the client's level of body fatness.

Weltman and colleagues (1987) developed an anthropometric equation to predict the FFM of obese (30-45% BF) men, 24 to 68 yr of age, using abdominal circumferences and body weight as predictors. Cross-validation of this equation yielded accurate estimates of average FFM and acceptable prediction error (SEE = 2.6 kg). In a similar study (Weltman et al. 1988) of obese women, cross-validation of the anthropometric equation predicting FFM indicated that this equation had only fair predictive accuracy (SEE = 3.2 kg). However, the accuracy of the equation estimating %BF in this sample was acceptable (SEE = 3.5% BF).

Although Teran et al. (1991) developed anthropometric equations to estimate the relative body

Table 13.1 Summary of Recommended Methods and Equations for Metabolic Diseases[a]

	Method/Equipment	Clinical population	Equation	Reference
Anthropometry	Tape measure	Obese men (24-68 yr)	1. $\%BF_{2-C} = 0.31457$ (abdominal C)[b] $- 0.10969$ (BW) $+ 10.8336$	Weltman et al. 1987
		Obese women (20-60 yr)	2. $\%BF_{2-C} = 0.11077$ (abdominal C)[b] $- 0.17666$ (HT) $+ 0.14354$ (BW) $+ 51.03301$	Weltman et al. 1988
BIA	RJL analyzer	Obese ($\geq$20% BF) men (17-62 yr)	3. FFM_{2-C} (kg) $= 0.00088580$ (HT^2) $- 0.02999$ (R) $+ 0.42688$ (BW) $- 0.07002$ (age) $+ 14.52435$	Segal et al. 1988
		Obese ($\geq$30% BF) women (17-62 yr)	4. FFM_{2-C} (kg) $= 0.00091186$ (HT^2) $- 0.01466$ (R) $+ 0.2999$ (BW) $- 0.07012$ (age) $+ 9.37938$	Segal et al. 1988

[a]Presently there are no disease-specific prediction equations for diabetes and thyroid diseases. For patients with type 2 diabetes who are also obese, you may try using Segal et al. (1988) fatness-specific BIA equations. For patients with type 1 diabetes, you may try using the Leiter et al. (1994) disease-specific BIA equation (see p. 191). However, please note that these equations need additional cross-validation to verify their applicability to individuals with these diseases.

[b]Abdominal C (cm) is the average of two abdominal circumferences measured (1) anteriorly midway between the xiphoid process of the sternum and the umbilicus and laterally between the lower end of the rib cage and iliac crests and (2) at the umbilical level.

Key: HT = height (cm); BW = body weight (kg); R = resistance (Ω); %BF = relative body fat; C = circumference in (cm).

fatness of women with obesity, these equations had unacceptable prediction errors (SEE = 3.9-4.3% BF). Also, the predictor variables in these equations include log transformations of SKF and circumference measures, as well as residual lung volume, making them impractical for field and clinical use.

In summary, we recommend using Weltman's (1987, 1988) fatness-specific anthropometric equations to obtain 2-C model estimates of body composition for clients who are obese. These equations are presented in table 13.1.

Diabetes Mellitus

This section defines diabetes mellitus, briefly describes differences between type 1 and type 2 diabetes, and presents the effects of this disease on body composition. Suitable reference methods, field methods, and prediction equations for this clinical population are summarized.

Definitions and Types of Diabetes Mellitus

The major clinical types of **diabetes mellitus (DM)** are type 1 and type 2 diabetes. **Type 1** or **insulin-dependent diabetes mellitus (IDDM)** is characterized by **hyperglycemia** (i.e., high blood glucose levels) caused by a lack of insulin production by the pancreas. This type of diabetes may occur at any age but most commonly develops in childhood, adolescence, or before age 30 yr (Beers & Berkow 1999). **Type 2** or **non-insulin-dependent diabetes mellitus (NIDDM)** is characterized by an impaired secretion of insulin in response to glucose, as well as insulin resistance. Type 2 diabetes usually is diagnosed in patients older than 30 yr but may also occur in children and adolescents. Approximately 90% to 95% of patients with diabetes have type 2 diabetes (Kriska et al. 1994).

Influence of Diabetes on Body Composition

Relatively few studies have described the body composition of patients with diabetes. Disease-related changes in body composition depend on the type of diabetes (type 1 or 2). In the past, it was commonly thought that type 1 diabetes was a catabolic disease, producing loss of protein and FFM. Recent studies, however, have indicated that the average FFM of individuals with type 1 diabetes is similar to or even slightly greater than that of healthy age- and gender-matched controls or reference population data (Gomez et al. 2001; Rosenfalck

et al. 1997, 2002; Svendsen & Hassager 1998). At onset of the disease, the FFM of individuals with type 1 diabetes was normal, even though body weight (BW) was 6.5 kg below ideal weight and FM was 25% lower than that of healthy individuals (Rosenfalck et al. 2002). These data suggest that uncontrolled type 1 diabetes produces a fat, rather than a protein, catabolic state.

The total body fat, abdominal fat, and %BF of individuals with type 1 diabetes tend to be less than values in healthy controls and individuals with type 2 diabetes (Gomez et al. 2001; Svendsen & Hassager 1998; Rosenfalck et al. 1995). As a result of insulin therapy for treatment of diabetes, BW, FM, and FFM increase. Rosenfalck et al. (2002) reported that during the first year of insulin therapy, average BW increased 4.3 kg, with a 13% increase in FM and a 5% increase in FFM of adults who had type 1 diabetes.

Type 1 diabetes and insulin therapy result in bone mineral loss (i.e., osteopenia) and osteoporosis (Levin et al. 1976; McNair et al. 1981; Munoz-Torres et al. 1996). Patients with type 1 diabetes have a lower bone mineral density compared to healthy individuals (Levin et al. 1976; Munoz-Torres et al. 1996). The rate of bone mineral loss (1.35% / yr) is greatest during the first 6 yr and tends to decrease with increasing duration of the disease (McNair et al. 1981).

Disease-related changes in body composition differ between individuals with type 2 diabetes and those with type 1 diabetes. The LTM and muscle mass of pre- and postmenopausal women with type 1 and type 2 diabetes do not differ significantly from values in age- and menopause-matched healthy controls; however, patients with type 2 diabetes are generally overweight or obese and have a greater degree of abdominal fat compared to individuals with type 1 diabetes (Svendsen & Hassager 1998). Also, compared to age- and gender-matched healthy adults, men and women with type 2 diabetes tend to have higher bone mass and bone mineral density (Rishaug et al. 1995). This finding is not surprising given that obesity is positively associated with bone mineral content and density and that approximately 85% of patients with type 2 diabetes are obese (Tsui et al. 1998).

Assessing Body Composition of Patients with Diabetes

The differential effects of type 1 and type 2 diabetes on body composition clearly suggest that multicomponent reference measures and disease-specific prediction equations need to be developed to accurately assess body composition of these clinical populations. To our knowledge, there are no comprehensive studies addressing the composition and density of the FFB in individuals with either type of diabetes. Studies are needed to quantify the amount of interindividual variability in mineral, water, and protein content of the FFB, controlling for factors (e.g., type and duration of disease, menopausal status, treatment regimens, and disease-related complications) that may potentially affect body composition.

We found only one study that evaluated reference methods and models for assessing body composition in this clinical population. Rosenfalck and colleagues (1995) compared FFM estimates from DXA to 2-C model estimates obtained from total body potassium (TBK) and TBW measures in a small sample (N = 11) of patients with type 1 diabetes and healthy controls (N = 13). There were no significant differences between average FFM_{DXA}, FFM_{TBK}, and FFM_{TBW} for either group. Also, the limits of agreement between DXA and the other two methods were similar in the type 1 diabetes and control groups, suggesting that DXA may be a valid method for assessing body composition of individuals with type 1 diabetes. However, Kistorp and Svendsen (1997) noted a lack of agreement between body composition measures obtained from two different types of DXA scanners. They observed that compared to the Lunar DPX instrument, the Hologic QDR-2000 software yielded higher %BF (10% greater) as well as lower LTM and bone mineral content (6% lower) in females with diabetes.

Few studies have assessed the predictive accuracy of body composition field methods and equations in diabetic populations. In each of these studies, reference body composition measures were obtained using DXA, and BIA was the only field method tested. Leiter and colleagues (1994, 1996) developed a whole-body BIA equation specifically for individuals with type 1 diabetes (19 females and 26 males, 13-39 yr of age). Instead of distally placing gel electrodes on the right hand and foot, as is typically done for whole-body BIA (see chapter 6), they used four rod electrodes placed proximally at the natural folds of the elbow joints and behind the knees. Four resistance values were obtained: right arm-right leg; right arm-left leg; left arm-left leg; and left arm-right leg. The minimum resistance value was used to develop the following BIA prediction model:

$$FFM\ (kg) = 4.63 + 0.205\ (HT^2/R) + 0.284\ (BW_1) - 0.0308\ (BW_2)$$

where HT^2/R = resistance index, BW_1 = body weight in kilograms; BW_2 = body weight in kilograms for females and zero for males.

The prediction error (SEE = 1.43 kg) and the estimated predictive accuracy of this equation when this model was applied to other samples (prediction of sum of squares statistic = 1.55 kg) were excellent. However, in light of the small sample size, this equation needs to be cross-validated on additional samples of individuals with type 1 diabetes before its clinical use can be recommended.

One other study examined the predictive accuracy of a leg-to-leg, segmental bioimpedance analyzer (Tanita TBF-105) for assessing $\%BF_{DXA}$ of 48 men and 48 women with type 2 diabetes (Tsui et al. 1998). The BIA equations, on average, significantly underestimated $\%BF_{DXA}$ of men by 3% BF and overestimated $\%BF_{DXA}$ of women by 1.5% BF; the prediction errors (SEE) were not reported. Further cross-validation of these equations is needed before their use can be recommended.

In light of the limited research dealing with body composition assessment for patients with diabetes, we cannot make any solid recommendations at this time. Stolarczyk and Heyward (1999), however, suggested that the applicability of the Segal et al. (1988) fatness-specific BIA equations, developed and cross-validated on multiple samples of both lean and obese individuals from diverse ethnic groups, should be tested on samples of adults with type 1 and type 2 diabetes. Future studies need to address the appropriateness of using 2-C models and prediction equations to estimate body composition of individuals with diabetes.

Thyroid Diseases

This section provides a brief definition and description of diseases associated with thyroid gland dysfunction. We also summarize a limited amount of information pertaining to changes in body composition resulting from these diseases and treatments. There is no available information about the validity and predictive accuracy of methods for assessing body composition of this clinical population.

Definitions of Thyroid Diseases

The thyroid gland and its hormones control general metabolic function, including the metabolism of carbohydrate, protein, fat, and minerals. **Hyperthyroidism** or **thyrotoxicosis** is a clinical condition characterized by increased synthesis and secretion of thyroid hormone. Diffuse toxic goiter (i.e., Graves' disease) and toxic nodular goiter produce hyperthyroidism. **Graves' disease,** which is the most common cause of hyperthyroidism, is an autoimmune disease (Beers & Berkow 1999). At the opposite extreme, **hypothyroidism** or **myxedema** is caused by thyroid hormone deficiency, resulting from autoimmune **thyroiditis** (i.e., inflammation of the thyroid gland), radioiodine therapy, or surgical removal of the thyroid gland. This disease is often associated with a firm goiter (Beers & Berkow 1999).

Influence of Thyroid Diseases on Body Composition

Much of what we know about body composition changes and thyroid disease has been inferred from the known metabolic effects of thyroid hormones. Two studies assessed the body composition of patients with thyroid disease using BIA (Miyakawa et al. 1999; Seppel et al. 1997). Compared to age- and gender-matched healthy individuals, patients with hyperthyroidism (Graves' disease and toxic nodular goiter) tend to have less BW, %BF, and body cell mass (BCM). In both studies, the reduction in BCM was attributed to a loss of muscle mass. However, significant bone mineral loss is also a consequence of hyperthyroidism (Cohn et al. 1973). In contrast, the BCM of patients with hypothyroidism was similar to that of healthy controls, even though these patients weighed more and were fatter than healthy individuals (Miyakawa et al. 1999; Seppel et al. 1997).

Lonn et al. (1998) used DXA and computerized tomography (CT) to monitor changes in body composition at diagnosis and after treatment (3 and 12 months) for hyperthyroidism. Following 12 months of treatment, BW, FM, and FFM increased significantly. At 3 and 12 months of treatment, skeletal muscle, as assessed by CT, increased; but there was no increase in subcutaneous adipose tissue until 12 months.

Kormas and colleagues (1998) assessed the body composition of postmenopausal women with hypothyroidism (benign goiter) both before and after surgery (i.e., total thyroidectomy). After the operation, the patients received thyroid hormone replacement. Compared to healthy age- and weight-matched controls, patients treated with thyroid hormone showed no significant changes in BW or body composition at 4 and 12 months postsurgery.

Additional research studies are warranted to lead to full understanding of changes in body composition resulting from thyroid disease and its treatment. Specifically, the FFB composition and applicability of field methods for assessing body composition of these clinical populations need to be addressed using multicomponent models and suitable methods to obtain reference measures of body composition.

Key Points

- Obesity is an excessive amount of body fat relative to BW.

- In epidemiological studies, obesity is defined as a BMI of 30 kg/m² or more.

- Abdominal obesity is characterized by excess subcutaneous and visceral fat in the abdominal region.

- Individuals with abdominal obesity have a relatively high risk of developing chronic diseases.

- There is a high degree of interindividual variability in the hydration of the FFM and density of the FFM of obese populations.

- The BIA and anthropometric (circumference) methods can be used in clinical settings to estimate body composition of persons who are obese, but the SKF and NIR methods are not recommended.

- Fatness-specific BIA equations are more accurate than generalized equations for assessing FFM of obese populations.

- Uncontrolled type 1 diabetes produces fat catabolism, which decreases BW and FM.

- Many individuals with type 1 diabetes have osteopenia or osteoporosis.

- Patients with type 2 diabetes tend to be overweight or obese and have a higher bone mass compared to healthy adults.

- Presently, there are no suitable field method prediction equations to estimate body composition of patients with type 1 or type 2 diabetes.

- Hyperthyroidism is characterized by an increased synthesis and secretion of thyroid hormones; hypothyroidism is caused by thyroid hormone deficiency.

- Patients with hyperthyroidism tend to lose BW and body fat, as well as skeletal muscle and bone mass.
- Patients with hypothyroidism tend to gain BW and body fat.
- Research is needed to quantify the FFB composition of patients with thyroid disease and to develop field method prediction equations applicable to this clinical population.

Key Terms

Learn the definition for each of the following key terms. Definitions of terms can be found in the Glossary, page 227.

abdominal obesity

diabetes mellitus (DM)

Graves' disease

hyperglycemia

hyperthyroidism

hypoglycemia

hypothyroidism

insulin-dependent diabetes mellitus (IDDM)

myxedema

non-insulin-dependent diabetes mellitus (NIDDM)

thyroiditis

thyrotoxicosis

type 1 diabetes

type 2 diabetes

Review Questions

1. Define obesity and abdominal obesity. What is the difference in health risk between upper-body and lower-body obesity?

2. Describe the changes in FFB composition of obese populations.

3. What methods have been used to obtain reference body composition measures for people who are obese?

4. Identify suitable field methods and prediction equations that can be used in clinical settings to estimate the body composition of clients who are obese.

5. Identify and define two types of diabetes mellitus. Describe how each type of diabetes affects body composition.

6. What reference and field methods can be used to assess body composition of individuals with diabetes?

7. What is the primary difference between hyperthyroidism and hypothyroidism? Describe how each of these thyroid diseases affects body composition.

8. Identify field methods, if any, that may be used to estimate body composition of patients with thyroid disease.

CHAPTER 14

BODY COMPOSITION AND WASTING DISEASES AND DISORDERS

Key Questions

- What are the most common wasting diseases and disorders and how do they develop?
- How do the various wasting diseases and disorders affect body composition?
- What reference methods are typically used to assess and quantify muscle wasting?
- What field methods can be used in clinical settings to assess the body composition of patients with wasting conditions?

Wasting can be defined as a state of nonvolitional weight loss that generally coincides with a clinical condition (Koch 1998). Weight loss is often indicative of infection and serious illness, but weight loss alone is not a sufficient measure of nutritional status and muscle wasting. Body cell mass (BCM), the metabolically active cellular components of the body, reflects the functional status of the patient and is a better indicator of survival from wasting diseases than body weight (Koch 1998; Kotler et al. 1989; Wanke et al. 2002). For this reason, body composition assessments that can help the practitioner identify losses in BCM and changes in the individual body components are needed. In addition to identifying and quantifying conditions of muscle wasting, body composition assessment is necessary for monitoring the efficacy of clinical interventions.

The muscle-wasting diseases and disorders addressed in this chapter include anorexia nervosa, HIV/AIDS, cancer, kidney failure and dialysis treatment, cirrhosis and other liver diseases, spinal cord injury, and other neuromuscular conditions such as multiple sclerosis and muscular dystrophy. Each condition is briefly defined, and the influence

of the disease or disorder on body composition is discussed. The most applicable methods and prediction equations are identified for assessing the body composition of patients with muscle wasting.

Anorexia Nervosa

This section defines anorexia nervosa and describes the effects of this psychological eating disorder on the body composition. It also summarizes studies on the applicability of various body composition methods and predictions for this population.

Definition of Anorexia Nervosa

Often classified as an eating disorder, **anorexia nervosa** is a psychological disorder characterized by a distorted body image, extreme fear of gaining weight, and refusal to maintain a minimally normal body weight. The stereotypical person with anorexia is an affluent or upper middle class adolescent or young adult female who is a high achiever or perfectionist. People with anorexia restrict calories and often exercise vigorously to control body weight even when emaciated, deny that there is a problem, and resist treatment. The diagnostic criteria for anorexia nervosa are as follows (American Psychiatric Association 1994):

- Body weight less than 85% of the minimally normal weight for age and height
- Intense fear of gaining weight or becoming fat, even though one is considered underweight by all medical criteria
- Distorted view or perception of one's body weight or shape
- Amenorrhea (absence of at least three consecutive menstrual cycles) in postmenarcheal females

Influence of Anorexia on Body Composition

Much of the weight lost by persons with anorexia is body fat, but a substantial amount of lean tissue is also lost. Studies have shown that compared to controls without eating disorders, persons with anorexia had an average reduction in body weight of 26% to 31%; body fat and LTM were reduced by 59% to 71% and 16% to 17%, respectively (Kooh et al. 1996; Scalfi et al. 1999, 2002). People with anorexia experience losses in all compartments of the body. Because of the decrease in FFM, TBW is less in people with anorexia than in age- and height-matched controls (Scalfi et al. 1997). Total body nitrogen (TBN), an index of protein, was also found to be less in anorexic patients (i.e., 73-75% of predicted values for age and height) (Kerruish et al. 2002). Additionally, the mineral component of the body is reduced in persons with anorexia. Total body bone mineral content (BMC) is about 25% less and the bone mineral density (BMD) of the femoral neck is reduced by 13% to 20% (Kooh et al. 1996; Mazess, Barden & Ohlrich 1990; Resch et al. 2000). This bone loss is not clinically significant in the first year of anorexia (Wong et al. 2001), but it can be difficult to recover bone once it is lost. Despite improved nutritional status in recovering anorexic adolescents, total body BMD and lumbar BMD initially decreased and did not start increasing until 21 months after treatment began (Jagielska et al. 2001). Hartman et al. (2000) reported that femur BMD remained significantly less even 21 yr after recovery from anorexia nervosa (0.92 g/cm^2) compared to values in those who had never had the disorder (1.07 g/cm^2).

Because the reduction in body fat is proportionally larger than the reduction in LTM during starvation, the percentage of each component of the FFB is increased when expressed relative to body weight in people with anorexia. For example,

although the absolute BCM is less for persons with anorexia, they have a greater BCM/body weight (38%) than healthy women (33%) (Moley et al. 1987). On the basis of multicomponent model estimates, Hannan and colleagues (1990) reported the following values for the relative composition of body weight of anorexia nervosa patients (BMI < 17 kg/m^2) and healthy controls (BMI = 23.9 kg/m^2): 19.4% versus 14.6% protein, 6.0% versus 4.4% mineral, 67.6% versus 52.1% water, and 7.0% versus 28.9% fat.

It appears that the FM and FFM of re-fed anorexia nervosa patients (BMI > 18.5 kg/m^2) can be recovered to levels similar to those of controls who never had an eating disorder (Polito et al. 1998; Scalfi et al. 1999, 2002). However, it is likely that the composition of the FFM will be altered. In a group of women who had been rehabilitated for >6 months following at least 1 yr of anorexia, Polito et al. (1998) reported that muscle mass, estimated from urinary creatinine levels, was still substantially lower than that of controls. Relative muscle mass was 61% of the FFM in controls but only 48% to 49% of the FFM in women with anorexia and those recovering from anorexia.

Assessing Body Composition of Anorexia Nervosa Patients

Most of the research on this clinical population has focused on the body composition of individuals with anorexia nervosa and on monitoring their recovery. Most studies have used dual-energy X-ray absorptiometry (DXA) for this purpose. The validity of field methods and the accuracy of prediction equations for anorexia nervosa patients have not been extensively studied.

SKFs

Kerruish and colleagues (2002) reported that triceps SKF thickness was a significant predictor of relative body fat (%BF$_{DXA}$) for adolescent females with anorexia nervosa. Using this measure, they derived the following equation: %BF = 2.14 (triceps SKF) − 1.91. The equation was developed using only 23 patients, however, and it was not cross-validated. Therefore, we cannot recommend its use at this time.

Bioelectrical Impedance Analysis

Scalfi et al. (1997) cross-validated several multifrequency bioelectrical impedance analysis (MFBIA) equations (Deurenberg et al. 1993; Kushner et al. 1992; Segal et al. 1991) on a sample of 19 anorexic women and 27 controls. Body weight alone was a better predictor of TBW (r^2 = .90, SEE = 1.6 L) than these BIA equations that included impedance index (HT²/Z). In contrast, the prediction error (SEE = 3.1% BF) for a single-frequency BIA equation, derived from a sample of 36 women with eating disorders and with BMI <16 kg/m², was less than that of an equation that included only anthropometric predictors (SEE = 4.2%) (Hannan et al. 1993). Although Hannan and colleagues did not cross-validate this equation in their study, it was found to be the most accurate equation for estimating the body composition of very lean female distance runners (Pichard, Kyle, Gremion et al. 1997) (see chapter 11). Thus, we recommend using the BIA equation of Hannan et al. (1993) (see table 14.1) to estimate the %BF of women with very low body fat.

HIV and AIDS

This section defines HIV and AIDS and describes how these diseases affect body composition. Body composition methods and prediction equations most applicable to this clinical population are presented.

Definitions of HIV and AIDS

Human immunodeficiency virus (HIV) is an infection spread by contact with infected body fluids. It progressively destroys white blood cells **(lymphocytes)**. HIV eventually leads to impaired immunity, severe immunosuppression known as **acquired immunodeficiency syndrome (AIDS)**, and opportunistic diseases (Beers & Berkow 1999). HIV attaches to a receptor protein called CD4+ and destroys these lymphocytes that help to coordinate the immune system. HIV patients are typically categorized by their CD4+ lymphocyte counts: ≥500 cells/μL, 200 to 499 cells/μL, or <200 cells/μL. AIDS is marked by a CD4+ count that falls below 200 cells/μL and the onset of opportunistic infections that do not normally cause disease in people with a healthy immune system (Beers & Berkow 1999).

Influence of HIV/AIDS on Body Composition

Patients with HIV have less body weight, BCM, FFM, and FM than noninfected controls (Kotler, Thea et al. 1999). Additionally, these body composition changes due to HIV may differ in men and women. Kotler et al. (1999) estimated that FFM accounts for about 51% of body weight loss in men, but only 18% of body weight loss in women. HIV-associated wasting has recently been defined as a patient meeting one of the following (Wanke et al. 2002):

- 10% unintentional weight loss over a year
- 7.5% unintentional weight loss over 6 months
- 5% loss of BCM within 6 months
- Males: BCM <35% of body weight and BMI <27 kg/m²
- Females: BCM <23% of body weight and BMI <27 kg/m²
- BMI <20 kg/m²

The density of the fat-free body (FFBd) has not been reported for HIV/AIDS patients, but data about the individual components of the FFB in this population are available. The relative hydration of the fat-free mass (TBW/FFM) in AIDS patients does not appear to be significantly altered from the value assumed for healthy people (~73%) (Wang, Kotler et al. 1992). Wang, Kotler et al. (1992) also noted that bone mineral loss is proportionally less than the FFM loss in AIDS patients; therefore, AIDS patients likely have an increased TBM/FFM ratio and an increased FFBd.

In recent years, antiretroviral therapy, such as use of protease inhibitors, has become a common treatment for HIV-infected patients. While this treatment appears to improve CD4+ counts and preserve immune function, it is often associated with **lipodystrophy,** also known as "fat redistribution syndrome." Lipodystrophic changes in body shape include peripheral lipoatrophy (a thinning of the cheeks and extremities), central lipohypertrophy (increased truncal fat), or a combination of both. Also, a small percentage of patients with lipodystrophy may experience an increased dorsocervical fat pad (i.e., "buffalo hump") (Gerrior et al. 2001). Presently, there are no definitive clinical criteria for lipodystrophy, and the severity of the syndrome is subjectively rated (Schwenk et al. 2001; Schwenk 2002). However, patients with lipodystrophy and a normal BMI are likely to have high waist-to-hip and waist-to-thigh ratios and below-average midarm circumferences and triceps SKF thicknesses (Gerrior et al. 2001; Schwenk et al. 2001). Also, when measured by magnetic resonance imaging (MRI) and DXA, HIV-infected patients with lipodystrophy had similar amounts of skeletal muscle and subcutaneous adipose tissue but greater visceral adipose tissue than HIV-infected patients without truncal enlargement (Engelson et al. 1999).

Assessing Body Composition of Patients With HIV/AIDS

Wasting is an important prognosticator of morbidity and mortality in HIV patients (Kotler et

Table 14.1 Summary of Recommended Methods and Equations for Muscle-Wasting Diseases and Disorders

	Method/Equipment	Clinical population	Equation	Reference
BIA	RJL 101	Anorexia nervosa (females, 15-44 yr, BMI <16 kg/m²)	1. %BF = 7.32 − 0.572 (HT²/R) + 0.664 (BW)	Hannan et al. 1993
	RJL 101A	HIV/AIDS (men without lipodystrophy, 40 yr*)	2. FFM (kg) = 0.50 ($HT^{1.48}/Z^{0.55}$ × 1.0/1.21) + 0.42 (BW) + 0.49	Kotler et al. 1996
	RJL 101A	HIV/AIDS (women without lipodystrophy, 40 yr*)	3. FFM (kg) = 0.88 ($HT^{1.97}/Z^{0.49}$ × 1.0/22.22) + 0.081 (BW) + 0.07	Kotler et al. 1996
	SEAC SFB3	HIV/AIDS (men without lipodystrophy, 37 ± 7 yr)	4. BCM (kg) = 0.360331 (HT²/Z) + 0.151123 (BW) − 2.95	Paton et al. 1998
	Holtain BCA	Dialysis (boys and girls, 4-22 yr)	5. TBW (L) = 0.144 (HT²/Z) + 0.40 (BW) + 1.99	Wuhl et al. 1996
	RJL 101A	Cirrhosis without ascites (men and women, 30-76 yr)	6. TBW (L) = 3.751 + 0.595 (HT²/R)	Zillikens & Conway 1991
	Analycor3 (dual frequency)	Spinal cord injury (men and women, 45.4 ± 12.8 yr)	7. TBW (L) = 2.896 + 0.366 (HT²/R)[a] + 0.137 (BW) + 2.485 (sex)[b]	Vache et al. 1998
	RJL 101A	Spinal cord injury (men and women, 18-73 yr)	8. FFM (kg) = 18.874 + 0.367 (HT²/R) + 0.253 (BW) − 0.081 (age) − 5.384 (sex)[c]	Kocina & Heyward 1997

[a]Resistance measured at 100 kHz; [b]Female = 0, male = 1; [c]Male = 0, female = 1.

Key: %BF = percent body fat; BCM = body cell mass (kg); TBW = total body water (L); HT = height (cm); R = resistance (Ω); Z = impedance (Ω); BW = body weight (kg), *average age.

al. 1989; Wanke et al. 2002). The lipodystrophy often associated with antiretroviral therapy may mask an underlying wasting syndrome (Wanke et al. 2002). Thus, simply monitoring body weight or BMI of HIV patients is not sufficient; body composition methods and techniques, capable of detecting alterations in the components of the body and quantifying BCM, are important in the treatment of HIV patients.

Wang, Kotler et al. (1992) compared the accuracy of several methods for predicting the body fat of men with AIDS. Not surprisingly, the laboratory method of TBW (tritium dilution) had the best agreement with criterion measures of %BF assessed by neutron activation analysis (NAA) (r = .97). Thus, the researchers derived an equation to predict %BF from TBW (R^2 = 0.95, SEE = 1.9% BF) for male AIDS patients; however, this equation was not cross-validated. The prediction equations derived from other methods (i.e., dual-photon absorptiometry, whole-body counting of total body potassium [TBK], SKF, and BIA) all had much higher prediction errors (>4.2% BF). Similarly, Corcoran et al. (2000) noted that FFM estimates from BIA and SKF do not correlate as well with TBK as the FFM assessed by DXA. Most researchers used TBW, TBK, DXA, or a combination of these methods to obtain reference measures for assessing the body composition of HIV-infected patients.

SKFs

Using total body electrical conductivity (TOBEC) as the reference method, five SKF equations, the RJL BIA manufacturer's equation, and a BMI equation were cross-validated in a sample of HIV-infected men (Risser et al. 1995). All of the equations had prediction errors that exceeded the acceptable limit for estimating FFM (SEE = 3.6-5.7 kg). Batterham et al. (1999) compared the predictive accuracy of seven SKF equations for estimating FFM_{DXA} of HIV patients. Although the SKF equation of Durnin and Womersley (1974) had the best predictive accuracy for HIV patients, the bias was significant (2.7 kg). Similarly, this SKF equation underestimated FFM_{DXA} in men and women (CE = –1.0 and –1.2 kg, respectively) with AIDS and no lipodystrophy (Corcoran et al. 2000). Body composition estimates from SKF and BIA also differed substantially in HIV-infected patients with lipodystrophy (Gerrior et al. 2001). Given these results, we do not recommend the SKF method for estimating the FFM of HIV-infected patients; however, there may be some value in using the SKF thicknesses at individual sites to track changes in fat redistribution with lipodystrophy.

BIA

Because of its ease of use in clinical settings and its potential to estimate BCM, BIA has been widely studied as a method of estimating the body composition of HIV-infected patients. Kotler and colleagues (1996) developed BIA equations using TBK, DXA, and TBW to obtain reference measures of body composition for a multiethnic sample of 134 HIV-infected patients and 198 healthy men and women. Reactance (Xc) was the best predictor of BCM, whereas impedance (Z) was the best predictor of FFM and TBW. Compared to using only body weight or BMI as predictors, inclusion of bioimpedance measures (i.e., Xc or Z) improved the accuracy of these equations for estimating BCM (SEE ~10%), FFM (SEE ~5%), and TBW (SEE ~8%).

The Kotler et al. (1996) equation was cross-validated in several BIA studies of HIV-infected patients (Batterham et al. 1999; Corcoran et al. 2000; Paton et al. 1998; Pichard et al. 1999). Pichard et al. (1999) recommended using the Kotler BIA equation to estimate the FFM of both healthy clients and individuals with AIDS. Likewise, Batterham et al. (1999) recommended using the BIA equations of either Kotler or Heitmann (1990) for estimating FFM of HIV-infected clients because the average FFM from these equations did not differ significantly from FFM_{DXA}. The Segal et al. (1988) fatness-specific BIA equations, however, had the best individual predictive accuracy. In contrast, Corcoran et al. (2000) reported that the BIA equation of Lukaski and Bolonchuk (1987) more closely estimated average FFM_{DXA} of AIDS patients than other BIA equations, including the Kotler equation.

The Kotler equation is based on a parallel circuit model (see chapter 6) and requires logarithmic calculations to estimate FFM (see table 14.1). Therefore, Paton et al. (1998) developed a BIA equation to estimate BCM of AIDS patients using the simpler, single-frequency serial BIA model. The predictive accuracy of their equation was similar to that of the Kotler equation. This equation, as well as the Kotler equation, is presented in table 14.1. Given that both of these equations were developed prior to the extensive use of antiretroviral therapy for AIDS patients, these equations may not be applicable to patients with lipodystrophy (Schwenk et al. 2001). Thus, we do not recommend using these equations in patients who have noticeable lipodystrophy.

Additionally, Arpadi and colleagues (1996) evaluated the predictive accuracy of eight BIA equations, developed for healthy children, for estimating either the TBW or FFM of prepubertal children (N = 20) with HIV. Given that the accuracy of these previously published equations was unacceptable, they derived a new equation for estimating FFM of children with HIV ($r^2 = 0.95$, SEE = 1.2 kg); however, this equation has not yet been cross-validated.

Recently, Wanke and colleagues (2002) reported that phase angle may be a useful predictive marker of survival in AIDS patients. **Phase angle** is arc tangent of the ratio of reactance to resistance (arctan Xc/R). Phase angle was the single best predictive marker of survival and had better prognostic value than measures commonly used to predict survival (e.g., BCM, body weight, BMI, and even CD4+ cell count) in AIDS patients.

Cancer

This section defines cancer and provides a brief description of the effects of this disease on body composition. Few studies have evaluated the predictive accuracy of body composition methods and equations in this clinical population.

Definition of Cancer

Cancer is simply the unregulated or uncontrolled growth of a cell. Cancer can develop in any tissue in the body, and it can grow and invade adjacent cells and spread (metastasize) throughout the body. Typical treatments for cancer include surgery, radiation, chemotherapy, or a combination of therapies.

Influence of Cancer on Body Composition

In a review of this topic, Tisdale (1999) noted that about one-half of all cancer patients experience cachexia (i.e., wasting of body fat and lean tissue). This wasting syndrome appears to be the result of tumor-induced catabolism rather than lack of caloric intake. Protein and BCM are preferentially spared during starvation; but in cancer patients, BCM is lost proportionally with body weight. For example, anorexia nervosa patients who had lost 31% of their body weight still had a relative BCM (BCM/BW) that was 30% greater than that of female cancer patients who had lost only 14% of their body weight (Moley et al. 1987). Additional alterations in body composition may be more specific to the type of cancer or treatment. For example, men with prostate cancer who were treated with androgen-deprivation therapy had less BMD and more obesity than healthy, age-matched men (Chen et al. 2002).

Assessing Body Composition of Cancer Patients

Despite the prevalence of cachexia in cancer patients and the likelihood that it is a prominent factor in the failure to survive, few studies have been done to validate body composition methods and prediction equations for this clinical population. Most of these studies examined the efficacy of the BIA method. In men with prostate cancer, average %BF, estimated by the Segal et al. (1988) fatness-specific BIA equations (22.5% BF) and the Deurenberg, van der Kooy et al. (1990) BIA equation (38.2% BF), differed significantly from %BF$_{DXA}$ (26.7% BF) (Smith et al. 2002). Several researchers reported that HT^2/R is highly correlated with TBW$_{D2O}$ in cancer patients (Fredrix et al. 1990; Simons et al. 1995); however, Simons et al. (1995) noted that BIA systematically overestimates TBW in underweight patients regardless of the BIA equation used. They developed equations to estimate the TBW of both underweight (<95% ideal body weight) and normal-weight cancer patients; but their sample size was small, and the equations were never cross-validated. Bauer and colleagues (2002) recently used the Simons BIA equations to estimate the TBW of pancreatic cancer patients, but no reference method was used to check the validity of these equations.

Kidney Failure and Dialysis

This section describes different methods of treating chronic kidney failure and the effects of this disease on body composition. Applicable methods and prediction equations for this clinical population are summarized.

Definitions of Kidney Failure and Dialysis Treatments

The primary function of the kidney is to filter metabolic waste products and excrete them from the body. Many diseases can contribute to kidney injury and eventually to **chronic kidney (renal) failure,** resulting in an accumulation of metabolic waste products in the blood due to a progressive decline in kidney function (Beers & Berkow 1999). The typical treatment for chronic kidney failure is **dialysis,** a process of removing excess water and toxins from the body. There are two methods of dialysis: hemodialysis and peritoneal dialysis. **Hemodialysis** consists of removing blood from the body, pumping it through a filtering machine, and returning the purified blood to the body. In **peritoneal dialysis,** a fluid mixture of glucose and salts is injected into the abdomen to attract the toxic waste and is then drained out of the body (Beers & Berkow 1999). Both methods adjust the body's fluid volume.

Influence of Kidney Failure and Dialysis on Body Composition

Factors such as the length of time the patient has had kidney disease and has been on dialysis, the type of dialysis, and how soon the body composition assessment is made after dialysis contribute to the variability in the body composition of patients with kidney failure. Using a variety of methods including NAA, whole-body counting, and isotope dilution, Stall et al. (2000) reported TBW/FFM values ranging from 75% to 77% and protein/FFM values ranging between 18% and 21% for groups of men and women on either hemodialysis or peritoneal dialysis. Hemodialysis patients were studied the day after treatment, and peritoneal dialysis patients were examined after peritoneal fluid was drained. Woodrow and colleagues (1996) reported that the relative hydration of the FFM (TBW/FFM) was not significantly different between controls and either hemodialysis or peritoneal dialysis patients when patients were measured within 1 hr of treatment. These researchers noted, however, that the average total body protein of both male and female hemodialysis patients was lower than that of the other groups.

Assessing Body Composition of Dialysis Patients

A single ideal method for assessing the body composition of the dialysis patient does not exist. Alterations in hydration level and BMD in this clinical population may affect the validity of 2-C model estimates of %BF from underwater weighing. Therefore, DXA and whole-body counting are typically used to obtain reference measures of bone status and BCM in this population (Lukaski 1997). Researchers reported moderately strong correlations (r ~.80) between field methods, such as SKF, BIA, and near-infrared interactance, and reference measures (i.e., DXA or TBK) for estimating the LTM of hemodialysis (Chen et al. 2000) and peritoneal dialysis (Lo et al. 1994) patients, suggesting that field methods may be applicable to this clinical population.

BIA

The efficacy of BIA will likely vary depending on the method of dialysis and the hydration status of the patient (Lukaski 1997). Dumler et al. (1992) reported that BIA resistance peaks at about 90 min into the hemodialysis procedure and does not reach a steady state until near the end of the treatment; however, the optimal time to perform BIA in relation to dialysis is still unknown (Chertow et al. 1995). Chertow et al. (1995) reported that although BIA was highly correlated to reference measures of TBW (r = .96) and BCM (r = .92) in hemodialysis patients, it significantly underestimated both variables by 3.5 kg and 5.8 kg, respectively. However, in a review of body composition assessment in children with chronic renal failure, Schaefer and colleagues (2000) recommended a BIA equation (see table 14.1) that they had previously developed (Wuhl et al. 1996) and cross-validated specifically for the pediatric dialysis patient. They noted that their population-specific equation estimates TBW with an error of 1.7 L (CV = 8.5%) compared to errors of ~2.8 L (CV = 14%) when equations for healthy children are applied to this clinical population.

Anthropometry

A height and weight equation, developed by Watson et al. (1980), was used to estimate the TBW of dialysis patients (Chertow et al. 1995; Johansson et al. 2001). Although this simple anthropometric equation estimated TBW reasonably well (r = .89, CE = −0.38 L, limits of agreement = −7.17 to 6.40 L), it tended to underestimate TBW in lean or overhydrated peritoneal dialysis patients and overestimate TBW in obese or underhydrated patients (Johansson et al. 2001). This research team developed a similar anthropometric equation to estimate TBW of dialysis patients, but it was not significantly better than the Watson et al. (1980) equation. They suggested using the Watson et al. (1980) equation to estimate average TBW of groups of dialysis patients, but cautioned that variability within individuals may be high.

Cirrhosis and Other Liver Diseases

This section defines cirrhosis and its classifications and describes changes in body composition due to this liver disease. The difficulty of assessing the body composition of patients with cirrhosis is addressed, and suitable field methods and prediction equations are summarized.

Definition of Cirrhosis

Injury to the liver often results in cirrhosis. **Cirrhosis** is the scarring of liver tissue, leaving it nonfunctional (Beers & Berkow 1999). This disease is often caused by alcohol abuse; however, various forms of hepatitis, bile duct obstruction, or other infections or autoimmune diseases can also lead to cirrhosis. Cirrhosis is usually progressive, and there is no cure. Patients with cirrhosis are typically classified into one of three categories (Childs A, Childs B, or Childs C) based on increasing abnormality of clinical and biochemical measures (Pugh et al. 1973). This classification is important because the body composition of individuals with cirrhosis often varies with the severity of the disease.

Influence of Cirrhosis on Body Composition

Both body fat and BCM progressively decrease with increasing severity of cirrhosis (Crawford et al. 1994; Morgan & Madden 1996). The decrease in BCM has been observed in both nonalcoholic (Crawford et al. 1994) and alcoholic individuals with cirrhosis (Prijatmoko et al. 1993). This wasting of BCM can be masked by abnormal fluid retention (i.e., **ascites**), which is common in patients with cirrhosis (Prijatmoko et al. 1993; Strauss et al. 2000). For example, the FFM estimated from a multicomponent model was not significantly different between healthy controls and persons with cirrhosis, but Prijatmoko et al. (1993) reported a significantly lower relative hydration of the FFM for controls (72.9%) compared to those with cirrhosis. The TBW/FFM increased with the severity of disease (Childs A = 77.7%, Childs B = 80.4%, Childs C = 82.6%) while the protein fraction decreased (controls = 20%, Childs A = 16%, Childs B = 13%, Childs C = 11%) (Prijatmoko et al. 1993). In addition to lower %BF, cirrhotic patients also had less BMC and BMD compared to healthy participants (Riggio et al. 1997). Decreases in bone mass in alcoholics, however, may be due to malnutrition rather than liver disease (Santolaria et al. 2000).

Assessing Body Composition of Patients With Cirrhosis

Morgan and Madden (1996) recommended using a 4-C model, modified to include extracellular water, to obtain reference measures of body composition for cirrhotic patients. Some research teams used a combination of NAA, DXA, and isotope dilution to assess the body composition of this clinical population (Prijatmoko et al. 1993; Strauss et al. 2000). Also, whole-body counters have been used to mea-

sure TBK for the estimation of BCM (Crawford et al. 1994; Lehnert et al. 2001).

Compared to a 4-C model, DXA provided better estimates (i.e., smaller CE and limits of agreement) of FM and FFM than TBK, SKF, single-frequency BIA, and MFBIA in men and women with cirrhosis (Strauss et al. 2000). Prijatmoko et al. (1993) also recommended using DXA and concluded that SKF and BIA methods are not precise enough to accurately reflect the changes in body composition that can occur with cirrhosis.

Although DXA may provide a reasonably accurate estimate of FFM, this method may be incapable of detecting fluid changes (i.e., ascites), which commonly occur in patients with cirrhosis. As mentioned previously, overhydration of a "normal" FFM can mask the wasting of protein and BCM. Thus, several researchers have studied the efficacy of using BIA to estimate TBW in this clinical population.

There is a good relationship between TBW_{BIA} and TBW_{D2O} for cirrhotic patients ($r \geq .89$), but the group and individual prediction errors for these patients are about double those for healthy clients (Lehnert et al. 2001; Zillikens, van den Berg, Wilson, Rietveld et al. 1992). Strauss et al. (2000) reported large 95% limits of agreement (ranging from about ±10 to 30 L) for males and females with cirrhosis. The accuracy of the BIA method varies with the amount of ascites. Most of the ascites in people with cirrhosis occurs in the trunk. Given that the traditional BIA method is insensitive to fluid changes in the trunk, Lehnert et al. (2001) theorized that this method would typically underestimate TBW in this clinical population. Zillikens and colleagues cross-validated selected BIA equations on healthy clients, as well as on cirrhotic patients with and without fluid overload (Zillikens, van den Berg, Wilson, Rietveld et al. 1992). The equations of Lukaski et al. (1985) and Zillikens and Conway (1991) both predicted reasonably well the average TBW of healthy clients and of persons who had cirrhosis without ascites, but significantly underestimated the TBW (–2.9 to –3.8 L) of patients with ascites. It is likely that any BIA equation that is accurate for a healthy clientele will also be appropriate for non-ascitic patients with cirrhosis. However, the BIA method is not recommended for estimating TBW in patients with ascites.

Spinal Cord Injury

This section defines spinal cord injury (SCI) and describes alterations in the body composition of SCI patients. The most suitable reference methods, field methods, and prediction equations for this clinical population are presented.

Definitions and Types of SCI

Severe SCI results in **paralysis** or a complete loss of neural communication to the innervated muscles, preventing any willed movement of the affected areas. The extent of the injury is dependent on the location of the spinal cord lesion. Individuals with SCI are typically classified as having either **paraplegia** (i.e., both legs and trunk paralyzed) or **quadriplegia** (i.e., all four extremities and trunk paralyzed).

Influence of SCI on Body Composition

Kocina (1997) reviewed body composition studies of adults with SCI and reported that this clinical population tends to have a greater FM and lesser LTM than able-bodied persons. Although BMI was not significantly different in a group of age- and height-matched persons with paraplegia compared to able-bodied controls, the SCI group had 16% less LTM and 47% more FM as measured by DXA (Jones et al. 1998). Similarly, in a study of eight identical twins, the average FM of the paralyzed twins was 4.8 kg greater per unit of BMI than that of the able-bodied twins (Spungen et al. 2000). The percent fat in the legs of the paraplegic twins was 17.5% greater than in the able-bodied twins. There was no significant difference in the LTM of the arms; but the LTM of the trunk, legs, and total body was reduced by an average of 3.0 kg, 10.0 kg, and 12.6 kg, respectively, in the paraplegic twins. Additionally, these researchers reported that loss of total body LTM is significantly associated with the duration of injury, independent of age.

Physical activity likely reduces the magnitude of LTM loss and FM gain. On average, physically active wheelchair athletes, who vigorously exercised for at least 120 min per week, had a lower %BF (15.6% BF) than sedentary individuals with SCI (23.3% BF) as measured by TOBEC (Olle et al. 1993). However, the body fat of SCI athletes is still likely to be greater than that of able-bodied individuals.

In addition to increased levels of body fat and decreased LTM, the three main constituents of the FFM—water, mineral, and protein—are also altered in SCI patients (Kocina 1997). Since body water is primarily associated with FFM and there is a loss of FFM with SCI, it is not surprising that TBW relative to body weight is less in SCI individuals than in able-bodied individuals; the higher the spinal cord lesion, the greater the decrease in FFM and TBW. Also, the distribution of water throughout the body is altered. Intracellular water, associated with potassium and BCM, decreases while extracellular water, indicative of edema, increases in this clinical population (Kocina 1997). Nuhlicek et al. (1988) suggested that the relative hydration of the FFM (TBW/FFM) of SCI patients may be similar to the reference value of the non-SCI population (~73%); however, Lockner et al. (1998) reported lower hydration levels for men (66.3% FFM) and women (69.4% FFM) with SCI.

The BMC and BMD of individuals with SCI are reduced (Kocina 1997). The magnitude of these losses is dependent on the level and duration of injury. Persons with quadriplegia have greater deficits than those with paraplegia, with the majority of bone mineral loss confined to the paralyzed extremities. Bone loss is greatest in the first 6 months following the spinal cord lesion and stabilizes at about two-thirds of the original value by 16 months postinjury (Garland et al. 1992). In persons with paraplegia who had had an SCI for at least 1 yr, total body BMC was 12% less and leg BMC was 30% less than the values for age- and height-matched controls (Jones et al. 1998). Average femoral neck BMD was only 74% that of the controls, placing those with paraplegia at risk for a fracture. When expressed as a percentage of FFM, the TBM of SCI women (7.0%) was similar to that of age-matched able-bodied women (7.3%), but the TBM/FFM of SCI men (5.5%) was significantly less than for able-bodied men (6.8%) (Lockner et al. 1998).

Because of loss of FFM and muscle atrophy with SCI, it is likely that protein is also decreased. Potassium levels, indicative of BCM, are reduced in proportion to the level and the duration of the SCI (Kocina 1997). However, when expressed as a percentage of the FFM, the protein fraction of men (28.2%) and women (23.6%) with SCI is substantially higher than that of able-bodied men (19.4%) and women (19.7%) (Lockner et al. 1998).

Using DXA and deuterium dilution methods, Lockner et al. (1998) estimated the FFBd of SCI women and men to be 1.114 g/cc and 1.116 g/cc, respectively. This is substantially greater than the reference value of 1.100 g/cc. This greater value is a consequence of the lower relative hydration (TBW/FFM) and concomitantly higher protein fraction (TBP/FFM) in the FFB of SCI patients compared to able-bodied individuals.

Assessing Body Composition of Patients With SCI

Because of the greater FM and reduced physical activity of SCI patients, this clinical population is at increased risk compared to others for cardiovascular disease and non-insulin-dependent diabetes mellitus (Kocina 1997). The physical inactivity and associated bone mineral loss also put these individuals at increased risk for osteoporosis and fractures (Kocina 1997). Thus, it is important to be able to assess and monitor changes in the body composition of this population.

Although hydrostatic weighing is the typical method for determining total Db in able-bodied clients, it may be impractical and unsafe for the SCI population. Kocina (1997) noted that it is difficult to transport SCI patients in and out of the tank and have them maintain the proper position in the hydrostatic weighing chair. Although the hydrodensitometry (HD) method has been reported to be reliable (r = .98) for measuring the Db of SCI clients (George et al. 1987), the validity of this method for this population has been questioned. The relationship between FFM_{TBW} and FFM_{HD} was good for able-bodied participants (r = .95, SEE = 2.8 kg) but poor for SCI men and women (r = .71, SEE = 6.0 kg) (George et al. 1988). The researchers theorized that difficulty in obtaining an accurate residual lung volume (RV) measurement in the SCI participants was the primary source of error. Bosch and Wells (1991) confirmed this premise with their findings of a significant difference in RV (0.99 L) measured on land versus under water for participants with quadriplegia, but no difference in RV between methods (on land vs. under water) for able-bodied participants. On average, Db will be overestimated (0.017 g/cc) and %BF underestimated (7.2% BF) for people with quadriplegia if RV is measured on land (Bosch & Wells 1991; George et al. 1988). For SCI clients with spinal cord lesions higher than the seventh thoracic level, RV should be measured while the person is submerged (Bosch & Wells 1991; George et al. 1988). Given the difficulty of performing the HD method on SCI patients, the Bod Pod (see chapter 3) may prove to be a better alternative for obtaining Db measures for this clinical population; however, this has not yet been studied.

DXA has been the reference method of choice for body composition researchers working with SCI clients, and it has been recommended as a practical method for assessing the body composition of this population (Jones et al. 1998). In addition to requiring little effort from the participant, DXA can measure regional body composition, important in the assessment of paralyzed limbs. Furthermore, DXA provides BMC and BMD data. As stated earlier, both of these measures are greatly diminished in individuals with SCI, placing them at increased risk for osteoporosis and fractures.

Recently, Maimoun et al. (2002) suggested that biochemical markers of bone turnover be used to complement DXA for detecting changes in bone mineral status during the initial phase of SCI. They found no significant difference in total, femoral, lumbar, or radial BMD in acute SCI patients (3 months postinjury) compared to age-matched controls as measured by DXA. However, biochemical markers of bone turnover revealed substantial resorption and demineralization, suggesting that this method is more sensitive than DXA during the early phase of injury.

Because of the change in the ratio of FM to LTM seen in paralyzed limbs compared to healthy body regions (Spungen et al. 2000), SKF equations derived from samples of able-bodied persons are not likely to be applicable to the SCI population (Kocina 1997). Desport et al. (2000) noted that the lack of mobility made it difficult or impossible to obtain certain SKF measures in some of their SCI patients.

BIA may be a more practical field method for the SCI population. Desport et al. (2000) reported that the Vache et al. (1998) BIA equation, which measures resistance at 100 kHz and was derived from a sample of healthy elderly men and women, accurately estimated the average TBW of 20 SCI patients (CE = 0.76 L). Also, the individual predictive accuracy of this equation was good (limits of agreement = −2.87 to 4.40 L). In comparison, other BIA prediction equations (i.e., Vache et al. [1998] equation with R measured at 50 kHz; Segal et al. [1991] equation; Analycor3 manufacturer's equation) had larger, significant prediction errors. Using DXA as the reference method, Kocina and Heyward (1997) derived a BIA equation for estimating FFM of SCI adults (N = 61) and cross-validated it on an additional sample of 30 paralyzed men and women. There was no significant difference between average FFM_{BIA} and FFM_{DXA} (0.2 kg), and the predictive accuracy of this equation was good (R^2 = 0.87, SEE = 3.2 kg, TE = 3.5 kg). On the basis of these studies, we recommend using either the Vache et al. (1998) equation or the Kocina and Heyward (1997) equation to estimate the TBW or FFM of your SCI clients (see table 14.1). Additionally, variations from the traditional BIA approach, such as using the parallel model to assess BCM or the segmental method and multifrequency approach (see chapter 6) to identify fluid shifts in paralyzed extremities, may prove useful. To date, no studies of these variations have been done for the SCI population.

Neuromuscular Diseases

This section briefly defines the neuromuscular diseases of multiple sclerosis and muscular dystrophy and describes how these diseases alter body composition. Presently, no disease-specific prediction equations have been developed or tested for these clinical populations.

Definitions of Neuromuscular Diseases

Neuromuscular disorders, such as multiple sclerosis and muscular dystrophy, can impair physical activity, thereby leading to muscle wasting. **Multiple sclerosis (MS)** involves the demyelination or damage and scarring of the myelin sheath of the nervous system (Beers & Berkow 1999). This obviously impairs the conduction of nerve impulses. **Muscular dystrophy (MD)** is an inherited muscle disease that leads to muscle weakness. The most common form of MD is **Duchenne's muscular dystrophy,** caused by a genetic defect carried on the X-chromosome, with signs of the disease usually becoming evident at about age 3 yr (Beers & Berkow 1999). Boys with this disease lack dystrophin, a protein necessary for muscle cell structure. These patients are typically in a wheelchair by age 12 yr, and most die by their early 20s.

Influence of Neuromuscular Disease on Body Composition

The extent to which one can remain mobile and physically active is likely to have the greatest influence on the MS patient's ability to maintain a normal body composition and resist muscle wasting. Nonambulatory MS patients had significantly less BMC and FFM than ambulatory MS patients and healthy controls, with no difference between the latter two groups (Formica et al. 1997). Likewise, Lambert et al. (2002) found no significant difference in %BF or %FFM between ambulatory MS patients and healthy clients.

Compared to healthy controls, boys with Duchenne's MD have more adipose tissue and intramuscular fat but less lean mass (Leroy-Willig et al. 1997; Pichiecchio et al. 2002). The water compartment also differs. When matched to healthy boys for age, height, and weight, Duchenne's patients had less TBW relative to body weight (51.8% vs. 58.5%) and a higher ECW/ICW ratio (1.15 vs. 0.70) (Bedogni et al. 1996).

Assessing Body Composition of Patients With MD

Several researchers have recommended using MRI to assess the body composition of patients with MD (Leroy-Willig et al. 1997; Pichiecchio et al. 2002). Unfortunately, field method equations derived from healthy children are not applicable to this clinical population. Although boys with MD have a higher %BF, subcutaneous FM as measured by MRI does not differ significantly from that of healthy boys (Leroy-Willig et al. 1997). Thus, it is not surprising that the commonly used SKF equation of Slaughter et al. (1988) for children significantly underestimated the average %BF (−16.1% BF) of boys with MD (Pichiecchio et al. 2002). Similarly, a BIA equation developed to predict TBW in the healthy controls significantly underestimated TBW in Duchenne's patients (Bedogni et al. 1996). The researchers noted that for the same resistance index (HT²/R), Duchenne's children tended to have higher TBW than healthy controls. Subsequently, they derived a population-specific BIA equation; however, this equation was developed using only 12 children with the disease and has not yet been cross-validated.

Key Points

- In patients with wasting syndromes, BCM is a better indicator of patient status than body weight.

- Anorexia nervosa is a psychological disorder in which the patient has an intense fear of gaining weight and refuses to eat, resulting in wasting.

- Most of the weight loss during anorexia is due to the loss of body fat; but FFM may also be lost. Loss of FFM may not be fully restored when body weight increases during recovery.

- HIV is an infection that results in an impaired immune system, greatly increasing one's susceptibility to opportunistic diseases. AIDS is the end stage of HIV, and wasting is a major prognostic marker for morbidity and mortality from this disease.

- Lipodystrophy is a fat redistribution syndrome characterized by a thinning in the face and extremities but increased fat in the trunk. This is a common side effect of antiretroviral therapy, a treatment for HIV.

- About one-half of all cancer patients experience cachexia, a wasting of body fat and lean tissue.

- In contrast to starvation, in which the majority of the weight loss is fat, cancer leads to loss of BCM in proportion to body weight.
- Altered hydration status is characteristic of chronic kidney failure, making it difficult to accurately assess body composition in this population.
- BCM progressively decreases with advancing liver disease, but water expansion (ascites) progressively increases, often masking the wasting associated with cirrhosis.
- SCI results in paralysis and decreases the absolute levels of bone mineral, water, and protein in the FFB.
- The loss of LTM in SCI patients is associated with the level and duration of injury.
- MS involves nerve demyelination, but patients who are able to remain ambulatory and active may be able to resist significant muscle wasting.
- MD, a genetic defect in the protein necessary for muscle cell structure, results in a decrease in LTM.
- SKF and BIA equations developed for healthy children are not applicable to boys with MD because they have a higher percentage of intramuscular fat compared to their healthy counterparts.
- For most of the wasting syndromes presented in this chapter, population-specific BIA equations have been developed and cross-validated to estimate either %BF, FFM, BCM, or TBW (see table 14.1). However, the utility of these equations is limited to those patients who are not experiencing "side effects" such as ascites or lipodystrophy from their wasting disease.

Key Terms

Learn the definition for each of the following key terms. Definitions of terms can be found in the Glossary, page 227.

acquired immunodeficiency syndrome (AIDS)	**lipodystrophy**
anorexia nervosa	**lymphocytes**
ascites	**multiple sclerosis (MS)**
cancer	**muscular dystrophy (MD)**
chronic kidney failure	**paralysis**
cirrhosis	**paraplegia**
dialysis	**peritoneal dialysis**
Duchenne's muscular dystrophy	**phase angle**
hemodialysis	**quadriplegia**
human immunodeficiency virus (HIV)	**wasting**

Review Questions

1. Why is body composition assessment, rather than simply tracking body weight, recommended for clients suspected of having a muscle-wasting disease or disorder?

2. Identify and define some of the most common wasting diseases and disorders.

3. Describe body composition changes in individuals with anorexia nervosa.

4. What field method and prediction equation are recommended for estimating the body composition of patients with anorexia?

5. What is lipodystrophy and why is it problematic when one is assessing the body composition of HIV/AIDS patients?

6. Identify suitable field methods and prediction equations for estimating the body composition of HIV/AIDS patients.

7. Describe the effects of cancer on body composition.

8. What factors are likely to have the greatest effect on body composition and its assessment in patients with chronic renal failure?

9. Explain how there can be considerable wasting of protein and BCM in a patient with cirrhosis but the FFM can remain "normal."

10. What factor must be considered when one is assessing the body composition of cirrhotic patients?

11. Describe the effects of SCI on FFB composition.

12. Identify a suitable field method and prediction equation to estimate the body composition of the SCI population.

13. Why are SKF and BIA prediction equations that were developed for healthy populations not likely to be accurate for patients with MD?

CHAPTER 15

ASSESSING
BODY COMPOSITION CHANGES

Key Questions

- What models and reference methods are recommended for assessing changes in body composition?
- Are field methods (i.e., skinfolds, anthropometry, bioelectrical impedance analysis, and near-infrared interactance) precise enough to adequately measure changes in body composition?
- What field methods, if any, can be used to monitor changes in body composition for clinical populations?

One of the primary reasons for conducting body composition assessments is to monitor or track the changes in a client's body composition over time. Factors such as increased physical activity, dietary modifications, pregnancy, and disease may alter body composition. Simply tracking body weight or BMI is not adequate, as a change in these variables does not indicate the amount of change in FM or FFM. Thus, one must assess body composition in order to modify weight goals, monitor the nutritional and health status of clients, and evaluate the effectiveness of diet and exercise regimens or clinical interventions. However, assessing changes in body composition is often more difficult and problematic than simply obtaining an accurate initial assessment. For example, it may not be possible to accurately measure changes in the body fat of individual clients within ±1.5 kg (i.e., 95% limits of agreement) even with use of multicomponent models (Jebb et al. 1993). This chapter addresses the validity of reference and field methods for assessing change in body composition.

Reference Methods

Multicomponent models provide the highest level of accuracy for making initial body composition assessments (see chapter 3) and tracking changes

in body composition. In a unique study that confined clients to a whole-body calorimeter and over- or underfed them for 12 days to induce body composition changes, Jebb et al. (1993) compared models and validated reference and field methods against fat balance estimated from the oxidation of fat. Three-component models had the smallest bias and best precision for estimating changes in body fat. The accuracy of 2-C reference methods (i.e., hydrodensitometry [HD] and TBW) and field methods (i.e., SKF and bioelectrical impedance [BIA]) was less. Heitmann et al. (1994) reported that the TBW method (2-C model) was more accurate for estimating changes in FFM_{4-C} of clients participating in a 12-week weight reduction program than SKF, BIA, BMI, and near-infrared interactance (NIR) methods.

In a study comparing the validity of body composition models for measuring changes in body composition of overweight women participating in a 16-week diet and exercise program, Evans et al. (1999) reported that 2-C and 3-C models accurately reflected the overall group changes in %BF as measured by a 4-C model. On average, the diet-only group lost 7 kg, and the diet-exercise group lost 4 kg; the average FFB density (FFBd) of the groups did not differ significantly from 1.100 g/cc. However, the authors noted that individual variation in FFBd during weight loss is substantial and recommended using a multicomponent model to obtain reference measures for assessing changes in the body composition of individuals. Also, given that many diseases affect the composition and density of the FFB (see chapters 12-14), multicomponent models should be used, whenever possible, to track changes in body composition.

In lieu of a multicomponent model approach, dual-energy X-ray absorptiometry (DXA) may be a suitable reference method for assessing body composition changes. For women with obesity who lost an average of 12 kg over 1 yr, DXA and whole-body potassium counting (TBK) yielded similar estimates of change in their FFM (1.3 and 0.9 kg, respectively) (Hendel et al. 1996). Also, given that DXA is not greatly affected by changes in hydration status (see chapter 3), it may be a useful method for monitoring changes in fat and FFM under conditions of fluid alteration. For example, large hydration changes caused only minimal changes in DXA estimates of fat and bone mineral following dialysis (Abrahamsen et al. 1996). In contrast, Nelson and colleagues (1996) noted that DXA is not sensitive enough to detect changes in the soft tissue of postmenopausal women following 1 yr of strength training. While DXA and neutron activation analysis showed no significant changes in FFM due to training, other methods (i.e., HD, TBW, urinary creatinine, TBK, and regional computed tomography) indicated significant increases in FFM. In contrast, Houtkooper and colleagues (2000) reported that DXA is more sensitive to increases in FFM with strength training than either HD or a multicomponent model.

Field Methods and Prediction Equations

The accuracy and precision of field methods to monitor and estimate changes in body composition are debatable. This section addresses the utility of commonly used field methods (SKF, anthropometry, BIA, and NIR) for tracking body composition change.

SKF Method

Several researchers noted that the SKF method is reasonably accurate for tracking changes in body composition that accompany exercise training (Broeder et al. 1997; Ross et al. 1989). Broeder et al. (1997) reported acceptable pre- and posttesting SEEs (<3.0% BF) and small mean differences between the SKF and HD methods for monitoring changes in FM and FFM of men participating in a 12-week endurance or resistance training program.

Acceptable prediction errors (SEE = 3.1 kg) were also reported for estimating the average FFM of overweight men both before and after a 10-week weight management program that included a combination of exercise training and a moderately restricted (1,400-1,800 kcal/day) diet (Ross et al. 1989). Additionally, although the absolute values differed, the increase in FM observed in recovering anorexia nervosa patients was similar for the SKF (8.9 kg) and DXA (8.3 kg) methods (Scalfi et al. 2002). The SKF method has been recommended over other field methods such as BIA for tracking body composition changes (Broeder et al. 1997; Jebb et al. 1993); however, keep in mind that the individual errors can be substantial (Friedl et al. 2001; Heitmann et al. 1994; Jebb et al. 1993).

Although the SKF method may be useful for estimating changes in lean and normal-weight clients on diet-exercise programs, it is not recommended for assessing body composition changes of clients who are obese. It is difficult to accurately measure SKF thicknesses of individuals who are obese (see chapter 13). Also, this method assumes that the ratio of internal to subcutaneous fat is constant within and between individuals. However, in clients with obesity who undergo rapid and substantial weight loss, there may be a disproportionate decrease in internal fat compared to subcutaneous fat (Scherf et al. 1986). Furthermore, relative decreases in SKF thickness during weight loss may not be the same at various sites (Ross, Leger et al. 1991). Thus, changes in SKF thickness are not highly correlated (r = .02-.38) with weight loss (Bray et al. 1978).

The SKF method is not recommended for assessing the body composition changes in clinical patients who may have a FFB composition that differs from the reference body. In healthy participants, the group and individual prediction errors for the SKF method in detecting change in subcutaneous adipose tissue, as measured by magnetic resonance imaging (MRI), were, respectively, 0.8 kg and –1.6 to 1.8 kg; however, this error doubled (SEE = 1.5 kg, 95% limits of agreement = –3.4 to 2.6 kg) when assessing change in the subcutaneous adipose tissue of HIV-infected patients (Andrade et al. 2002). Furthermore, the sum of SKF thicknesses, typically an indicator of fat, was reduced by 5% following hemodialysis in which fluid, but no fat, was removed (Abrahamsen et al. 1996).

The SKF method should not be used to monitor changes in body composition during pregnancy. Using a 4-C model, Paxton et al. (1998) reported an increase in the relative hydration of the FFM (TBW/FFM = 73.8-75.7%) and a decrease in FFBd (1.100-1.091 g/cc) between weeks 14 and 37 of gestation; commonly used anthropometric and SKF equations based on the 2-C model overestimated the increase in body fat (~0.9 kg error) that occurs with pregnancy. Therefore, the authors derived an anthropometric prediction equation to estimate the fat change (Δfat) that occurs during this time of pregnancy [Δfat in kilograms = 0.84 (body weight in kilograms) – 6.49]. Adding thigh SKF did not significantly improve this equation (r^2 = .71 vs. .73).

Anthropometric Method

There is little research documenting the use of anthropometric equations for assessing changes in body composition. Compared to SKFs, circumference measures are more highly related (r = 0.40-0.83) to weight loss (Bray et al. 1978). Friedl et al. (2001) evaluated the accuracy of several anthropometric and SKF equations in estimating changes in body composition of women engaging in an 8-week endurance training program. They reported that the predictive accuracy of the U.S. Navy circumference equation (Hodgdon & Beckett 1984), which includes waist and hip circumferences, is similar to that of the Durnin and Womersley (1974) SKF equation and better than other anthropometric equations in estimating body composition changes. All of the equations evaluated in the study, however, had poor sensitivity (12-55%), indicating the potential for large individual error when they are used to assess change.

Although individual circumferences reflect weight loss, the waist-to-hip ratio (WHR) does not change in response to rapid weight loss in individuals with upper-body obesity (WHR > 0.80 for women, WHR > 0.95 for men). Therefore, WHR should not be used in clinical settings to assess changes in adipose tissue distribution during acute weight loss (Ross et al. 1991). In addition, BMI does not adequately reflect changes in FFM during weight loss. Van der Kooy et al. (1992) reported that Deurenberg, Weststrate et al. (1991) age- and gender-specific BMI equations significantly overestimated FFM loss in obese women and men consuming a hypocaloric diet. Also, Friedl et al. (2001) reported a mean increase in BMI (0.2 kg/m^2), which would translate into an estimated increase in body fat with use of BMI equations, even though the participants in their study lost an average of 1.2 kg of body fat after 8 weeks of basic military training. The increase in BMI was due to increases in body weight (0.8 kg) and FFM (2.0 kg).

BIA Method

There is much debate about the validity of the BIA method for accurately estimating changes

in FFM. In chapter 6 we stressed the necessity of maintaining normal hydration levels; BIA estimates of body composition will be inaccurate when hydration status is altered. Following an exercise protocol that reduced body weight by 4% to 5% from sweating, BIA underestimated TBW loss by –1.4 L (O'Brien et al. 1999), and acute changes in hydration status (hypohydration, rehydration, and hyperhydration) significantly changed the BIA estimate of %BF (Demura, Yamaji, Goshi, & Nagasawa 2002; Saunders et al. 1998). Also, the BIA method is not suitable for monitoring acute changes in TBW that occur with some clinical treatments such as dialysis (Abrahamsen et al. 1996; Dumler et al. 1992) or **paracentesis** (surgical puncture of the abdominal cavity for the aspiration of peritoneal fluid). Most of the fluid removed with paracentesis is from the abdomen. Given that most of the total body resistance measured by single-frequency BIA comes from the arms and legs and not the trunk, this method underestimates TBW loss by paracentesis; the degree of inaccuracy increases with the amount of ascitic fluid removed (McCullough et al. 1991; Zillikens, van den Berg, Wilson, & Swart 1992).

Many studies have been done to determine if the BIA method is sensitive enough to detect body composition change as a result of weight loss interventions. Compared to HD, leg-to-leg bioimpedance equations (Tanita, model TBF 105) accurately estimated average decreases in FM in groups of obese women participating in either a diet-only, exercise-only, or diet-exercise program for 12 weeks; however, the individual accuracy of the equations was not reported (Utter et al. 1999). The validity of the BIA method for assessing change in FFM with weight loss may be dependent on the prediction equation (Hendel et al. 1996; Kushner et al. 1990; Paton et al. 1997). Ross et al. (1989) reported that the BIA equations of Lukaski (1987) and Segal et al. (1988) accurately estimated the average FFM of slightly obese men before and after weight loss, and Kushner et al. (1990) noted that their BIA equation (Kushner & Schoeller 1986) accurately estimated changes in FFM and TBW in obese women on very low caloric diets. However, many other researchers have concluded that most BIA equations do not adequately reflect changes in FFM during weight loss (Deurenberg, Weststrate et al. 1989; van der Kooy et al. 1992; Vazquez & Janosky 1991). Hendel et al. (1996) evaluated the accuracy of 11 BIA equations for assessing the change in FFM in obese women who lost an average of 12 kg over a year. Although the limits of agreement were large for most of the equations, the authors noted that the change in resistance may be indicative of change in FFM. However, Forbes et al. (1992) reported that losses in FFM appear to be more highly related to changes in body weight (r = .69) than total body resistance (r = .56). In fact, resistance may increase, decrease, or stay the same during weight loss, depending on the ratio of change in FFM and body weight (ΔFFM/ΔBW).

Because of its ease of use, BIA has also been considered as a method for tracking change in clinical populations. In HIV patients, changes in FFM greater than 3% were detected 95% of the time, and changes greater than 5% were detected 100% of the time using BIA (Kotler et al. 1996). This research group also reported that single-frequency BIA and a parallel transformation of resistance and reactance (see chapter 6) could be used to accurately detect a change in body cell mass (measured by TBK; r = .78, SEE = 1.2 kg or about 4.4%) in individuals infected with HIV (Kotler, Rosenbaum et al. 1999).

In patients recovering from eating disorders, the accuracy of a BIA equation developed for anorexia nervosa was tested. The error of this population-specific BIA equation (SEE = 2.6% BF) was not much better than that of simple anthropometric measurements (SEE = 2.7%) in predicting the %BF$_{DXA}$ of patients recovering from eating disorders (Hannan et al. 1993). Additionally, changes in body fat measured by the SKF and BIA methods were poorly related (r = 0.31) in anorexia nervosa patients undergoing refeeding (Birmingham et al. 1996). Because of fluid shifts during recovery from anorexia nervosa, whole-body, single-frequency impedance should not be used to assess body composition changes in this population. Phase angle, however, may be useful in monitoring such changes. Phase angle is significantly related to gains in body weight (r = .66), BMI (r = .75), FFM (r = .53), and FM (r = .62) (Scalfi et al. 1999). Whole-body phase angle increases with weight recovery in anorexic patients (Polito et al. 1998; Scalfi et al. 1999). Also, multifrequency BIA (MFBIA) may be a valuable tool for monitoring body composition changes during recovery from anorexia nervosa. Using MFBIA to estimate water compartments, Wotton, Trocki et al. (2000) noted that changes in TBK are related to intracellular water but not extracellular water in recovering anorexics.

NIR

Kalantar-Zadeh and colleagues (2001) suggested using the NIR method to assess body composition changes in dialysis patients because of its reliability and significant relationship with anthropometric changes over time; however, no reference body composition methods were used in their study. Other researchers have not been supportive of using NIR as a method for tracking change. Heitmann et al. (1994) reported that the prediction error for the NIR method in measuring changes in FFM of participants in a weight loss program was similar to that for other field methods but too large to be of clinical use. Furthermore, although the average FFM estimated from TBW increased by 0.4 kg, the average change in FFM estimated by NIR was negative (–0.4 kg). Similarly, Broeder et al. (1997) compared NIR and HD estimates of FFM change in men engaging in a 12-week resistance training program. The NIR method indicated a decline in the FFM (–0.5 kg), whereas the HD method showed an increase in average FFM (2.5 kg). Additionally, the change in FFM as estimated by NIR was not significantly related to the change in FFM from HD.

Summary

Because the composition of the FFB is likely to be altered if there is a significant change in body weight, we recommend using a multicomponent model whenever possible to obtain reference measures for tracking changes in body composition. Field methods may be able to provide a reasonable average estimate of body composition change for some groups; however, the SKF method should not be used with the obese population. Also, the BIA method should not be used in clinical populations with altered hydration levels.

In short, it appears that field methods are not precise enough to offer accurate estimates of individual changes in body composition. Still, we recommend taking anthropometric measures of your clients periodically to create an anthropometric profile (see chapter 5). These measures can be used to track changes in a client's body profile over time. Finally, to maximize reliability and sensitivity when tracking change in body composition, we recommend that the same equipment, testing procedures, and technician be used each time the client is assessed.

Key Points

- Multicomponent models offer the greatest accuracy for monitoring changes in body composition.

- DXA is a potential alternative to the multicomponent model for tracking change in FFM.

- The SKF method may provide reasonable group estimates of body composition change for average to lean participants in diet or exercise programs.

- The SKF method should not be used to monitor changes in the body composition of clients who are obese or patients with an altered FFB composition.

- Anthropometric indices such as BMI and WHR do not adequately reflect changes in FM or FFM.

- The BIA method should not be used to assess body composition change when the fluid distribution or hydration status of the client is acutely altered.

- Using alternative BIA approaches (e.g., parallel model) or monitoring less commonly used bioimpedance measures (i.e., phase angle) may increase the utility of the BIA method as a tool for assessing body composition changes in clinical populations.

- There is no research to support the use of NIR to monitor the body composition change of clients on diet and exercise programs.

- There is the potential for substantial individual error when using field methods to track body composition change.

- SKFs and circumferences can be used to monitor changes in your client's anthropometric profile.

Key Term

Learn the definition for the following key term. Definitions of terms can be found in the Glossary, page 227.

paracentesis

Review Questions

1. Which body composition models and reference methods are recommended for assessing change in body composition?

2. Why is the SKF method not recommended for monitoring body composition changes of persons with obesity who are participating in weight loss programs?

3. Explain why simple measures such as body weight and BMI can be misleading if they are used to assess changes in body composition.

4. Explain how SKFs and anthropometry can best be used to track body composition change.

5. What is the greatest limitation of using the BIA method to monitor body composition change?

6. Describe how the traditional, single-frequency BIA method may be modified to track changes in body composition of clinical populations.

DERIVATION OF CONSTANTS FOR 2-C MODEL CONVERSION FORMULAS

Calculation of FFB density (FFB$_d$)

For molecular models, the fat-free body (FFB) consists of total body water, mineral, and protein. The following table lists the assumed proportions and densities of these FFB components for the Siri (1961) and Brozek et al. (1963) 2-C models:

FFB component	% of FFB	Density (g/cc)
Water	73.8	0.9937
Protein	19.4	1.34
Mineral	6.8	3.038
Overall	100.0	1.100

Given these assumed proportions with their respective densities, the overall FFB density (FFB$_d$) is calculated as follows:

$$1.00 / \text{FFB}_d = \%\text{water}/d_{water} + \%\text{protein}/d_{protein} + \%\text{mineral}/d_{mineral}$$

[Eq. 1]

To use this equation:

1. Convert the percentage (%) to a decimal (e.g., 6.8% = 0.068).

2. Substitute values into the equation and divide each decimal by its respective density.

$$1.00 / \text{FFB}_d = 0.738/0.9937 + 0.194/1.34 + 0.068/3.038$$

$$1.00 / \text{FFB}_d = 0.74278 + 0.1448 + 0.0224$$

$$1.00 / \text{FFB}_d = 0.9099$$

3. Solve for FFB$_d$ by dividing 1.00 by the sum.

$$\text{FFB}_d = 1.00/0.9099 = 1.099 \text{ or } 1.10 \text{ g/cc}$$

Siri 2-C Model Conversion Formula

To derive 2-C model formulas for converting Db to %BF, you can use the following generic equation provided you know the densities of the FFB (FFB$_d$) and fat (fat$_d$) components. For the Siri 2-C model conversion formula the assumed densities for the FFB and fat components are, respectively, 1.10 g/cc and 0.900 g/cc.

$$\%\text{BF} = 1 / \text{Db} [\text{FFB}_d \times \text{fat}_d / (\text{FFB}_d - \text{fat}_d)] - [\text{fat}_d / (\text{FFB}_d - \text{fat}_d)]$$

[Eq. 2]

Substituting assumed densities into the equation yields:

$$\%\text{BF} = 1 / \text{Db} [1.10 \times 0.900 / (1.10 - 0.900)] - [0.900 / (1.10 - 0.900)]$$

$$\%\text{BF} = 1 / \text{Db} [0.9900 / 0.20] - [0.900 / 0.20]$$

$$\%\text{BF} = 1 / \text{Db} [4.95] - 4.50$$

$$\%\text{BF} = [4.95 / \text{Db} - 4.50] \times 100$$

2-C Model Population-Specific Conversion Formulas

Using multicomponent model methods (e.g., HD, DXA, and D$_2$O) to measure total body density (HD), total body water (D$_2$O), and total body mineral (DXA), the relative proportions of the three components of the FFB (i.e., water, mineral, and protein) can be calculated (protein is estimated by subtracting the %water and %mineral from 100). Average values of %FFB$_{water}$, %FFB$_{mineral}$, and %FFB$_{protein}$ can be substituted into Eq. 1 to estimate the average overall FFB$_d$ of specific population subgroups. This value, along with assumed value for the density of fat (i.e., 0.9007 g/cc) can be substituted into Eq. 2 to generate a population-specific conversion formula for a specific group. We derived many of the 2-C model population-specific conversion formulas in table 1.4 (see page 9) using this method.

For example, the average $FFB_d \cong 1.106$ g/cc and $fat_d = 0.9007$ g/cc for black men. Substituting these values into Eq. 2 yields a 2-C model population-specific conversion formula for black men:

$$\%BF = 1 / Db \, [1.106 \times 0.9007 / (1.106 - 0.9007)]$$
$$- 0.9007 / (1.106 - 0.9007)$$

$$\%BF = [4.86 / Db - 4.39] \times 100$$

APPENDIX B

FIELD METHOD AND PREDICTION EQUATION FINDERS

B.1 Field Method and Prediction Equation Finder for Healthy Adults

Ethnicity	Gender	Age	SKF	Anthro	BIA	NIR	Table
African American	Male	18-61	Jackson				10.1, Eq. 1, p. 140
	Female	18-55	Jackson				10.1, Eq. 2, p. 140
	Male	17-62			Segal		10.1, Eq. 5, 6, 7, p. 140
	Female	17-62			Segal		10.1, Eq. 8, 9, 10, p. 140
American Indian	Female	18-60	Hicks			Hicks	10.2, Eq. 1, 7, p. 144
	Female	17-62			Segal or Stolarczyk		10.2, Eq. 3, 4, 5, 6, p. 144
Asian	Male	24-43	Eston				10.3, Eq. 1, p. 148
	Male	18-27			Demura		10.3, Eq. 4, p. 148
	Female	18-54			Nakadomo		10.3, Eq. 2, p. 148
	Obese female	18-68			Tanaka		10.3, Eq. 3, p. 148
Caucasian	Male	18-61	Jackson		Segal		10.4, Eq. 1, 3, 4, 5, p. 151
	Male	18-40		Wilmore & Behnke			10.4, Eq. 10, p. 151
	Female	18-62	Jackson		Segal		10.4, Eq. 2, 6, 7, 8, p. 151
	Female	20-72				Heyward	10.4, Eq. 9, p. 151
	Female	15-79		Tran & Weltman			10.4, Eq. 11, p. 151
Hispanic	Male	18-62			Segal		10.5, Eq. 2, 3, 4, p. 154
	Female	18-62	Jackson		Segal		10.5, Eq. 1, 5, 6, 7, p. 154

B.2   Field Method and Prediction Equation Finder for Children and Older Adults

	Ethnicity	Gender	Age	SKF	Anthro	BIA	NIR	Table
CHILDREN	African American	Male	8-17	Slaughter				8.3, Eq. 1, 3, 5, p. 118
		Female	8-17	Slaughter				8.3, Eq. 2, 4, 6, p. 118
	American Indian	Male and female	7-11	Lohman				10.2, Eq. 2, p. 144
	Asian	Male	9-14			Kim		8.3, Eq. 9, p. 118
		Female	9-15			Watanabe		8.3, Eq. 10, p. 118
	Caucasian	Male	8-17	Slaughter				8.3, Eq. 1, 3, 5, p. 118
		Female	8-17	Slaughter				8.3, Eq. 2, 4, 6, p. 118
		Male and female	5-10			Kushner		8.3, Eq. 7, p. 118
		Male and female	10-19			Houtkooper		8.5, Eq. 8, p. 118
OLDER ADULTS	Caucasian	Male and female	22-94			Kyle		9.3, Eq. 2, p. 132
		Male and female	65-94			Baumgartner		9.3, Eq. 1, p. 132
		Male	15-78		Tran & Weltman			9.3, Eq. 1, p. 132
		Female	15-79		Tran & Weltman			9.3, Eq. 3, p. 132

B.3   Field Method and Prediction Equation Finder for Athletes

Sport	Gender	Age	SKF	Anthro	BIA	NIR	Table
All sports	Male	>18	Jackson				11.4, Eq. 1, p. 168
	Male	14-19	Lohman				11.4, Eq. 3, p. 168
	Female	>18	Jackson		Fornetti	Fornetti	11.4, Eq. 2, 4, 7, p. 168
	Female	11-19	Jackson				11.4, Eq. 2, p. 168
Gymnastics	Female	13-17	Jackson		Van Loan		11.4, Eq. 2, 6, p. 168
Elite distance runners	Female	16-37			Hannan		11.4, Eq. 5, p. 168
Wrestling	Male	14-22	Lohman				11.4, Eq. 3, p. 168
	Male	14-18				Oppliger	11.4, Eq. 8, p. 168

B.4　Field Method and Prediction Equation Finder for Clinical Populations

	Disease	Gender	Age	SKF	Anthro	BIA	NIR	Table and equation
CARDIOPULMONARY DISEASE	Cardiac	Male	46-82		Tran & Weltman			12.1, Eq. 1, p. 181
	Heart and lung transplants	Male and female	49*			Kyle		12.1, Eq. 2, p. 181
	Chronic obstructive pulmonary disease	Male and female	45-86			Kyle		12.1, Eq. 3, p. 181
	Cystic fibrosis	Male	27*			Kyle		12.1, Eq. 4, p. 181
		Female	34*			Kotler		12.1, Eq. 5, p. 181
		Male	10-18	Durnin & Rahaman or Slaughter				12.1, Eq. 6, 8, p. 181
		Female	10-18	Durnin & Rahaman or Slaughter				12.1, Eq. 7, 9, p. 181
METABOLIC	Obesity	Male	24-68		Weltman			13.1, Eq. 1, p. 189
		Male	17-62			Segal		13.1, Eq. 3, p. 189
		Female	20-60		Weltman			13.1, Eq. 2, p. 189
		Female	17-62			Segal		13.1, Eq. 4, p. 189
WASTING	Anorexia nervosa	Female	15-44			Hannan		14.1, Eq. 1, p. 198
	HIV/AIDS	Male	40*			Kotler		14.1, Eq. 2, p. 198
		Female	40*			Kotler		14.1, Eq. 3, p. 198
		Male	37*			Paton		14.1, Eq. 4, p. 198
	Cirrhosis	Male and female	30-76			Zillikens & Conway		14.1, Eq. 6, p. 198
	Kidney/ Dialysis	Male and female	4-22			Wuhl		14.1, Eq. 5, p. 198
	Spinal cord injury	Male and female	45.4*			Vache		14.1, Eq. 7, p. 198
		Male and female	18-73			Kocina & Heyward		14.1, Eq. 8, p. 198

*Average age.

SOURCES FOR BODY COMPOSITION EQUIPMENT

Product	Supplier's Address
AIR DISPLACEMENT PLETHYSMOGRAPH	
Bod Pod Body Composition System	LMI, Inc. 1980 Oliveri Rd. Ste. C Concord, CA 94520 800-426-3763 www.bodpod.com
ANTHROPOMETERS	
Spreading calipers Sliding calipers Standard skeletal anthropometer	Pfister Import-Export, Inc. 450 Barell Ave. Carlstadt, NJ 07072 202-939-4606 Rosscraft Industries 14732 16A Ave. Surrey, BC Canada V4A 5M7 604-531-5049 tep2000.com/Rosscraft.htm
Anthropometric tape measure	Country Technology, Inc. P.O. Box 87 Gays Mills, WI 54631 608-735-4718 www.fitnessmart.com
BIOIMPEDANCE ANALYZERS	
American Weights & Measures (formerly Valhalla Scientific)	American Weights & Measures, Inc. 16501 Zumaque St. Rancho Santa Fe, CA 92067 800-395-4565 http://amerweights.com
Bio-Analogics	Bio-Analogics 7909 SW Cirrus Dr. Beaverton, OR 97008 800-327-7953 www.bioanalogics.com
Biodynamics	Biodynamics Corp. 3511 NE 45th St. #2 Seattle, WA 98105 800-869-6987 www.biodyncorp.com

Product	Supplier's Address
BIOIMPEDANCE ANALYZERS (continued)	
Biospace	Biospace Co., Ltd. 363 Yangjae Dong Seocho Gu, Seoul 137-898, Korea +82-2-501-3939 www.biospace.co.kr
Bodystat	Bodystat USA Inc. 2 Adalia Ave., Ste. 401 Tampa, FL 33606 813-258-3570 www.bodystat.com
Data-Input	Data Input GmbH Trakehner St. 5 60487 Frankfurt, Germany Tel. +49 69-970 840-0 www.b-i-a.de
Holtain	Holtain, Ltd. Crosswell, Crymych Pembrokeshire, SA41 3UF, United Kingdom +44 0 1239-891656 www.fullbore.co.uk/holtain/medical
Impedimed (distributors of SEAC BIA analyzers)	Impedimed Pty Ltd P.O. Box 2121 Mansfield, Queensland 4122, Australia +61 07 3849 3444
Maltron	Maltron International Ltd. P.O. Box 15 Rayleigh, Essex, SS6 9SN, United Kingdom +44 0 1268 778251 www.maltronint.com
OMRON	OMRON Healthcare, Inc. 300 Lakeview Pkwy. Vernon Hills, IL 60061 847-680-6200 www.omronhealthcare.com
RJL	RJL Systems 33955 Harper Ave. Clinton Twp., MI 48035 800-528-4513 www.rjlsystems.com
Tanita	Tanita Corp. 2625 S. Clearbrook Dr. Arlington Heights, IL 60005 800-TANITA-8 www.tanita.com
Valhalla	See American Weights & Measures
Xitron Hydra ECF/ICF	Xitron Technologies, Inc. 9770-A Carroll Centre Rd. San Diego, CA 92126 858-530-8099 www.xitron-tech.com

Product	Supplier's Address
CALIBRATION INSTRUMENTS/SUPPLIES	
Skinfold calibration blocks 15 mm	Creative Health Products 7621 East Joy Rd. Ann Arbor, MI 48105 800-742-4478 www.chponline.com
Standard calibration weights	Ohaus Scale Corp. 29 Hanover Rd. Florham, NJ 07932 800-672-7722 www.ohaus.com
Vernier caliper	L.S. Starrett Co. 121 Crescent St. Athol. MA 01331 978-249-3551 www.lsstarrett.com
DUAL-ENERGY X-RAY ABSORPTIOMETERS	
Hologic	Hologic, Inc. 35 Crosby Dr. Bedford, MA 01730 781-999-7300 www.hologic.com
Norland	Norland Medical Systems, Inc. W6340 Hackbarth Rd. Fort Atkinson, WI 53538 800-563-9504 www.norland.com
Lunar	GE Lunar Medical Systems P.O. Box 414 Milwaukee, WI 53201 608-274-2663 www.gemedicalsystems.com
NEAR-INFRARED INTERACTANCE	
Futrex NIR analyzers	Zelcore, Inc. 130 Western Maryland Pkwy. Hagerstown, MD 21740 800-576-0295 www.zelcore.com
SCALES	
Chatillon underwater weighing scale Detecto balance beam scale Health-O-Meter balance beam scale Health-O-Meter digital scale Seca digital scale	Creative Health Products 7621 East Joy Rd. Ann Arbor, MI 48105 800-742-4478 www.chponline.com

Product	Supplier's Address
SKINFOLD CALIPERS	
Adipometer	Ross Products Division Abbot Laboratories 625 Cleveland Ave. Columbus, OH 43216 800-344-9739 www.ross.com
Body Caliper	The Caliper Company 7 Millside Lane Mill Valley, CA 94941 800-655-4960 www.bodycaliper.com
Fat-Control Fat-O-Meter Lafayette Skyndex Slim Guide	Creative Health Products 7621 East Joy Rd. Ann Arbor, MI 48105 800-742-4478 www.chponline.com
Harpenden	Quinton Instruments 3303 Monte Villa Pkwy. Bothell, WA 98021 800-426-0347 www.quinton.com
Holtain	Pfister Import-Export, Inc. 450 Barell Ave. Carlstadt, NJ 07072 201-939-4606
Lange	Cambridge Scientific Products 26 New St. Cambridge, MA 21613 888-354-8908 www.cambridgescientific.com
McGaw	McGaw, Inc. P.O. Box 19791 Irvine, CA 92713 714-660-2055 www.mcgaw.com
STADIOMETERS	
Harpenden stadiometer Holtain stadiometer	Pfister Import-Export, Inc. 450 Barell Ave. Carlstadt, NJ 07072 201-939-4606

Abbreviations and Symbols

Symbol	Term
	GENERAL
%BF	Relative body fat; percent body fat
ADP	Air displacement plethysmography
AIDS	Acquired immunodeficiency syndrome
BCM (kg)	Body cell mass
BIA	Bioelectrical impedance analysis
BIS	Bioelectrical impedance spectroscopy
BMC (g)	Bone mineral content
BMD (g/cm^2)	Bone mineral density
BSA	Body surface area
BM (kg)	Body mass
BV (L)	Body volume
BW (kg)	Body weight
CAD	Coronary artery disease
COPD	Chronic obstructive pulmonary disease
CF	Cystic fibrosis
CT	Computerized tomography
d (g/cc)	Density
Db (g/cc)	Total body density
D$_2$O	Deuterium oxide
DM	Diabetes mellitus
DPA	Dual-photon absorptiometry
DXA	Dual-energy X-ray absorptiometry
ECW	Extracellular water
FFB or FFM (kg)	Fat-free body or fat-free mass
FFBd (g/cc)	Fat-free body density
FM (kg)	Fat mass
FRC (L)	Functional residual capacity
GV (L)	Gastrointestinal volume
HD	Hydrodensitometry
HIV	Human immunodeficiency virus
HT (cm)	Height

Symbol	Term
HW	Hydrostatic weighing
^{3}H$_2$O	Tritium oxide
H$_2$^{18}O	Oxygen-18
ICW	Intracellular water
IDDM	Insulin-dependent diabetes mellitus
^{40}K	Potassium isotope
LBM (kg)	Lean body mass
LTM	Lean tissue mass
MC	Multicomponent model
MD	Muscular dystrophy
MRI	Magnetic resonance imaging
MS	Multiple sclerosis
MW (kg)	Minimal body weight
NAA	Neutron activation analysis
NIDDM	Non-insulin-dependent diabetes mellitus
NIR	Near-infrared interactance
RV (L)	Residual lung volume
SCI	Spinal cord injury
SKF (mm)	Skinfold
STM (kg)	Soft-tissue mass
TBBM (kg)	Total body bone mineral
TBK (kg)	Total body potassium
TBM (kg)	Total body mineral
TBW (L)	Total body water
TGV (L)	Thoracic gas volume
TLC (L)	Total lung capacity
UWW	Underwater weighing
2-C	Two-component model
3-C	Three-component model
4-C	Four-component model
6-C	Six-component model
	SKINFOLD METHOD
ΣSKF (mm)	Sum of skinfolds
Σ2SKF (mm)	Sum of two skinfolds
Σ3SKF (mm)	Sum of three skinfolds
Σ4SKF (mm)	Sum of four skinfolds
Σ7SKF (mm)	Sum of seven skinfolds

BIA METHOD	
MFBIA	Multifrequency bioelectrical impedance analysis
SBIA	Segmental bioelectrical impedance analysis
HT^2/R	Resistance index
HT^2/Z	Impedance index
ρ	Specific resistivity
R (Ω)	Resistance
ϕ	Phase angle
X_C (Ω)	Reactance
Z (Ω)	Impedance
NIR METHOD	
OD	Optical density
ΔOD_2	[$OD_{2Standard} - OD_{2Biceps}$]
ANTHROPOMETRY	
Ab C (cm)	Average abdominal circumference
Arm C (cm)	Arm circumference
Bi-iliac D (cm)	Bi-iliac diameter
BMI (kg/m^2)	Body mass index
C (cm)	Circumference
D (cm)	Bony diameter or bony width
Forearm C (cm)	Forearm circumference
Hip C (cm)	Hip circumference
Knee C (cm)	Knee circumference
Neck C (cm)	Neck circumference
Thigh C (cm)	Proximal thigh circumference
SAD (cm)	Sagittal abdominal diameter
WHR	Waist-to-hip ratio
MEASUREMENT UNITS	
cc	cubic centimeter
cm	centimeter
g	gram
g/cc	grams/cubic centimeter

in.	inch
Kcal	kilocalorie
kg	kilogram
L	liter
lb	pound
kHz	kilo-Hertz
MHz	Mega-Hertz
m	meter
mm	millimeter
nm	nanometer
yr	year
°C	degrees in Celsius
μA	microampere
μL	microliter
Ω	ohm
STATISTICAL TERMS	
Σ	Sum of scores
a	y-intercept of regression line
b	Slope of regression line
CE	Constant error or bias
CI	Confidence interval
CV	Coefficient of variation
LA	95% limits of agreement
M	Mean or average
N	Sample size
R_{mc}	Multiple correlation coefficient
$r_{x,y}$	Pearson product-moment correlation coefficient
$r_{x,y}^2$	Coefficient of determination
$r_{y,y'}$	Validity coefficient
SEE	Standard error of estimate
TE	Total error

Glossary

abdominal obesity–Excessive subcutaneous and visceral fat in the abdominal region; upper-body obesity.

acquired immunodeficiency syndrome (AIDS)–Severe immunosuppression caused by HIV, making individual susceptible to opportunistic infections that do not normally cause disease in people with healthy immune systems.

adiabatic conditions–Occurring without gain or loss of heat.

adipose tissue–Fat (~83%) plus its supporting structures (~2% protein and ~15% water).

air displacement plethysmography (ADP)–Densitometric method for estimating body volume using air displacement and pressure-volume relationships.

amenorrhea–Absence of three or more consecutive menstrual cycles after menarche.

android obesity–Type of obesity in which excess fat is localized on the trunk and abdomen; upper-body obesity; apple-shaped body.

angina pectoris–Chest pain.

anorexia nervosa–A psychological disorder characterized by distorted body image, extreme fear of gaining weight, and refusal to maintain normal body weight; an eating disorder characterized by food restriction, prolonged fasting, and use of diet pills, diuretics, and laxatives.

anthropometric profile–Somatogram graphically depicting the pattern of muscle and fat distribution of the body.

anthropometry–Measurement of body size and proportions including skinfold thicknesses, circumferences, bony widths and lengths, stature, and body weight.

Archimedes' principle–Principle stating that weight loss under water is directly proportional to the volume of water displaced by the body's volume.

arrhythmias–Abnormal heart rhythms.

ascites–Effusion and accumulation of body water in the abdominal cavity, common in patients with cirrhosis.

athletic amenorrhea–Exercise-induced cessation of menstrual periods (three or fewer periods per year).

attenuation–Weakening or lessening of an amount.

bias–Average difference between the measured and predicted values for the cross-validation group; constant error.

bioelectrical impedance analysis (BIA)–Body composition method used to estimate total body water or fat-free mass through measurement of the conductance of low-level electrical current through the body.

bioelectrical impedance spectroscopy (BIS)–A BIA approach that uses multiple frequencies to assess extracellular, intracellular, and total body water; see multifrequency BIA.

bivariate regression–Statistical method used to predict one variable (Y) from another variable (X).

Bland and Altman method–A statistical approach used to assess the degree of agreement between methods by calculating the 95% limits of agreement and confidence intervals.

Bod Pod–Large, egg-shaped fiberglass chamber used to measure body volume from air displacement; air displacement plethysmograph.

body cell mass (BCM)–Metabolically active tissues of the body.

body density (Db)–Overall density of the fat, water, mineral, and protein components of the human body; total body mass expressed relative to total body volume.

body mass–Measure of size of the body; body weight.

body mass index (BMI)–Crude index of obesity; the ratio of body weight to height squared.

body surface area–Amount of surface area of the body estimated from client's height and body weight.

body volume (BV)–Measure of body size estimated by water or air displacement.

bone-free lean tissue mass (LTM)–In DXA, a measure of the bone-free lean tissues of the body.

bone mineral density (BMD)–Bone (osseous) mineral content expressed relative to the cross-sectional area of the bone, that is, in g/cm^2.

bootstrap technique–An internal cross-validation approach in which data for each subject in a sample are excluded one at a time and regression analysis is performed to generate prediction equation; the value for each omitted subject is predicted, and a PRESS residual score is calculated for each subject; the accuracy of the prediction equation is evaluated by measuring the sum of squares of all PRESS residuals; also known as PRESS technique.

Boyle's law–Isothermal gas law stating that volume and pressure are inversely related.

cachexia–Wasting syndrome resulting in loss of body weight and lean body tissues.

cadaver analysis–Reference method used to determine the water, mineral, lipid, and protein content of the human body.

calibration block–Block with a fixed, known width (e.g., 10, 15, and 20 mm) used to check the measurement accuracy of skinfold calipers and small sliding calipers.

cancer–Unregulated or uncontrolled growth of a cell that can develop in any tissue of the body, invade adjacent cells, and spread throughout the body.

capacitance–Storage of voltage for a brief moment in time.

central adiposity–Fat deposition in the abdominal region.

chronic bronchitis–Persistent inflammation of mucous membranes of the bronchi.

chronic kidney failure—A progressive decline in kidney function, resulting in the accumulation of metabolic waste products in the blood; typically treated by dialysis.

chronic obstructive pulmonary disease (COPD)—Condition in which airflow to lungs is obstructed.

circumference—A measure of the girth of body segments.

cirrhosis—A scarring of liver tissue, leaving it nonfunctional; disease caused by alcohol abuse, hepatitis, bile duct obstruction, or autoimmune diseases.

coefficient of determination (r_{xy}^2)—Correlation coefficient squared; amount of variance shared by two variables.

Cole model—Procedure using resistance values ranging from zero to infinite frequencies to analyze multi-frequency bioimpedance data.

computerized tomography (CT)—Radiographic method that measures the differences in attenuation of X-ray beams as they pass through the body to create a computer-generated image of the scanned area; used for regional assessment of bone, adipose tissue, and lean tissue.

confidence interval—Statistic used to assess the precision of the body composition estimates for the whole population; the width of the confidence intervals is an estimate of the degree of variability in scores for the entire population, with a small interval indicating less variability.

constant error (CE)—Average difference between the measured and predicted values for the cross-validation group; bias.

coronary artery disease (CAD)—Disease caused by deposition of atherosclerotic lesions on inner lining of arteries supplying the heart muscle.

correlation—Statistical technique used to determine relationship between two variables.

criterion method—Gold standard or reference method; typically a direct measure of a component used to validate other tests.

cross-validation—Statistical method used to establish the accuracy of prediction equations and to establish the validity of methods.

cubic regression model—A statistical model used to derive a third-degree polynomial (X^3) equation with two bends in the regression line.

curvilinear regression—A type of polynomial regression in which one or more of the predictor variables in the regression model is raised to a certain power, for example quadratic (X^2) or cubic (X^3).

cystic fibrosis (CF)—Genetic disease of the exocrine glands, primarily affecting the gastrointestinal and respiratory systems.

densitometry—Measurement of total body density; hydrodensitometry and air displacement plethysmography are densitometric methods.

diabetes mellitus (DM)—Disease characterized by hyperglycemia caused by lack of insulin production or insulin resistance.

dialysis—A procedure that removes excess water and toxins from the body.

difference score—Difference between actual and predicted scores.

dilution principle—The volume of a solvent is equal to the amount of a compound added to the solvent divided by the concentration of the compound in that solvent.

double cross-validation technique—A cross-validation approach in which the sample is randomly divided into two equal groups, and data from each group are used to derive two prediction equations; each equation is then applied to the other group to test its predictive accuracy.

dual-energy X-ray absorptiometry (DXA)—Method used in clinical and research settings to estimate bone density and the bone mineral, fat, and mineral-free lean tissues of the body from X-ray attenuation.

dual-photon absorptiometry (DPA)—Method used to assess total body bone mineral and bone mineral density via measurement of photon beams from a radionuclide source.

Duchenne's muscular dystrophy—Common form of muscular dystrophy in boys caused by a genetic defect carried on the X-chromosome, resulting in a lack of dystrophin, a protein necessary for muscle cell structure.

ectomorphy—A measure of proportionality between body weight and height; the third component of an anthropometric somatotype.

emphysema—Condition characterized by permanent, abnormal enlargement of air spaces in lungs and destruction of terminal bronchiole walls.

endomorphy—A measure of body fatness; the first component of an anthropometric somatotype.

essential lipids—Compound lipids (phospholipids) needed for cell membrane formation; approximately 10% of the total body lipid pool.

eumenorrhea—Normal menstrual cycle with 10 to 13 menstrual periods per year.

external cross-validation—A cross-validation approach in which a prediction equation is applied to a sample that is independent from the one used to develop the original equation.

extracellular water (ECW)—Plasma water, interstitial water, transcellular water, and water found in bone, cartilage, and dense connective tissues.

fat-free body (FFB)—All residual lipid-free chemicals and tissues including water, muscle, bone, connective tissue, and internal organs.

fat-free body density (FFBd)—Overall density of the fat-free body calculated from the proportions and respective densities of the water, mineral, and protein components of the body.

fat-free mass (FFM)—A measure of the size of the fat-free body.

fat mass (FM)—All extractable lipids from adipose and other tissues in the body.

female athlete triad—Pattern of disordered eating, amenorrhea, and premature osteoporosis characteristic of some female athletes.

field methods—Methods suitable for estimating body composition in practical settings, for example skinfolds, bioimpedance analysis, and anthropometry.

FIT index (FIT)—Crude measure of physical activity level based on subjective ratings of the frequency, intensity, and time (duration) of aerobic activity; scores range from 1 to 100.

four-component model (4-C)—Body composition model that divides the body into its molecular components: water, mineral, protein, and fat.

functional residual capacity (FRC)—Amount of air remaining in the lungs at the end of a normal expiration; sum of the residual lung volume and expiratory reserve volume.

generalized prediction equation—Prediction equation used to estimate body composition of a heterogeneous group who varies greatly in age, gender, ethnicity, level of body fatness, and physical activity level; usually based on a quadratic regression model.

gold standard—Method used in research settings to obtain reference measures of body composition (e.g., hydrodensitometry, neutron activation analysis).

Graves' disease—Diffuse toxic goiter resulting in overactive thyroid gland or hyperthyroidism.

group predictive accuracy—A measure of the accuracy of a method or prediction equation in estimating the average body composition of an entire sample; for example, constant error and SEE are statistics used to describe group predictive accuracy.

gynoid obesity—Type of obesity in which excess fat is localized in the lower body; lower-body obesity; pear-shaped body.

healthy body weight—A body weight that does not increase an individual's risk of disease; BMI of 18.5 to 24.9 kg/m^2.

heart and lung transplantations—Surgical procedures that restore cardiac and pulmonary function for patients with end-stage heart failure or respiratory diseases.

heart failure—Symptomatic myocardial dysfunction resulting in excessive fluid accumulation in lungs, abdomen, and peripheral tissues and sometimes cachexia.

hemodialysis—A procedure that removes blood from the body, pumps blood through a filtering machine, and returns purified blood to the body; treatment for chronic kidney failure.

hierarchical regression—An approach used in multiple regression analysis in which the order of entry of predictor variables into the analysis is predetermined and specified by the researcher based on theory.

homoscedasticity—Assumption in regression analysis that variances of Y scores for each value of X are equally spread.

human immunodeficiency virus (HIV)—An infection, spread by contact with infected body fluids, that progressively destroys white blood cells.

hydrodensitometry (HD)—Body composition method used to estimate body volume via measurement of weight loss when the body is totally submerged under the water; hydrostatic weighing; underwater weighing.

hydrometry—Measurement of body water using isotope dilution methods.

hydrostatic weighing (HW)—Body composition method used to estimate body volume via measurement of weight loss when the body is totally submerged under water; hydrodensitometry; underwater weighing.

hyperglycemia—High blood glucose levels.

hyperthyroidism—Clinical condition characterized by increased synthesis and secretion of thyroid hormones; thyrotoxicosis.

hypoglycemia—Low blood glucose levels.

hypothyroidism—Condition caused by thyroid hormone deficiency; myxedema.

impedance (Z)—Measure of the opposition to the flow of electrical current through the body; composed of two vectors, resistance and reactance.

individual predictive accuracy—A measure of the accuracy of a method or prediction equation in estimating the body composition of individuals within a group; the 95% limits of agreement (Bland and Altman method) is one statistic used to evaluate individual predictive accuracy.

insulin-dependent diabetes mellitus (IDDM)—Disease characterized by hyperglycemia resulting from lack of insulin production by pancreas; type 1 diabetes.

internal cross-validation—A cross-validation approach in which one sample is divided into groups (see double cross-validation and jackknife technique) or subjects are excluded one at time (see bootstrap technique) to generate and assess predictive accuracy of equations.

intra-abdominal fat—Visceral fat in the abdominal cavity.

intracellular water (ICW)—Water found in the cytosol of every tissue in the body; calculated as the difference between measured total body water and extracellular water.

intratechnician reliability—Measure of the internal consistency or stability of scores measured by the same technician.

isothermal conditions—Constant temperature.

jackknife technique—An internal cross-validation approach in which total sample is divided into groups of equal size to generate a prediction equation and the sum of squares of residuals for each group is used to determine accuracy of the equation.

laboratory methods—Methods used to provide reference or criterion measures of body composition, for example dual-energy X-ray absorptiometry, densitometry, and neutron activation analysis.

law of propagation of errors—Sources of measurement error are independent and additive; the total measurement error reflects the degree of precision that can be expected when different methods are used to assess various components of body composition.

lean body mass (LBM)—Fat-free mass plus essential lipids.

leptin—Energy-regulating protein secreted by adipocytes.

limits of agreement—A statistical method used to assess the degree of agreement between methods; also known as the Bland and Altman method.

linear regression model–Statistical model depicting the linear relationship between the dependent variable and one or more predictor variables.

line of best fit–Regression line depicting the relationship between the reference values and predictor variables in an equation.

line of identity–Straight line with a slope equal to 1 and an intercept equal to 0; used in a scatterplot to illustrate the differences in the measured and predicted scores of the cross-validation sample.

lipodystrophy–Side effect of antiretroviral therapy in HIV patients that changes one's body shape due to the redistribution of body fat from the face and extremities (i.e., peripheral lipodystrophy), or to the trunk (i.e., central lipodystrophy), or both; also known as fat redistribution syndrome.

lower-body BIA method–Method that uses foot-to-foot impedance measures.

lower-body obesity–Type of obesity in which excess body fat is localized in the lower body; gynoid obesity; pear-shaped body.

lymphocytes–White blood cells.

magnetic resonance imaging (MRI)–Technique used to create computerized cross-sectional images of the human body from radio frequency signals emitted by hydrogen nuclei.

measurement error–Amount of error relating to the degree of precision of a measuring instrument.

mesomorphy–A measure of muscularity; the second component of an anthropometric somatotype.

minimal body fat–Lower limit of body fat needed for normal physiological and metabolic functions.

minimal body weight (MW)–Body weight corresponding to 5% to 7% BF for males and 12% to 14% BF for females.

multicomponent model–Body composition model that accounts for interindividual variation in the water, mineral, and protein content of the fat-free body.

multifrequency BIA (MFBIA)–A BIA approach in which a wide range of frequencies (1 kHz to 1 MHz) are used to estimate extracellular, intracellular, and total body water; also known as bioelectrical impedance spectroscopy.

multiple correlation coefficient (R_{mc})–Correlation between the reference measure and predictor variables in the body composition equation.

multiple regression–Statistical method used to derive body composition prediction equations.

multiple sclerosis (MS)–Neuromuscular disease characterized by demyelination or damage and scarring of the myelin sheath of the nervous system, thereby impairing the conduction of nerve impulses.

muscular dystrophy (MD)–An inherited muscle disease leading to muscle weakness; in boys, commonly caused by lack of dystrophin, a protein necessary for muscle cell structure (i.e., Duchenne's MD).

myocardial infarction–Heart attack.

myocardium–Heart muscle.

myxedema–Hypothyroidism caused by thyroid hormone deficiency.

near-infrared interactance (NIR)–Body composition method used to estimate percent body fat or total body density via measurement of the reflectance of near-infrared light at the measurement site.

neck circumference–A measure of neck girth; indirect index of upper-body fat and obesity.

neutron activation analysis (NAA)–Method used for direct, in vivo chemical analysis of the body at the atomic level; used to obtain reference measures of body composition from total body content of major elements (i.e., calcium, sodium, chlorine, phosphorus, nitrogen, hydrogen, oxygen, and carbon).

nomogram–Graph usually consisting of three parallel scales graduated for different variables so that when two values are plotted and connected by a straight line, the value of the third variable can be read directly at the point intersected by the line; used in body composition to calculate variables such as %BF, waist-to-hip ratio, and body mass index.

nonessential lipids–Triglycerides found primarily in adipose tissue; approximately 90% of the total body lipid pool.

non-insulin-dependent diabetes mellitus (NIDDM)–Disease characterized by impaired secretion of insulin in response to glucose and insulin resistance; type 2 diabetes.

obesity–Excessive amount of total body fat relative to body weight; BMI of 30 kg/m^2 or more.

objectivity–Intertechnician reliability; ability of test to yield similar scores for given individual when same test is administered by different technicians.

optical density (OD)–A measure of the amount of near-infrared light reflected by the body's tissues at specific wavelengths, e.g., 940 and 950 nm.

osteopenia–Loss of bone mineral mass.

osteoporosis–A disease associated with loss of bone mineral content due to aging, amenorrhea, malnutrition, or excessive exercise.

overweight–BMI of 25.0 to 29.9 kg/m^2.

paracentesis–Surgical puncture of abdominal cavity for aspiration of peritoneal fluid.

parallel model–BIA model used to estimate intracellular water or body cell mass; the reciprocal of the series model.

paralysis–Complete loss of neural communication to innervated muscles, preventing any willed movement of the affected areas.

paraplegia–Type of spinal cord injury resulting in paralysis of both legs and the trunk.

Pearson product-moment correlation coefficient (r_{xy})–Statistical test that quantifies the degree of relationship between two continuous variables.

percent body fat (%BF)–Fat mass expressed relative to body mass; relative body fat.

peritoneal dialysis–A procedure in which a mixture of glucose and salts is injected into the abdomen to attract toxic waste and then is drained from the body; treatment for chronic kidney failure.

phase angle–A bioimpedance measure calculated as the arc tangent of the ratio of reactance to resistance.

Poisson's law–Law describing the relationship between pressure (P) and volume (V) under adiabatic conditions; $P_1/P_2 = (V_2/V_1)\gamma$, where γ is ratio of specific heat of a gas at constant pressure to that at constant volume; $\gamma \cong 1.4$ for air.

polynomial regression–A statistical approach used to describe nonlinear relationships among variables; see quadratic regression model and curvilinear regression.

population-specific conversion formula–Formula used to convert Db to %BF for individuals from a specific population subgroup; formulas vary depending on the age, ethnicity, gender, level of body fatness, and physical activity level of the individual.

population-specific equations–Prediction equations used to estimate body composition of individuals from a specific homogeneous group, for example children, athletes, American Indians.

prediction equation–Mathematical formula derived from multiple regression analysis and used to estimate body composition measures (e.g., %BF, Db, and FFM).

prediction error–Measure of the predictive inaccuracy of a prediction equation in estimating reference measures of body composition, that is, the standard error of estimate and total error.

predictor variables–Variables used to estimate the reference measure (e.g., %BF, FFM, Db) of body composition.

PRESS technique–An internal cross-validation approach in which data for each subject in sample are excluded one at a time and regression analysis is performed to generate prediction equation; the value for each omitted subject is predicted and a PRESS residual score is calculated for each subject; the accuracy of the prediction equation is evaluated through measurement of the sum of squares of all PRESS residuals; also known as bootstrap technique.

pure error–Total error of prediction equation.

quadratic regression model–Statistical model depicting a curvilinear relationship between the dependent variable and one or more predictor variables.

quadriplegia–Type of spinal cord injury resulting in paralysis of all four extremities and the trunk.

reactance (Xc)–Measure of opposition to current flow through the body due to the capacitance of cell membranes; a vector of impedance.

reference body–Theoretically and empirically derived body whose fat-free body is 73.8% water, 6.8% mineral, and 19.4% protein.

reference method–Gold standard or criterion method used in research settings to develop and cross-validate body composition models, methods, and prediction equations; typically a direct measure of body composition component.

regional fat distribution–The distribution of fat in the upper and lower body.

regression–Statistical method used to derive and cross-validate prediction equations.

regression line–Line of best fit depicting relationship between the dependent variable and one or more predictor variables.

regression weight–Constant for each predictor variable in a body composition prediction equation; each predictor variable is multiplied by its constant, and the products are summed to yield an estimate of the body composition measure being predicted.

relative body fat (%BF)–Fat mass expressed as a percentage of total body mass; percent body fat.

reliability–Ability of test to yield consistent and stable scores across trials and over time.

residual lung volume (RV)–Volume of air remaining in the lungs following a maximal expiration.

residual score–Difference between the actual and predicted scores (Y – Y′).

resistance (R)–Measure of pure opposition to the flow of electrical current through the body; a vector of impedance; the lesser the resistance, the greater the current flow.

resistance index (HT^2/R)–Predictor variable in some BIA equations that is calculated by dividing height squared by total body resistance.

sagittal abdominal diameter (SAD)–Indirect measure of central adiposity and visceral fat.

sarcopenia–Loss of skeletal muscle mass usually associated with aging.

scatterplot–Graph used to illustrate the relationship between two variables.

segmental BIA (SBIA)–BIA approach in which the resistance of each body segment is measured separately; useful in patients with altered fluid distribution.

series model–BIA model which assumes that the body consists of resistors (body segments) connected in series.

six-component model–Body composition model that divides the body into six major elements: water, nitrogen, calcium, sodium, potassium, and chloride; requires direct analysis of the chemical composition of the body in vivo.

skeletal anthropometer–Instrument used to measure bony widths and body breadths.

skeletal diameter–A measure of the width of bones.

skinfold (SKF)–Measure of the thickness of two layers of skin and the underlying subcutaneous fat.

sliding or spreading caliper–Small anthropometer used to measure the breadth of small body segments with maximum measurement of 30 cm.

slope (b)–In regression analysis, a measure of the change in Y values with respect to the change in X values.

soft-tissue mass (STM)–In DXA, the sum of the bone-free lean tissue mass and fat mass.

somatogram–Anthropometric profile that graphically depicts the body's pattern of muscle and fat distribution.

somatotype–Anthropometric method used to describe an individual's body type in terms of endomorphic (fat), mesomorphic (muscle), and ectomorphic (stature) components.

specific resistivity (ρ)–Constant physical property analogous to specific gravity; the reciprocal of conductance; assumed to be a constant value in most BIA models.

spreading caliper–Anthropometer shaped to allow measurement between anatomical landmarks that would be difficult to measure using a standard anthropometer, for example wrist and ankle breadths.

stadiometer–Instrument used to measure stature or standing height.

standard error of estimate (SEE)–Measure of prediction error; quantifies the average deviation of individual data points around the line of best fit.

stepwise regression–An approach to multiple regression analysis in which the order of entry of predictor variables into the analysis is based on statistical criteria.

subcutaneous fat–Adipose tissue stored underneath the skin.

sum of skinfolds (ΣSKF)–Sum of two or more skinfold thicknesses.

tare weight–Weight of the underwater weighing apparatus used to determine the net underwater weight of a client.

thoracic gas volume (TGV)–Volume of air in the lungs and thorax.

three-component model (3-C)–Body composition model that divides the body into fat, water or mineral, and residual components.

thyroiditis–Inflammation of thyroid gland.

thyrotoxicosis–Hyperthyroidism caused by increased synthesis and secretion of thyroid hormones.

total body bone mineral (TBBM)–Measure of the osseous (bone) mineral content of the body.

total body mineral (TBM)–Measure of the osseous (bone) and non-osseous (cell) mineral content of the body.

total body potassium (TBK)–Amount of potassium in the body stored intracellularly; used to estimate body cell mass.

total body water (TBW)–Measure of the intracellular and extracellular fluid compartments of the body.

total error (TE)–Average deviation of individual scores of the cross-validation sample from the line of identity.

total error of measurement (TEM)–Quantifies the degree of precision to which variables can be measured; the standard deviation of the total measurement error; total measurement error reflects the degree of precision that can be expected when different methods are used to assess various components of body composition.

total lung capacity (TLC)–Total amount of air that can be held in the lungs; sum of the vital capacity and residual lung volume.

two-component model (2-C)–Body composition model that divides the body into fat and fat-free body components.

type 1 diabetes–See insulin-dependent diabetes mellitus.

type 2 diabetes–See non-insulin-dependent diabetes mellitus.

underwater weighing (UWW)–Densitometric method used to estimate body volume from water displacement; hydrostatic weighing; hydrodensitometry.

underweight–BMI <18.5 kg/m^2.

upper-body BIA method–Method that uses hand-to-hand impedance measures to estimate fat-free mass.

upper-body obesity–Type of obesity in which excess fat is localized to the upper body; android obesity; apple-shaped body.

validity–Ability of a test to accurately measure, with minimal error, a specific component.

validity coefficient ($r_{y,y'}$)–Correlation between the reference measure and predicted scores.

Vernier caliper–High-precision instrument with a small, movable, graduated scale running parallel with a fixed, graduated scale of a ruler, providing highly precise and accurate linear measurements; used in body composition to calibrate skinfold calipers and small sliding calipers.

visceral fat–Adipose tissue within and around the organs in the thoracic and abdominal cavities.

waist circumference–Measure of central adiposity and upper-body obesity; waist girth.

waist-to-hip ratio (WHR)–Waist circumference divided by the hip circumference; used as a measure of upper-body obesity and visceral fat.

wasting–State of nonvolitional weight loss generally coinciding with a clinical condition.

whole-body BIA method–Method that uses traditional wrist-to-ankle impedance measure.

whole-body counter–Device consisting of gamma ray detector, shielded room, and data acquisition system that measures gamma rays emitted by a naturally occurring potassium isotope (^{40}K) in the body; used to estimate total body potassium, a reference measure of body cell mass.

y-intercept (a)–For regression equation, the value of Y when X is equal to zero; point at which regression line intersects the Y axis in a scatterplot.

References

AACVPR Outcomes Committee. 1995. Outcome measurement in cardiac and pulmonary rehabilitation. *Journal of Cardiopulmonary Rehabilitation* 15: 394-405.

Abrahamsen, B., Hansen, T.B., Hogsberg, I.M., Pedersen, F.B., and Beck-Nielsen, H. 1996. Impact of hemodialysis on dual X-ray absorptiometry, bioelectrical impedance measurements, and anthropometry. *American Journal of Clinical Nutrition* 63: 80-86.

Ainsworth, B.E., Stolarczyk, L.M., Heyward, V.H., Berry, C.B., Irwin, M.L., and Mussulman, L.M. 1997. Predictive accuracy of bioimpedance in estimating fat-free mass of African-American women. *Medicine & Science in Sports & Exercise* 29: 781-787.

Akers, R., and Buskirk, E.R. 1969. An underwater weighing system utilizing "force cube" transducers. *Journal of Applied Physiology* 26: 649-652.

Albu, S., Lichtman, S., Heymsfield, S.B., Wang, J., Pierson, R.N., and Pi-Sunyer, F.X. 1989. Reassessment of body composition models in morbidly obese. *Federation of American Societies for Experimental Biology Journal* 3: A336.

Alfredson, H., Nordstrom, P., and Lorentzon, R. 1997. Bone mass in female volleyball players: A comparison of total and regional bone mass in female volleyball players and nonactive females. *Calcified Tissue International* 60: 338-342.

Aloia, J.F., Vaswani, A., Ma, R., and Flaster, E. 1997. Comparison of body composition in black and white premenopausal women. *Journal of Clinical Medicine* 129: 294-299.

American College of Sports Medicine. 1996. Weight loss in wrestlers: Position stand. *Medicine & Science in Sports & Exercise* 28: ix-xii.

American College of Sports Medicine. 1997. The female athlete triad: Position stand. *Medicine & Science in Sports & Exercise* 29: i-ix.

American Psychiatric Association. 1994. *Diagnostic and statistical manual of mental disorders (DSM-IV),* 4th ed. Washington, DC: Author.

Andersen, R.E. 2000. The spread of the childhood obesity epidemic. *Canadian Medical Association Journal* 163(11): 1461-1462.

Andrade, S., Lan, S.J.J., Engelson, E.S., Agin, D., Wang, J., Heymsfield, S.B., and Kotler, D.P. 2002. Use of a Durnin-Womersley formula to estimate change in subcutaneous fat content in HIV-infected subjects. *American Journal of Clinical Nutrition* 75: 587-592.

Anker, S.D., Clark, A.L., Teixeira, M.M., Hellewell, P.G., and Coats, A.J. 1999. Loss of bone mineral in patients with cachexia due to chronic heart failure. *American Journal of Cardiology* 83: 612-615.

Arngrimsson, S.A., Evans, E.M., Saunders, M.J., Ogburn, C.L., Lewis, R.D., and Cureton, K.J. 2000. Validation of body composition estimates in male and female distance runners using estimates from a four-component model. *American Journal of Human Biology* 12: 301-314.

Arpadi, S.M., Wang, J., Cuff, P.A., Thornton, J., Horlick, M., Kotler, D.P., and Pierson, R.N. 1996. Application of bioimpedance analysis for estimating body composition in prepubertal children infected with human immunodeficiency virus type 1. *Journal of Pediatrics* 129: 755-757.

Azcue, M., Fried, M., and Pencharz, P.B. 1993. Use of bioelectrical impedance analysis to measure total body water in patients with cystic fibrosis. *Journal of Pediatric Gastroenterology and Nutrition* 16: 440-445.

Bale, P., Doust, J., and Dawson, D. 1996. Gymnasts, distance runners, anorexics body composition and menstrual status. *Journal of Sports Medicine and Physical Fitness* 36: 49-53.

Batterham, M.J., Garsia, R., and Greenop, P. 1999. Measurement of body composition in people with HIV/AIDS: A comparison of bioelectrical impedance and skinfold anthropometry with dual-energy X-ray absorptiometry. *Journal of the American Dietetic Association* 99: 1109-1111.

Bauer, J., Capra, S., Davies, P.S.W., Ash, S., and Davidson, W. 2002. Estimation of total body water from bioelectrical impedance analysis in patients with pancreatic cancer—agreement between three methods of prediction. *Journal of Human Nutrition and Dietetics* 15: 185-188.

Baumgartner, R.N., Chumlea, W.C., and Roche, A.F. 1988. Bioelectric impedance phase angle and body composition. *American Journal of Clinical Nutrition* 48: 16-23.

Baumgartner, R.N., Heymsfield, S.B., Lichtman, S., Wang, J., and Pierson, R.N. 1991. Body composition in elderly people: Effect of criterion estimates on predictive equations. *American Journal of Clinical Nutrition* 53: 1345-1353.

Baumgartner, R.N., Heymsfield, S.B., and Roche, A.F. 1995. Human body composition and the epidemiology of chronic disease. *Obesity Research* 3: 73-95.

Baun, W.B., and Baun, M.R. 1981. A nomogram for the estimate of percent body fat from generalized equations. *Research Quarterly for Exercise and Sport* 52: 380-384.

Beals, K.A., and Manore, M.M. 1994. The prevalence and consequences of subclinical eating disorders in female athletes. *International Journal of Sports Nutrition* 4: 175-179.

Bedogni, G., Merlini, L., Ballestrazzi, A., Severi, S., and Battistini, N. 1996. Multifrequency bioelectric impedance measurements for predicting body water compartments in Duchenne muscular dystrophy. *Neuromuscular Disorders* 6: 55-60.

Beers, M.H., and Berkow, R. 1999. *The Merck manual,* 17th ed. West Point, PA: Merck & Co.

Behnke, A.R. 1961. Qualitative assessment of body build. *Journal of Applied Physiology* 16: 960-968.

Behnke, A.R., Feen, B.G., and Welham, W.C. 1942. The specific gravity of healthy men. Body weight and volume as an index of obesity. *Journal of the American Medical Association* 118: 495-498.

Behnke, A.R., Osserman, E.F., and Welham, W.C. 1953. Lean body mass. *Archives of Internal Medicine* 91: 585-601.

Behnke, A.R., and Wilmore, J.H. 1974. *Evaluation and regulation of body build and composition.* Englewood Cliffs, NJ: Prentice Hall.

Ben-Noun, L., Sohar, E., and Laor, A. 2001. Neck circumference as a simple screening measure for identifying overweight and obese patients. *Obesity Research* 9: 470-477.

Bergsma-Kadijk, J.A., Baumeister, B., and Deurenberg, P. 1996. Measurement of body fat in young and elderly women: Comparison between a four-compartment model and widely used reference methods. *British Journal of Nutrition* 75: 649-657.

Biaggi, R.R., Vollman, M.W., Nies, M.A., Brener, C.E., Flakoll, P.J., Levenhagen, D.K., Sun, M., Karabulut, Z., and Chen, K.Y. 1999. Comparison of air-displacement plethysmography with hydrostatic weighing and bioelectrical impedance analysis for the assessment of body composition in healthy adults. *American Journal of Clinical Nutrition* 69: 898-903.

Birmingham, C.L., Jones, P.J.H., Orphanidou, C., Bakan, R., Cleator, I.G.M., Goldner, E.M., and Phang, P.T. 1996. The reliability of bioelectrical impedance analysis for measuring changes in the body composition of patients with anorexia nervosa. *International Journal of Eating Disorders* 19: 311-315.

Blair, S.B., and Brodney, S. 1999. Effects of physical inactivity and obesity on morbidity and mortality: Current evidence and research issues. *Medicine & Science in Sports & Exercise* 31: S646-S662.

Bland, J.M., and Altman, D.G. 1986. Statistical methods for assessing agreement between two methods of clinical measurement. *The Lancet* 12: 307-310.

Boileau, R.A. 1996. Body composition assessment in children and youths. In *The child and adolescent athlete. Encyclopaedia of sports medicine,* ed. O. Bar-Or, 6: 523-537. Cambridge, MA: Blackwell Science.

Boileau, R.A., Lohman, T.G., and Slaughter, M.H. 1985. Exercise and body composition in children and youth. *Scandinavian Journal of Sports Medicine* 7: 17-27.

Boileau, R.A., Lohman, T.G., Slaughter, M.H., Ball, T.E., Going, S.B., and Hendrix, M.K. 1984. Hydration of the fat-free body in children during maturation. *Human Biology* 56: 651-666.

Boileau, R.A., Wilmore, J.H., Lohman, T.G., Slaughter, M.H., and Riner, W.F. 1981. Estimation of body density from skinfold thicknesses, body circumferences and skeletal widths in boys aged 8 to 11 years: Comparison of two samples. *Human Biology* 53: 575-592.

Bolton, M.P., Ward, L.C., Khan, A., Campbell, I., Nightingale, P., Dewit, O., and Elia, M. 1998. Sources of error in bioimpedance spectroscopy. *Physiological Measurement* 19: 235-245.

Bonge, D., and Donnelly, J.E. 1989. Trials to criteria for hydrostatic weighing at residual volume. *Research Quarterly for Exercise and Sport* 60: 176-179.

Borowitz, D., and Conboy, K. 1994. Are bioelectric impedance measurements valid in patients with cystic fibrosis? *Journal of Pediatric Gastroenterology and Nutrition* 18: 453-456.

Bosch, P.R., and Wells, C.L. 1991. Effect of immersion on residual volume of able-bodied and spinal cord injured males. *Medicine & Science in Sports & Exercise* 23: 384-388.

Bottaro, M.F., Heyward, V.H., Bezerra, R.F.A., and Wagner, D.R. 2002. Skinfold method vs. dual-energy X-ray absorptiometry to assess body composition in normal and obese women. *Journal of Exercise Physiology*$_{online}$ 5(2): 11-18.

Bouchard, C. 1994. Genetics of obesity: Overview and research direction. In *Genetics of obesity,* ed. C. Bouchard, 223-233. Boca Raton, FL: CRC Press.

Brandon, L.J. 1998. Comparison of existing skinfold equations for estimating body fat in African American and white women. *American Journal of Clinical Nutrition* 67: 1155-1161.

Brandon, L.J., and Bond, V. 1999. Are practical methods of evaluating body fat in African-American women accurate? *American Journal of Health Promotion* 13: 200-202.

Bray, G.A. 1978. Definitions, measurements and classifications of the syndromes of obesity. *International Journal of Obesity and Related Metabolic Disorders* 2: 99-113.

Bray, G.A., and Gray, D.S. 1988a. Anthropometric measurements in the obese. In *Anthropometric standardization reference manual,* ed. T.G. Lohman, A.F. Roche, and R. Martorell, 131-136. Champaign, IL: Human Kinetics.

Bray, G.A., and Gray, D.S. 1988b. Obesity. Part I–Pathogenesis. *Western Journal of Medicine* 149: 429-441.

Bray, G.A., Greenway, F.L., Molitch, M.E., Dahms, W.T., Atkinson, R.L., and Hamilton, K. 1978. Use of anthropometric measures to assess weight loss. *American Journal of Clinical Nutrition* 31: 769-773.

Brodowicz, G.R., Mansfield, R.A., McClung, M.R., and Althoff, S.A. 1994. Measurement of body composition in the elderly: Dual energy X-ray absorptiometry, underwater weighing, bioelectrical impedance analysis, and anthropometry. *Gerontology* 40: 332-339.

Broeder, C.E., Burrhus, K.A., Svanevik, L.S., Volpe, J., and Wilmore, J.H. 1997. Assessing body composition before and after resistance or endurance training. *Medicine & Science in Sports & Exercise* 29: 705-712.

Brook, C.G.D. 1971. Determination of body composition of children from skinfold measurements. *Archives of Disease in Childhood* 46: 182-184.

Brooke-Wavell, K., Jones, P.R., Norgan, N.G., and Hardman, A.E. 1995. Evaluation of near infra-red interactance for assessment of subcutaneous and total body fat. *European Journal of Clinical Nutrition* 49: 57-65.

Brownell, K.D., Rodin, J., and Wilmore, J.H., eds. 1992. *Eating, body weight and performance in athletes: Disorders of modern society.* Philadelphia: Lea & Febiger.

Brozek, J., Grande, F., Anderson, J.T., and Keys, A. 1963. Densitometric analysis of body composition: Revision of some quantitative assumptions. *Annals of the New York Academy of Sciences* 110: 113-140.

Brozek, J., and Keys, A. 1951. Evaluation of leanness-fatness in man: Norms and interrelationships. *British Journal of Nutrition* 5: 194-206.

Bunt, J.C., Going, S.B., Lohman, T.G., Heinrich, C.H., Perry, C.D., and Pamenter, R.W. 1990. Variation in bone mineral content and estimated body fat in young adult females. *Medicine & Science in Sports & Exercise* 22: 564-569.

Bunt, J.C., Lohman, T.G., and Boileau, R.A. 1989. Impact of total body water fluctuation on estimation of body fat from body density. *Medicine & Science in Sports & Exercise* 21: 96-100.

Burgert, S.L., and Anderson, C.F. 1979. A comparison of triceps skinfold values as measured by the plastic McGaw caliper and the Lange caliper. *American Journal of Clinical Nutrition* 32: 1531-1533.

Buskirk, E.R. 1961. Underwater weighing and body density: A review of procedures. In *Techniques for measuring body composition,* ed. J. Brozek and A. Henschel, 90-105. Washington, DC: National Academy of Sciences, National Research Council.

Cable, A., Nieman, D.C., Austin, M., Hogen, E., and Utter, A.C. 2001. Validity of leg-to-leg bioelectrical impedance measurement in males. *Journal of Sports Medicine and Physical Fitness* 41: 411-414.

Callaway, C.W., Chumlea, W.C., Bouchard, C., Himes, J.H., Lohman, T.G., Martin, A.D., Mitchell, C.D., Mueller, W.H., Roche, A.F., and Seefeldt, V.D. 1988. Circumferences. In *Anthropometric standardization reference manual,* ed. T.G. Lohman, A.F. Roche, and R. Martorell, 39-54. Champaign, IL: Human Kinetics.

Canadian Academy of Sport Medicine. 2001. Abandoning routine body composition assessment: A strategy to reduce disordered eating among female athletes and dancers: Position statement. *Clinical Journal of Sport Medicine* 11: 280.

Carey, D. 2000. The validity of anthropometric regression equations in predicting percent body fat in collegiate wrestlers. *Journal of Sports Medicine and Physical Fitness* 40: 254-259.

Carter, J.E.L. 1982. *Physical structure of Olympic athletes: Part I. Medicine and sport* (Vol. 16). New York: S. Karger.

Casas, Y.G., Schiller, B.C., DeSouza, C.A., and Seals, D.R. 2001. Total and regional body composition across age in healthy Hispanic and white women of similar socioeconomic status. *American Journal of Clinical Nutrition* 73: 13-18.

Cassady, S.L., Nielsen, D.H., Janz, K.F., Wu, Y., Cook, J.S., and Hansen, J.R. 1993. Validity of near infrared body composition analysis in children and adolescents. *Medicine & Science in Sports & Exercise* 25: 1185-1191.

Cataldo, D., and Heyward, V. 2000. Pinch an inch: A comparison of several high-quality and plastic skinfold calipers. *ACSM's Health & Fitness Journal* 4(3): 12-16.

Caton, J.R., Mole, P.A., Adams, W.C., and Heustis, D.S. 1988. Body composition analysis by bioelectrical impedance: Effect of skin temperature. *Medicine & Science in Sports & Exercise* 20: 489-491.

Centers for Disease Control. 1998. Prevalence of diagnosed diabetes among American Indians/Alaska Natives–United States, 1996. *Morbidity and Mortality Weekly Report* 47(2): 901-904.

Centers for Disease Control. 2000. Prevalence of selected cardiovascular disease risk factors among American Indians and Alaska Natives–United States, 1997. *Morbidity and Mortality Weekly Report* 49(21): 461-465.

Centers for Disease Control. 2003. Early release of selected estimates based on data from the 2002 National Health Interview Survey. On-line report dated July 18, 2003 at www.cdc.gov/nchs.

Chen, Y.C., Chen, H.H., Yeh, J.C., Chen, S.Y., and Lin, K.H. 2000. Each of anthropometry, bioelectrical impedance analysis and dual-energy X-ray absorptiometry methods can be used to assess lean body mass in hemodialysis patients. *Nephron* 84: 374-375.

Chen, Z., Maricic, M., Nguyen, P., Ahmann, F.R., Bruhn, R., and Dalkin, B.L. 2002. Low bone density and high percentage of body fat among men who were treated with androgen deprivation therapy for prostate carcinoma. *Cancer* 95: 2136-2144.

Chertow, G.M., Lowrie, E.G., Wilmore, D.W., Gonzalez, J., Lew, N.L., Ling, J., Leboff, M.S., Gottlieb, M.N., Huang, W., Zebrowski, B., College, J., and Lazarus, J.M. 1995. Nutritional assessment with bioelectrical impedance analysis in maintenance hemodialysis patients. *Journal of the American Society of Nephrology* 6: 75-81.

Chumlea, W.C., and Baumgartner, R.N. 1989. Status of anthropometry and body composition data in elderly subjects. *American Journal of Clinical Nutrition* 50: 1158-1166.

Chumlea, W.C., Baumgartner, R.N., and Roche, A.F. 1988. Specific resistivity used to estimate fat-free mass from segmental body measures of bioelectric impedance. *American Journal of Clinical Nutrition* 48: 7-15.

Clark, R.R., Kuta, J.M., and Sullivan, J.C. 1993. Prediction of percent body fat in adult males using dual energy X-ray absorptiometry, skinfolds, and hydrostatic weighing. *Medicine & Science in Sports & Exercise* 25: 528-535.

Clark, R.R., Kuta, J.M., Sullivan, J.C., Bedford, W.M., Penner, J.D., and Studesville, E.A. 1993. A comparison of methods to predict minimal weight in high school wrestlers. *Medicine & Science in Sports & Exercise* 25: 151-158.

Clarys, J.P., Martin, A.D., Drinkwater, D.T., and Marfell-Jones, M.J. 1987. The skinfold: Myth and reality. *Journal of Sports Sciences* 5: 3-33.

Clasey, J.L., Hartman, M.L., Kanaley, J., Wideman, L., Teates, C.D., Bouchard, C., and Weltman, A. 1997. Body composition by DEXA in older adults: Accuracy and influence of scan mode. *Medicine & Science in Sports & Exercise* 29: 560-567.

Clasey, J.L., Kanaley, J.A., Wideman, L., Heymsfield, S.B., Teates, C.D., Gutgesell, M.E., Thorner, M.O., Hartman,

M.L., and Weltman, A. 1999. Validity of methods of body composition assessment in young and older men and women. *Journal of Applied Physiology* 86:1728-1738.

Cohn, S.H., Roginsky, M.S., Aloia, J.F., Ellis, K.J., and Shukla, K.K. 1973. Alterations in elemental body composition in thyroid disorders. *Journal of Clinical Endocrinology and Metabolism* 36: 742-749.

Collins, M.A., Millard-Stafford, M.L., Sparling, P.B., Snow, T.K., Rosskopf, L.B., Webb, S.A., and Omer, J. 1999. Evaluation of the Bod Pod for assessing body fat in collegiate football players. *Medicine & Science in Sports & Exercise* 31: 1350-1356.

Colville, B.C., Heyward, V.H., and Sandoval, W.M. 1989. Comparison of two methods for estimating body composition of bodybuilders. *Journal of Applied Sport Science Research* 3: 57-61.

Conway, J.M., and Norris, K.H. 1987. Noninvasive body composition in humans by near infrared interactance. In *In vivo body composition studies,* ed. K.J. Ellis, S. Yasumura, and W.D. Morgan, 163-170. Brookhaven, NY: Institute of Physical Sciences in Medicine.

Conway, J.M., Norris, K.H., and Bodwell, C.E. 1984. A new approach for the estimation of body composition: Infrared interactance. *American Journal of Clinical Nutrition* 40: 1123-1130.

Corcoran, C., Anderson, E.J., Burrows, B., Stanley, T., Walsh, M., Poulos, A.M., and Grinspoon, S. 2000. Comparison of total body potassium with other techniques for measuring lean body mass in men and women with AIDS wasting. *American Journal of Clinical Nutrition* 72: 1053-1058.

Cornish, B.H., Jacobs, A., Thomas, B.J., and Ward, L.C. 1999. Optimizing electrode sites for segmental bioimpedance measurements. *Physiological Measurement* 20: 241-250.

Cornish, B.H., and Ward, L.C. 1998. Data analysis in multiple-frequency bioelectrical impedance analysis. *Physiological Measurement* 19: 275-283.

Cote, K.D., and Adams, W.C. 1993. Effect of bone density on body composition estimates in young adult black and white women. *Medicine & Science in Sports & Exercise* 25: 290-296.

Crawford, D.H., Shepherd, R.W., Halliday, J.W., Cooksley, G.W., Golding, S.D., Cheng, W.S., and Powell, L.W. 1994. Body composition in nonalcoholic cirrhosis: The effect of disease etiology and severity on nutritional compartments. *Gastroenterology* 106: 1611-1617.

Curb, J.D., and Marcus, E.B. 1991. Body fat and obesity in Japanese Americans. *American Journal of Clinical Nutrition* 53: 1552S-1555S.

Davis, P.G., Van Loan, M., Holly, R.G., Krstich, K., and Phinney, S.D. 1989. Near infrared interactance vs. hydrostatic weighing to measure body composition in lean, normal, and obese women. *Medicine & Science in Sports & Exercise* 21: S100 [abstract].

De Lorenzo, A., Bertini, I., Candeloro, N., Iacopino, L., Andreoli, A., Van Loan, M. 1998. Comparison of different techniques to measure body composition in moderately active adolescents. *British Journal of Sports Medicine* 32: 215-219.

De Lorenzo, A., Bertini, I., Iacopino, L., Pagliato, E., Testolin, C., and Testolin, G. 2000. Body composition measurement in highly trained male athletes. *Journal of Sports Medicine and Physical Fitness* 40: 178-183.

de Meer, K., Gulmans, V.A.M., Westerterp, K.R., Houwen, R.H.J., and Berger, R. 1999. Skinfold measurements in children with cystic fibrosis: Monitoring fat-free mass and exercise effects. *European Journal of Pediatrics* 158: 800-806.

Demerath, E.W., Guo, S.S., Chumlea, W.C., Towne, B., Roche, A.F., and Siervogel, R.M. 2002. Comparison of percent body fat estimates using air displacement plethysmography and hydrodensitometry in adults and children. *International Journal of Obesity and Related Metabolic Disorders* 26:389-397.

Dempster, P., and Aitkens, S. 1995. A new air displacement method for the determination of human body composition. *Medicine & Science in Sports & Exercise* 27: 1692-1697.

Demura, S., Yamaji, S., Goshi, F., Kobayashi, H., Sato, S., and Nagasawa, Y. 2002. The validity and reliability of relative body fat estimates and the construction of new prediction equations for young Japanese adult males. *Journal of Sports Sciences* 20: 153-164.

Demura, S., Yamaji, S., Goshi, F., and Nagasawa, Y. 2002. The influence of transient change of total body water on relative body fats based on three bioelectrical impedance analyses methods. *Journal of Sports Medicine and Physical Fitness* 42: 38-44.

DeOnis, M., and Blossner, M. 2000. Prevalence and trends of overweight among preschool children in developing countries. *American Journal of Clinical Nutrition* 72: 1032-1039.

Desport, J.C., Preux, P.M., Guinvarch, S., Rousset, P., Salle, J.Y., Daviet, J.C., Dudognon, P., Munoz, M., and Ritz, P. 2000. Total body water and percentage fat mass measurements using bioelectrical impedance analysis and anthropometry in spinal cord-injured patients. *Clinical Nutrition* 19: 185-190.

Despres, J.P. 1993. Abdominal obesity as important component of insulin-resistance syndrome. *Nutrition* 9: 452-459.

Deurenberg, P. 2001. Universal cut-off BMI points for obesity are not appropriate. *British Journal of Nutrition* 85: 135-136.

Deurenberg, P., Andreoli, A., Borg, P., Kukkonen-Harjula, K., de Lorenzo, A., van Marken Lichtenbelt, W.D., Testolin, G., Vigano, R., and Vollaard, N. 2001. The validity of predicted body fat percentage from body mass index and from impedance in samples of five European populations. *European Journal of Clinical Nutrition* 55: 973-979.

Deurenberg, P., and Deurenberg-Yap, M. 2001. Differences in body-composition assumptions across ethnic groups: Practical consequences. *Current Opinion in Clinical Nutrition and Metabolic Care* 4: 377-383.

Deurenberg, P., and Deurenberg-Yap, M. 2002. Validation of skinfold thickness and hand-held impedance measurements for estimation of body fat percentage among Singaporean Chinese, Malay and Indian subjects. *Asia Pacific Journal of Clinical Nutrition* 11: 1-7.

Deurenberg, P., Deurenberg-Yap, M., and Guricci, S. 2002. Asians are different from Caucasians and from each other in their body mass index/body fat percent relationship. *Obesity Reviews* 3: 141-146.

Deurenberg, P., Deurenberg-Yap, M., Wang, J., Lin, F.P., and Schmidt, G. 1999. The impact of body build on the relationship between body mass index and percent body fat. *International Journal of Obesity and Related Metabolic Disorders* 23: 537-542.

Deurenberg, P., Deurenberg-Yap, M., Wang, J., Lin, F.P., and Schmidt, G. 2000. Prediction of percentage body fat from anthropometry and bioelectrical impedance in Singaporean and Beijing Chinese. *Asia Pacific Journal of Clinical Nutrition* 9: 93-98.

Deurenberg, P., Kusters, C.S., and Smit, H.E. 1990. Assessment of body composition by bioelectrical impedance in children and young adults is strongly age-dependent. *European Journal of Clinical Nutrition* 44: 261-268.

Deurenberg, P., Leenan, R., van der Kooy, K., and Hautvast, J.G. 1989. In obese subjects the body fat percentage calculated with Siri's formula is an overestimation. *European Journal of Clinical Nutrition* 43: 569-575.

Deurenberg, P., Pieters, J.J.L., and Hautvast, J.G.A.J. 1990. The assessment of the body fat percentage by skinfold thickness measurements in childhood and young adolescence. *British Journal of Nutrition* 63: 293-303.

Deurenberg, P., Schouten, F.J., Andreoli, A., and DeLorenzo, A. 1993. Assessment of changes in extra-cellular water and total body water using multi-frequency bio-electrical impedance. *Basic Life Science* 60: 129-132.

Deurenberg, P., Tagliabue, A., and Schouten, F.J.M. 1995. Multi-frequency impedance for prediction of extra-cellular water and total body water. *British Journal of Nutrition* 73: 349-358.

Deurenberg, P., van der Kooy, K., Evers, P., and Hulshof, T. 1990. Assessment of body composition by bioelectrical impedance in a population aged greater than 60 y. *American Journal of Clinical Nutrition* 51: 3-6.

Deurenberg, P., van der Kooy, K., Hulshop, T., and Evers, P. 1989. Body mass index as a measure of body fatness in the elderly. *European Journal of Clinical Nutrition* 43: 231-236.

Deurenberg, P., van der Kooy, K., and Leenan, R. 1989. Differences in body impedance when measured with different instruments. *European Journal of Clinical Nutrition* 43: 885-886.

Deurenberg, P., van der Kooy, K., Leenan, R., Weststrate, J.A., and Seidell, J.C. 1991. Sex and age-specific population prediction formulas for estimating body composition from bioelectrical impedance: A cross-validation study. *International Journal of Obesity and Related Metabolic Disorders* 15: 17-25.

Deurenberg, P., Weststrate, J.A., and Hautvast, J.G. 1989. Changes in fat-free mass during weight loss measured by bioelectrical impedance and by densitometry. *American Journal of Clinical Nutrition* 49: 33-36.

Deurenberg, P., Weststrate, J.A., Paymans, I., and van der Kooy, K. 1988. Factors affecting bioelectrical impedance measurements in humans. *European Journal of Clinical Nutrition* 42: 1017-1022.

Deurenberg, P., Weststrate, J.A., and Seidell, J. 1991. Body mass index as a measure of body fatness: Age- and sex-specific prediction formulas. *British Journal of Nutrition* 65: 105-114.

Deurenberg, P., and Yap, M. 1999. The assessment of obesity: Methods for measuring body fat and global prevalence of obesity. *Baillieres Best Practice and Research in Clinical Endocrinology and Metabolism* 13: 1-11.

Deurenberg, P., Yap, M., and van Staveren, W.A. 1998. Body mass index and percent body fat: A meta-analysis among different ethnic groups. *International Journal of Obesity and Related Metabolic Disorders* 22: 1164-1171.

Deurenberg-Yap, M., Schmidt, G., van Staveren, W.A., and Deurenberg, P. 2000. The paradox of low body mass index and high body fat percentage among Chinese, Malays and Indians in Singapore. *International Journal of Obesity and Related Metabolic Disorders* 24: 1011-1017.

Deurenberg-Yap, M., Schmidt, G., van Staveren, W.A., Hautvast, J.G.A.J., and Deurenberg, P. 2001. Body fat measurement among Singaporean Chinese, Malays and Indians: A comparative study using a four-compartment model and different two-compartment models. *British Journal of Nutrition* 85: 491-498.

Dewit, O., Fuller, N.J., Fewtrell, M.S., Elia, M., and Wells, J.C.K. 2000. Whole body air displacement plethysmography compared with hydrodensitometry for body composition analysis. *Archives of Disease in Childhood* 82: 159-164.

Donnelly, J.R., Brown, T.E., Israel, R.G., Smith-Sintek, S., O'Brien, K.F., and Caslavka, B. 1988. Hydrostatic weighing without head submersion: Description of a method. *Medicine & Science in Sports & Exercise* 20: 66-69.

Dook, J.E., James, C., Henderson, N.K., and Price, R.I. 1997. Exercise and bone mineral density in mature female athletes. *Medicine & Science in Sports & Exercise* 29: 291-296.

Douphrate, D.I., Green, J.S., Heffner, K.D., Berman, W.I., East, C., Robinzine, K., and Verstraete, R. 1994. Evaluation of three near-infrared instruments for body composition assessment in an aged cardiac patient population. *Journal of Cardiopulmonary Rehabilitation* 14: 399-405.

DuBois, D., and DuBois, E.F. 1916. A formula to estimate the approximate surface area if height and weight be known. *Archives of Internal Medicine* 17: 863-871.

Dumler, F., Schmidt, R., Kilates, C., Faber, M., Lubkowski, T., and Frinak, S. 1992. Use of bioelectrical impedance for the nutritional assessment of chronic hemodialysis patients. *Mineral and Electrolyte Metabolism* 18: 284-287.

Duppe, H., Gardsell, P., Johnell, O., and Ornstein, E. 1996. Bone mineral density in female junior, senior and former football players. *Osteoporosis International* 6: 437-441.

Durnin, J.V.G.A., and Rahaman, M.M. 1967. The assessment of the amount of fat in the human body from measurements of skinfold thickness. *British Journal of Nutrition* 21: 681-689.

Durnin, J.V.G.A., and Womersley, J. 1974. Body fat assessed from total body density and its estimation from skinfold thickness: Measurements on 481 men and women aged from 17 to 72 years. *British Journal of Nutrition* 32: 77-97.

Eaton, A.W., Israel, R.G., O'Brien, K.F., Hortobagyi, T., and McCammon, M.R. 1993. Comparison of four methods to assess body composition in women. *European Journal of Clinical Nutrition* 47: 353-360.

Eckerson, J.M., Evetovich, T.K., Stout, J.R., Housh, T.J., Johnson, G.O., Housh, D.J., Ebersole, K.T., and Smith, D.B. 1997. Validity of bioelectrical impedance equations for estimating fat-free weight in high school female gymnasts. *Medicine & Science in Sports & Exercise* 29: 962-968.

Eckerson, J.M., Housh, T.J., and Johnson, G.O. 1992. Validity of bioelectrical impedance equations for estimating fat-free weight in lean males. *Medicine & Science in Sports & Exercise* 24: 1298-1302.

Eckerson, J.M., Stout, J.R., Evetovich, T.K., Housh, T.J., Johnson, G.O., and Worrell, N. 1998. Validity of self-assessment techniques for estimating percent fat in men and women. *Journal of Strength and Conditioning Research* 12: 243-247.

Economos, C.D., Nelson, M.E., Fiatarone Singh, M.A., Kehayias, J.J., Dallal, G.E., Heymsfield, S.B., Wang, J., Yasumura, S., Ma, R., and Pierson, R.N. 1999. Bone mineral measurements: A comparison of delayed gamma neutron activation, dual-energy X-ray absorptiometry and direct chemical analysis. *Osteoporosis International* 10: 200-206.

Edelman, I.S., Olney, J.M., James, A.H., Brooks, L., and Moore, F.D. 1952. Body composition: Studies in the human being by the dilution principle. *Science* 115: 447-454.

Edwards, D.A., Hammond, W.H., Healy, M.J., Tanner, J.M., and Whitehouse, R.H. 1955. Design and accuracy of calipers for measuring subcutaneous tissue thickness. *British Journal of Nutrition* 9: 133-143.

Elia, M., Parkinson, S.A., and Diaz, E. 1990. Evaluation of near infra-red interactance as a method of predicting body composition. *European Journal of Clinical Nutrition* 44: 113-121.

Ellis, K.J. 1997. Body composition of a young, multiethnic, male population. *American Journal of Clinical Nutrition* 66: 1323-1331.

Ellis, K.J. 2000. Human body composition: In vivo methods. *Physiological Reviews* 80: 649-680.

Ellis, K.J. 2001. Selected body composition methods can be used in field studies. *Journal of Nutrition* 131: 1589S-1595S.

Ellis, K.J., Abrams, S.A., and Wong, W.W. 1997. Body composition of a young, multiethnic female population. *American Journal of Clinical Nutrition* 65: 724-731.

Ellis, K.J., Bell, S.J., Chertow, G.M., Chumlea, W.C., Knox, T.A., Kotler, D.P., Lukaski, H.C., and Schoeller, D.A. 1999. Bioelectrical impedance methods in clinical research: A follow-up to the NIH technology assessment conference. *Nutrition* 15: 874-880.

Ellis, K.J., Shypailo, R.J., Hergenroeder, A., Perez, M., and Abrams, S. 1996. Total body calcium and bone mineral content: Comparison of dual-energy X-ray absorptiometry with neutron activation analysis. *Journal of Bone Mineral Research* 11: 843-848.

Elsen, R., Siu, M.L., Pineda, O., and Solomons, N.W. 1987. Sources of variability in bioelectrical impedance determinations in adults. In *In vivo body composition studies,* ed. K.J. Ellis, S. Yasamura, and W.D. Morgan, 184-188. London: Institute of Physical Sciences in Medicine.

Engelen, M.P.K.J., Schols, A.M.W.J., Heidendal, G.A.K., and Wouters, E.F.M. 1998. Dual-energy X-ray absorptiometry in the clinical evaluation of body composition and bone mineral density in patients with chronic obstructive pulmonary disease. *American Journal of Clinical Nutrition* 68: 1298-1303.

Engelson, E.S., Kotler, D.P., Tan, Y.X., Agin, D., Wang, J., Pierson, R.N., and Heymsfield, S.B. 1999. Fat distribution in HIV-infected patients reporting truncal enlargement quantified by whole-body magnetic resonance imaging. *American Journal of Clinical Nutrition* 69: 1162-1169.

Eston, R.G., Cruz, A., Fu, F., and Fung, L.M. 1993. Fat-free mass estimation by bioelectrical impedance and anthropometric techniques in Chinese children. *Journal of Sports Sciences* 11: 241-247.

Eston, R.G., Fu, F., and Fung, L. 1995. Validity of conventional anthropometric techniques for predicting body composition in healthy Chinese adults. *British Journal of Sports Medicine* 29: 52-56.

Evans, E.M., Arngrimsson, S.A., and Cureton, K.J. 2001. Body composition estimates from multicomponent models using BIA to determine body water. *Medicine & Science in Sports & Exercise* 33: 839-845.

Evans, E.M., Prior, B.M., Arngrimsson, S.A., Modlesky, C.M., and Cureton, K.J. 2001. Relation of bone mineral density and content to mineral content and density of the fat-free mass. *Journal of Applied Physiology* 91: 2166-2172.

Evans, E.M., Saunders, M.J., Spano, M.A., Arngrimsson, S.A., Lewis, R.D., and Cureton, K.J. 1999. Effects of diet and exercise on the density and composition of the fat-free mass in obese women. *Medicine & Science in Sports & Exercise* 31: 1778-1787.

Evans, P.E., Israel, R.G., Flickinger, E.G., O'Brien, K.F., and Donnelly, J.E. 1989. Hydrostatic weighing without head submersion in morbidly obese females. *American Journal of Clinical Nutrition* 50: 400-403.

Ferland, M., Despres, J.P., Tremblay, A., Pinault, S., Nadeau, A., Moorjani, S., Lupien, P.J., Theriault, G., and Bouchard, C. 1989. Assessment of adipose distribution by computed axial tomography in obese women: Association with body density and anthropometric measurements. *British Journal of Nutrition* 61: 139-148.

Fields, D.A., and Goran, M.I. 2000. Body composition techniques and the four-compartment model in children. *Journal of Applied Physiology* 89: 613-620.

Fields, D.A., Goran, M.I., and McCrory, M.A. 2002. Body-composition assessment via air-displacement plethysmography in adults and children: A review. *American Journal of Clinical Nutrition* 75: 453-467.

Fields, D.A., Hunter, G.R., and Goran, M.I. 2000. Validation of the Bod Pod with hydrostatic weighing: Influence of body clothing. *International Journal of Obesity and Related Metabolic Disorders* 24: 200-205.

Fields, D.A., Wilson, G.D., Gladden, L.B., Hunter, G.R., Pascoe, D.D., and Goran, M.I. 2001. Comparison of the Bod Pod with the four-compartment model in adult females. *Medicine & Science in Sports & Exercise* 33: 1605-1610.

Fleck, S.J. 1983. Body composition of elite American athletes. *American Journal of Sports Medicine* 11: 398-403.

Flegal, K.M., Carroll, M.D., Kuczmarski, R.J., and Johnson, C.L. 1998. Overweight and obesity in the United States: Prevalence and trends, 1960-1994. *International Journal of Obesity and Related Metabolic Disorders* 22: 39-47.

Flegal, K.M., Carroll, M.D., Ogden, C.L., and Johnson, C.L. 2002. Prevalence and trends in obesity among US adults, 1999-2000. *Journal of the American Medical Association* 288(14): 1723-1727

Fogelholm, G.M., Sievanan, H.T., van Marken Lichtenbelt, W.D., and Westerterp, K.R. 1997. Assessment of fat-mass loss during weight reduction in obese women. *Metabolism* 446: 968-975.

Fogelholm, M., Sievanen, H., Kukkonen-Harjula, K., Oja, P., and Vuori, I. 1993. Effects of meal and its electrolytes on bioelectrical impedance. In *Human body composition: In vivo methods, models and assessment,* ed. K.J. Ellis and J.D. Eastman, 331-332. New York: Plenum Press.

Fogelholm, M., and van Marken Lichtenbelt, W. 1997. Comparison of body composition methods: A literature analysis. *European Journal of Clinical Nutrition* 51: 495-503.

Fohlin, L. 1977. Body composition, cardiovascular and renal function in adolescent patients with anorexia nervosa. *Acta Paediatrica Scandinavica* 268 (Suppl.): 7-20.

Folsom, A.R., Burke, G.L., Byers, C.L., Hutchinson, R.G., Heiss, G., Flack, J.M., Jacobs, D.R., and Caan, B. 1991. Implications of obesity for cardiovascular disease in blacks: The CARDIA and ARIC studies. *American Journal of Clinical Nutrition* 53: 1604S-1611S.

Fomon, S.J., Haschke, F., Ziegler, E.E., and Nelson, S.E. 1982. Body composition of reference children from birth to age 10 years. *American Journal of Clinical Nutrition* 35: 1169-1175.

Forbes, G.B. 2001. On the matter of ethnic differences in body composition. *American Journal of Clinical Nutrition* 74: 555.

Forbes, G.B., Simon, W., and Amatruda, J.M. 1992. Is bioimpedance a good predictor of body-composition change? *American Journal of Clinical Nutrition* 56: 4-6.

Forbes, R.M., Cooper, A.R., and Mitchell, H.H. 1953. The composition of the adult human body as determined by chemical analysis. *Journal of Biological Chemistry* 203: 359-366.

Forbes, R.M., Mitchell, H.H., and Cooper, A.R. 1956. Further studies on the gross composition and mineral elements of the adult human body. *Journal of Biological Chemistry* 223: 969-975.

Formica, C.A., Cosman, F., Nieves, J., Herbert, J., and Lindsay, R. 1997. Reduced bone mass and fat-free mass in women with multiple sclerosis: Effects of ambulatory status and glucocorticoid use. *Calcified Tissue International* 61: 129-133.

Fornetti, W.C., Pivarnik, J.M., Foley, J.M., and Fiechtner, J.J. 1999. Reliability and validity of body composition measures in female athletes. *Journal of Applied Physiology* 87: 1114-1122.

Forsyth, H.L., and Sinning, W.E. 1973. The anthropometric estimation of body density and lean body weight of male athletes. *Medicine and Science in Sports* 5: 174-180.

Franzen, R. 1929. *Physical measures of growth and nutrition.* New York: American Child Health Association.

Fredrix, E.W., Saris, W.H., Soeters, P.B., Wouters, E.F., Kester, A.D., von Meyenfeldt, M.F., and Westerterp, K.R. 1990. Estimation of body composition by bioelectrical impedance in cancer patients. *European Journal of Clinical Nutrition* 44: 749-752.

Friedl, K.E., DeLuca, J.P., Marchitelli, L.J., and Vogel, J.A. 1992. Reliability of body-fat estimations from a four-compartment model by using density, body water, and bone mineral measurements. *American Journal of Clinical Nutrition* 55: 764-770.

Friedl, K.E., Westphal, K.A., Marchitelli, L.J., Patton, J.F., Chumlea, W.C., and Guo, S.S. 2001. Evaluation of anthropometric equations to assess body-composition changes in young women. *American Journal of Clinical Nutrition* 73: 268-275.

Frisancho, A.R. 1984. New standard of weight and body composition by frame size and height for assessment of nutritional status of adults and the elderly. *American Journal of Clinical Nutrition* 40: 808-819.

Frisancho, A.R., and Flegel, P.N. 1983. Elbow breadth as a measure of frame size for US males and females. *American Journal of Clinical Nutrition* 37: 311-314.

Fruth, S.J., and Worrell, T.W. 1995. Factors associated with menstrual irregularities and decreased bone mineral density in female athletes. *Journal of Orthopaedic & Sports Physical Therapy* 22: 26-38.

Fuller, N.J., Dewit, O., and Wells, J.C.K. 2001. The potential of near infra-red interactance for predicting body composition in children. *European Journal of Clinical Nutrition* 55: 967-972.

Fuller, N.J., and Elia, M. 1989. Potential use of bioelectrical impedance of "whole body" and of body segments for the assessment of body composition: A comparison with densitometry and anthropometry. *European Journal of Clinical Nutrition* 43: 779-791.

Fuller, N.J., Jebb, S.A., Laskey, M.A., Coward, W.A., and Elia, M. 1992. Four-component model for the assessment of body composition in humans: Comparison with alternative methods, and evaluation of the density and hydration of fat-free mass. *Clinical Science* 82: 687-693.

Fuller, N.J., Sawyer, M.B., and Elia, M. 1994. Comparative evaluation of body composition methods and predictions, and calculation of density and hydration fraction of fat-free mass, in obese women. *International Journal of Obesity and Related Metabolic Disorders* 18: 503-512.

Futrex, Inc. 1988. *Futrex-5000 research manual.* Gaithersburg, MD: Author.

Gallagher, D., Visser, M., Sepulveda, D., Pierson, R.N., Harris, T., and Heymsfield, S.B. 1996. How useful is body mass index for comparison of body fatness across age, sex, and ethnic groups? *American Journal of Epidemiology* 143: 228-239.

Gallagher, M.R., Walker, K.Z., and O'Dea, K. 1998. The influence of a breakfast meal on the assessment of body composition using bioelectrical impedance. *European Journal of Clinical Nutrition* 52: 94-97.

Garland, D.E., Stewart, C.A., Adkins, R.H., Hu, S.S., Rosen, C., Liotta, F.J., and Weinstein, D.A. 1992. Osteoporosis after spinal cord injury. *Journal of Orthopaedic Research* 10: 371-378.

Garrow, J.S., and Webster, J. 1985. Quetelet's index (W/H^2) as a measure of fatness. *International Journal of Obesity and Related Metabolic Disorders* 9: 147-153.

Genton, L., Hans, D., Kyle, U.G., and Pichard, C. 2002. Dual-energy X-ray absorptiometry and body composition: Differences between devices and comparison with reference methods. *Nutrition* 18: 66-70.

Genton, L., Karsegard, V.L., Kyle, U.G., Hans, D.B., Michel, J.P., and Pichard, C. 2001. Comparison of four bioelectrical impedance analysis formulas in healthy elderly subjects. *Gerontology* 47: 315-323.

George, C.M., Wells, C.L., and Dugan, N.L. 1988. Validity of hydrodensitometry for determination of body composition in spinal injured subjects. *Human Biology* 60: 771-780.

George, C.M., Wells, C.L., Dugan, N.L., and Hardison, R. 1987. Hydrostatic weights of patients with spinal injury. Reliability of measurements in standard sit-in and Hubbard tanks. *Physical Therapy* 67: 921-925.

Gerrior, J., Kantaros, J., Coakley, E., Albrecht, M., and Wanke, C. 2001. The fat redistribution syndrome in patients infected with HIV: Measurements of body shape abnormalities. *Journal of the American Dietetic Association* 101: 1175-1180.

Gibson, A.L., Heyward, V.H., and Mermier, C.M. 2000. Predictive accuracy of Omron body logic analyzer in estimating relative body fat of adults. *International Journal of Sport Nutrition and Exercise Metabolism* 10: 216-227.

Gleichauf, C.N., and Rose, D.A. 1989. The menstrual cycle's effect on the reliability of bioimpedance measurements for assessing body composition. *American Journal of Clinical Nutrition* 50: 903-907.

Going, S.B. 1996. Densitometry. In *Human body composition,* ed. A.F. Roche, S.B. Heymsfield, and T.G. Lohman, 3-24. Champaign, IL: Human Kinetics.

Going, S.B., Massett, M.P., Hall, M.C., Bare, L.A., Root, P.A., Williams, D.P., and Lohman, T.G. 1993. Detection of small changes in body composition by dual-energy X-ray absorptiometry. *American Journal of Clinical Nutrition* 57: 845-850.

Goldman, R.F., and Buskirk, E.R. 1961. Body volume measurement by under-water weighing: Description of a method. In *Techniques for measuring body composition,* ed. J. Brozek and A. Henschel, 78-89. Washington, DC: National Academy of Sciences, National Research Council.

Gomez, J.M., Maravall, F.J., Soler, J., and Fernandez-Castaner, M. 2001. Body composition assessment in type 1 diabetes mellitus patients over 15 years old. *Hormone and Metabolism Research* 33: 670-673.

Gonzalez, C., Evans, J.A., Smye, S.W., and Holland, P. 1999. Variables affecting bioimpedance analysis measurements of body water. *Medical & Biological Engineering & Computing* 37: 106-107.

Gonzalez, C.H., Evans, J.A., Smye, S.W., and Holland, P. 2002. Total body water measurement using bioelectrical impedance analysis, isotope dilution and total body potassium: A scoring system to facilitate intercomparison. *European Journal of Clinical Nutrition* 56: 326-337.

Goran, M.I., Allison, D.B., and Poehlman, E.T. 1995. Issues relating to normalization of body fat content in men and women. *International Journal of Obesity and Related Metabolic Disorders* 19: 638-643.

Goran, M.I., Driscoll, P., Johnson, R., Nagy, T.R., and Hunter, G. 1996. Cross-calibration of body-composition techniques against dual-energy X-ray absorptiometry in young children. *American Journal of Clinical Nutrition* 63: 299-305.

Goran, M.I., Toth, M.J., and Poehlman, E.T. 1998. Assessment of research-based body composition techniques in healthy elderly men and women using the 4-component model as a criterion method. *International Journal of Obesity and Related Metabolic Disorders* 22: 135-142.

Gordon, C.C., Chumlea, W.C., and Roche, A.F. 1988. Stature, recumbent length, and weight. In *Anthropometric standardization reference manual,* ed. T.G. Lohman, A.F. Roche, and R. Martorell, 3-8. Champaign, IL: Human Kinetics.

Gore, C.J., Woolford, S.M., and Carlyon, R.G. 1995. Calibrating skinfold calipers. *Journal of Sports Sciences* 13: 355-360.

Gotfredsen, A., Borg, J., Christiansen, C., and Mazess, R.B. 1984. Total body bone mineral in vivo by dual photon absorptiometry. I. Measurement procedures. *Clinical Physiology* 4: 343-355.

Graves, J.E., Pollock, M.L., Colvin, A.B., Van Loan, M., and Lohman, T.G. 1989. Comparison of different bioelectrical impedance analyzers in the prediction of body composition. *American Journal of Human Biology* 1: 603-611.

Gray, D.S., Bray, G.A., Bauer, M., Kaplan, K., Gemayel, N., Wood, R., Greenway, R., and Kirk, S. 1990. Skinfold thickness measurements in obese subjects. *American Journal of Clinical Nutrition* 51: 571-577.

Gray, D.S., Bray, G.A., Gemayel, N., and Kaplan, K. 1989. Effect of obesity on bioelectrical impedance. *American Journal of Clinical Nutrition* 50: 255-260.

Gray, D.S., and Fujioka, K. 1991. Use of relative weight and body mass index for the determination of adiposity. *Journal of Clinical Epidemiology* 44: 545-550.

Gruber, J.J., Pollock, M.L., Graves, J.E., Colvin, A.B., and Braith, R.W. 1990. Comparison of Harpenden and Lange calipers in predicting body composition. *Research Quarterly for Exercise and Sport* 61: 184-190.

Grundy, S., Blackburn, G., Higgins, M., Lauer, R., Perri, M., and Ryan, D. 1999. Roundtable consensus statement. Physical activity in the prevention and treatment of obesity and its comorbidities: Evidence report of independent panel to assess the role of physical activity in the treatment of obesity and its comorbidities. *Medicine & Science in Sports & Exercise* 31 (Suppl.): S502-S508.

Gudivaka, R., Schoeller, D., and Kushner, R.F. 1996. Effect of skin temperature on multifrequency bioelectrical impedance analysis. *Journal of Applied Physiology* 81: 838-845.

Gudivaka, R., Schoeller, D.A., Kushner, R.F., and Bolt, M.J.G. 1999. Single- and multifrequency models for bioelectrical impedance analysis of body water compartments. *Journal of Applied Physiology* 87: 1087-1096.

Gulmans, V.A.M., de Meer, K., Binkhorst, R.A., Helders, P.J.M., and Saris, W.H.M. 1997. Reference values for maximum work capacity in relation to body composition in healthy Dutch children. *European Respiratory Journal* 10: 94-97.

Guo, S.S., and Chumlea, W.C. 1996. Statistical methods for the development and testing of predictive equations. In *Human body composition,* ed. A.F. Roche, S.B. Heymsfield, and T.G. Lohman, 191-202. Champaign, IL: Human Kinetics.

Guo, S., Roche, A.F., and Houtkooper, L. 1989. Fat-free mass in children and young adults predicted from bioelectric impedance and anthropometric variables. *American Journal of Clinical Nutrition* 50: 435-443.

Gutin, B., Litaker, M., Islam, S., Manos, T., Smith, C., and Treiber, F. 1996. Body-composition measurement in 9-11-y-old children by dual-energy X-ray absorptiometry, skinfold-thickness measurements, and bioimpedance analysis. *American Journal of Clinical Nutrition* 63: 287-292.

Haarbo, J., Gotfredsen, A., Hassajer, C., and Christiansen, C. 1991. Validation of body composition by dual energy X-ray absorptiometry (DEXA). *Clinical Physiology* 11: 331-341.

Habash, D. 2002. Tactile and interpersonal techniques for fatfold anthropometry. Unpublished paper.

Han, T.S., Carter, R., Currall, J.E.P., and Lean, M.E.J. 1996. The influence of fat free mass on prediction of densitometric body composition by bioelectrical impedance analysis and by anthropometry. *European Journal of Clinical Nutrition* 50: 542-548.

Han, T.S., McNeill, G., Seidell, J.C., and Lean, M.E. 1997. Predicting intra-abdominal fatness from anthropometric measures: The influence of stature. *International Journal of Obesity and Related Metabolic Disorders* 21: 587-593.

Hannan, W.J., Cowen, S., Freeman, C., Mackie, A., and Shapiro, C.M. 1990. Assessment of body composition in anorexic patients. In *Advances in in vivo body composition studies,* ed. S. Yasumura, 149-154. New York: Plenum Press.

Hannan, W.J., Cowen, S.J., Freeman, C.P., and Wrate, R.M. 1993. Can bioelectrical impedance improve the prediction of body fat in patients with eating disorders? *European Journal of Clinical Nutrition* 47: 741-746.

Hansen, N.J., Lohman, T.G., Going, S.B., Hall, M.C., Pamenter, R.W., Bare, L.A., Boyden, T.W., and Houtkooper, L.B. 1993. Prediction of body composition in premenopausal females from dual-energy x-ray absorptiometry. *Journal of Applied Physiology* 75: 1637-1641.

Harris, M.I., Cowie, C.C., Stern, M.P., Boyko, E.J., Reiber, G.E., and Bennett, P.H. 1995. *Diabetes in America,* 2nd ed. Washington, DC: U.S. Department of Health and Human Services, Public Health Service, National Institutes of Health. DHHS (NIH) Pub. No. 95-1468.

Harrison, G.G., Buskirk, E.R., Carter, J.E.L., Johnston, F.E., Lohman, T.G., Pollock, M.L., Roche, A.F., and Wilmore, J.H. 1988. Skinfold thicknesses and measurement technique. In *Anthropometric standardization reference manual,* ed. T.G. Lohman, A.F. Roche, and R. Martorell, 55-70. Champaign, IL: Human Kinetics.

Harsha, D.W., Frerichs, R.R., and Berenson, G.S. 1978. Densitometry and anthropometry of black and white children. *Human Biology* 50: 261-280.

Hartman, D., Crisp, A., Rooney, B., Rackow, C., Atkinson, R., and Patel, S. 2000. Bone density of women who have recovered from anorexia nervosa. *International Journal of Eating Disorders* 28: 107-112.

Haschke, F., 1983. Body composition of adolescent males. Part II. Body composition of the male reference adolescent. *Acta Paediatrica Scandinavica* 307 (Suppl.): 11-23.

Hawkins, J.D. 1983. An analysis of selected skinfold measuring instruments. *Journal of Health, Physical Education, Recreation and Dance* 54(1): 25-27.

Hayes, P.A., Sowood, P.J., Belyavin, A., Cohen, J.B., and Smith, F.W. 1988. Sub-cutaneous fat thickness measured by magnetic resonance imaging, ultrasound, and calipers. *Medicine & Science in Sports & Exercise* 20: 303-309.

He, M., Li, E.T.S., and Kung, A.W.C. 1999. Dual-energy X-ray absorptiometry for body composition estimation in Chinese women. *European Journal of Clinical Nutrition* 53: 933-937.

Heath, E.M., Adams, T.D., Daines, M.M., and Hunt, S.C. 1998. Bioelectrical impedance and hydrostatic weighing with and without head submersion in persons who are morbidly obese. *Journal of the American Dietetic Association* 98: 869-875.

Heitmann, B.L. 1990. Prediction of body water and fat in adult Danes from measurement of electrical impedance. A validation study. *International Journal of Obesity and Related Metabolic Disorders* 14: 789-802.

Heitmann, B.L., Kondrup, J., Engelhart, M., Kristensen, J.H., Podenphant, J., Hoie, L.H., and Andersen, V. 1994. Changes in fat free mass in overweight patients with rheumatoid arthritis on a weight reducing regimen. A comparison of eight different body composition methods. *International Journal of Obesity and Related Metabolic Disorders* 18: 812-819.

Hendel, H.W., Gotfredsen, A., Hojgaard, L., Andersen, T., and Hilsted, J. 1996. Change in fat-free mass assessed by bioelectrical impedance, total body potassium and dual energy X-ray absorptiometry during prolonged weight loss. *Scandinavian Journal of Clinical and Laboratory Investigation* 56: 671-679.

Henderson, R.C., and Madsen, C.D. 1999. Bone mineral content and body composition in children and young adults with cystic fibrosis. *Pediatric Pulmonology* 27: 80-84.

Henderson, R.C., and Specter, B.B. 1994. Kyphosis and fractures in children and young adults with cystic fibrosis. *Journal of Pediatrics* 125: 208-212.

Hergenroeder, A.C., Brown, B., and Klish, W.J. 1993. Anthropometric measurements and estimating body composition in ballet dancers. *Medicine & Science in Sports & Exercise* 25: 145-150.

Heymsfield, S.B., Lichtman, S., Baumgartner, R.N., Wang, J., Kamen, Y., Aliprantis, A., and Pierson, R.N. 1990. Body composition in humans: Comparison of two improved four-component models that differ in expense, technical complexity, and radiation exposure. *American Journal of Clinical Nutrition* 52: 52-58.

Heymsfield, S.B., Wang, J., Lichtman, S., Kamen, Y., Kehayias, J., and Pierson, R.N. 1989. Body composition in elderly subjects: A critical appraisal of clinical methodology. *American Journal of Clinical Nutrition* 50: 1167-1175.

Heymsfield, S.B., Wang, Z.M., and Withers, R.T. 1996. Multicomponent molecular level models of body composition analysis. In *Human body composition,* ed. A.F. Roche, S.B. Heymsfield, and T.G. Lohman, 129-148. Champaign, IL: Human Kinetics.

Heyward, V.H. 2001. ASEP methods recommendation: Body composition assessment. *Journal of Exercise Physiology$_{online}$* 4(4): 1-12.

Heyward, V.H., Cook, K.L., Hicks, V.L., Jenkins, K.A., Quatrochi, J.A., and Wilson, W. 1992. Predictive accuracy of three field methods for estimating relative body fatness of nonobese and obese women. *International Journal of Sport Nutrition* 2: 75-86.

Heyward, V.H., Hicks, V., Reano, L., and Stolarczyk, L. 1996. Comparison of dual-energy X-ray absorptiometry and four-component model estimates of body fat in American Indian men. Presented at International In Vivo Body Composition Symposium, Malmo, Sweden, September.

Heyward, V.H., Hicks, V., Reano, L., and Stolarczyk, L. 1998. Comparison of dual-energy X-ray absorptiometry and four-component model estimates of body fat in American Indian men. *Applied Radiation and Isotopes* 49: 625-626.

Heyward, V.H., Jenkins, K.A., Cook, K.L., Hicks, V.L., Quatrochi, J.A., Wilson, W., and Going, S. 1992. Validity of single-site and multi-site models of estimating body composition of women using near-infrared interactance. *American Journal of Human Biology* 4: 579-593.

Heyward, V.H., Jenkins, K.A., Mermier, C.M., and Stolarczyk, L. 1993. Sources of variability for optical density measures. *Medicine & Science in Sports & Exercise* 25: S60 [abstract].

Heyward, V.H., Stolarczyk, L.M., Goodman, J.A., Grant, D., Kessler, K., Kocina, P., and Wilmerding, V. 1995. Comparison of two component and multi-component models in estimating body composition of Hispanic women. *Medicine & Science in Sports & Exercise* 27: S118 [abstract].

Heyward, V.H., Wilson, W.L., and Stolarczyk, L.M. 1994. Predictive accuracy of BIA equations for estimating fat-free mass of American Indian, Black, and Hispanic men. *Medicine & Science in Sports & Exercise* 26: S202 [abstract].

Hicks, V.L., Heyward, V.H., Baumgartner, R.N., Flores, A.J., Stolarczyk, L.M., and Wotruba, E.A. 1993. Body composition of native-American women estimated by dual-energy X-ray absorptiometry and hydrodensitometry. In *Human body composition: In vivo methods, models and assessment,* ed. K.J. Ellis and J.D. Eastman, 89-92. New York: Plenum Press.

Hicks, V.L., Stolarczyk, L.M., Heyward, V.H., and Baumgartner, R.N. 2000. Validation of near-infrared interactance and skinfold methods for estimating body composition of American Indian women. *Medicine & Science in Sports & Exercise* 32: 531-539.

Hicks, V.L., Wilson, W.L., and Heyward, V.H. 1992. Accuracy of bioelectrical impedance in estimating body composition of American Indians. *Medicine & Science in Sports & Exercise* 24: S6 [abstract].

Higgins, P.B., Fields, D.A., Hunter, G.R., and Gower, B.A. 2001. Effect of scalp and facial hair on air displacement plethysmography estimates of percentage of body fat. *Obesity Research* 9: 326-330.

Hill, J.O., and Melanson, E.L. 1999. Overview of the determinants of overweight and obesity: Current evidence and research issues. *Medicine & Science in Sports & Exercise* 31: S515-S521.

Himes, J.H., and Bouchard, C. 1985. Do the new Metropolitan Life Insurance weight-height tables correctly assess body frame and body fat relationships? *American Journal of Public Health* 75: 1076-1079.

Himes, J.H., and Frisancho, R.A. 1988. Estimating frame size. In *Anthropometric standardization reference manual,* ed. T.G. Lohman, A.F. Roche, and R. Martorell, 121-124. Champaign, IL: Human Kinetics.

Ho, C.P., Kim, R.W., Schaffler, M.B., and Sartoris, D.J. 1990. Accuracy of dual-energy radiographic absorptiometry of the lumbar spine: Cadaver study. *Radiology* 176: 171-173.

Hodgdon, J.A. 1992. Body composition in the military services: Standards and methods. In *Body composition and physical performance: Applications for the military services,* ed. B.M. Marriott and J. Grumstrup-Scott, 57-70. Washington, DC: National Academy Press.

Hodgdon, J.A., and Beckett, M.B. 1984. Prediction of %BF for U.S. Navy women from body circumferences and height. San Diego: Naval Health Research Center. Technical Report No. T17-88.

Hoffer, E.C., Meador, C.K., and Simpson, D.C. 1969. Correlation of whole-body impedance with total body water volume. *Journal of Applied Physiology* 27: 531-534.

Horswill, C.A., Lohman, T.G., Slaughter, M.H., Boileau, R.A., and Wilmore, J.H. 1990. Estimation of minimal weight of adolescent males using multicomponent models. *Medicine & Science in Sports & Exercise* 22: 528-532.

Hortobagyi, T., Israel, R.G., Houmard, J.A., McCammon, M.R., and O'Brien, K.F. 1992. Comparison of body composition assessment by hydrodensitometry, skinfolds, and multiple site near-infrared spectrophotometry. *European Journal of Clinical Nutrition* 46: 205-211.

Hortobagyi, T., Israel, R.G., Houmard, J.A., O'Brien, K.F., Johns, R.A., and Wells, J.M. 1992. Comparison of four methods to assess body composition in black and white athletes. *International Journal of Sport Nutrition* 2: 60-74.

Houmard, J.A., Israel, R.G., McCammon, M.R., O'Brien, K.F., Omer, J., and Zamora, B.S. 1991. Validity of a near-infrared device for estimating body composition in a college football team. *Journal of Applied Sport Science Research* 5: 53-59.

Housh, T.J., Johnson, G.O., Housh, D.J., Eckerson, J.M., and Stout, J.R. 1996. Validity of skinfold estimates of percent fat in high school female gymnasts. *Medicine & Science in Sports & Exercise* 28: 1331-1335.

Houtkooper, L.B., Going, S.B., Lohman, T.G., Roche, A.F., and Van Loan, M. 1992. Bioelectrical impedance estimation of fat-free body mass in children and youth: A cross-validation study. *Journal of Applied Physiology* 72: 366-373.

Houtkooper, L.B., Going, S.B., Sproul, J., Blew, R.M., and Lohman, T.G. 2000. Comparison of methods for assessing body-composition changes over 1 y in postmenopausal women. *American Journal of Clinical Nutrition* 72: 401-406.

Houtkooper, L.B., Lohman, T.G., Going, S.B., and Hall, M.C. 1989. Validity of bioelectric impedance for body composition assessment in children. *Journal of Applied Physiology* 66: 814-821.

Houtkooper, L.B., Lohman, T.G., Going, S.B., and Howell, W.H. 1996. Why bioelectrical impedance analysis should be used for estimating adiposity. *American Journal of Clinical Nutrition* 64: 436S-448S.

Houtkooper, L.B., Mullins, V.A., Going, S.B., Brown, C.H., and Lohman, T.G. 2001. Body composition profiles of elite American heptathletes. *International Journal of Sport Nutrition and Exercise Metabolism* 11: 162-173.

Human Kinetics, 1995. *Practical body composition kit.* Champaign, IL: Author [includes guide, plastic calipers, tape measure, software, and video].

Human Kinetics. 1999. *Assessing body composition.* Champaign, IL: Author [self-study distance education course].

Irwin, M.L., Ainsworth, B.E., Stolarczyk, L.M., and Heyward, V.H. 1998. Predictive accuracy of skinfold equations for estimating body density of African-American women. *Medicine & Science in Sports & Exercise* 30: 1654-1658.

Israel, R.G., Houmard, J.A., O'Brien, K.F., McCammon, M.R., Zamora, B.S., and Eaton, A.W. 1989. Validity of near-infrared spectrophotometry device for estimating human body composition. *Research Quarterly for Exercise and Sport* 60: 379-383.

Jackson, A.S. 1984. Research design and analysis of data procedures for predicting body density. *Medicine & Science in Sports & Exercise* 16: 616-620.

Jackson, A.S. 1989. Application of regression analysis to exercise science. In *Measurement concepts in physical education and exercise science,* ed. M.J. Safrit and T.M. Wood, 181-205. Champaign, IL: Human Kinetics.

Jackson, A.S., and Pollock, M.L. 1976. Factor analysis and multivariate scaling of anthropometric variables for the assessment of body composition. *Medicine & Science in Sports & Exercise* 8: 196-203.

Jackson, A.S., and Pollock, M.L. 1978. Generalized equations for predicting body density of men. *British Journal of Nutrition* 40: 487-504.

Jackson, A.S., and Pollock, M.L. 1985. Practical assessment of body composition. *Physician and Sportsmedicine* 13: 76-90.

Jackson, A.S., Pollock, M.L., Graves, J.E., and Mahar, M.T. 1988. Reliability and validity of bioelectrical impedance in determining body composition. *Journal of Applied Physiology* 64: 529-534.

Jackson, A.S., Pollock, M.L., and Ward, A. 1980. Generalized equations for predicting body density of women. *Medicine & Science in Sports & Exercise* 12: 175-182.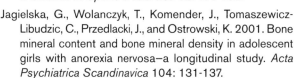

Jagielska, G., Wolanczyk, T., Komender, J., Tomaszewicz-Libudzic, C., Przedlacki, J., and Ostrowski, K. 2001. Bone mineral content and bone mineral density in adolescent girls with anorexia nervosa—a longitudinal study. *Acta Psychiatrica Scandinavica* 104: 131-137.

Jakicic, J.M., Wing, R.R., and Lang, W. 1998. Bioelectrical impedance analysis to assess body composition in obese adult women: The effect of ethnicity. *International Journal of Obesity and Related Metabolic Disorders* 22: 243-249.

Janssen, I., Heymsfield, S.B., Allison, D.B., Kotler, D.P., and Ross, R. 2002. Body mass index and waist circumference independently contribute to the prediction of nonabdominal, abdominal subcutaneous, and visceral fat. *American Journal of Clinical Nutrition* 75: 683-688.

Janz, K.F., Nielsen, D.H., Cassady, S.L., Cook, J.S., Wu, Y., and Hansen, J.R. 1993. Cross-validation of the Slaughter skinfold equations for children and adolescents. *Medicine & Science in Sports & Exercise* 25: 1070-1076.

Jebb, S.A., Cole, T.J., Doman, D., Murgatroyd, P.R., and Prentice, A.M. 2000. Evaluation of the novel Tanita body-fat analyzer to measure body composition by comparison with a four-component model. *British Journal of Nutrition* 83: 115-122.

Jebb, S.A., Murgatroyd, P.R., Goldberg, G.R., Prentice, A.M., and Coward, W.A. 1993. In vivo measurement of changes in body composition: Description of methods and their validation against 12-d continuous whole-body calorimetry. *American Journal of Clinical Nutrition* 58: 455-462.

Jenkins, K.A., Heyward, V.H., Cook, K.L., Hicks, V.L., Quatrochi, J.A., Wilson, W.L., and Colville, B.C. 1994. Predictive accuracy of bioelectrical impedance equations for women. *American Journal of Human Biology* 6: 293-303.

Johansson, A.C., Samuelsson, O., Attman, P.O., Bosaeus, I., and Haraldsson, B. 2001. Limitations in anthropometric calculations of total body water in patients on peritoneal dialysis. *Journal of the American Society of Nephrology* 12: 568-573.

Johnson, J., and Dawson-Hughes, B. 1991. Precision and stability of dual-energy X-ray absorptiometry measurements. *Calcified Tissue International* 49: 174-178.

Johnston, J.L., Leong, M.S., Checkland, E.G., Zuberbahler, P.C., Conger, P.R., and Quinney, H.A. 1988. Body fat assessed from body density and estimated from skinfold thickness in normal children and children with cystic fibrosis. *American Journal of Clinical Nutrition* 48: 1362-1366.

Jones, L.M., Goulding, A., and Gerrard, D.F. 1998. DEXA: A practical and accurate tool to demonstrate total and regional bone loss, lean tissue loss and fat mass gain in paraplegia. *Spinal Cord* 36: 637-640.

Kahn, H.S. 1991. A major error in nomograms for estimating body mass index. *American Journal of Clinical Nutrition* 54: 435-437.

Kalantar-Zadeh, K., Block, G., Kelly, M.P., Schroepfer, C., Rodriguez, R.A., and Humphreys, M.H. 2001. Near infrared interactance for longitudinal assessment of nutrition in dialysis patients. *Journal of Renal Nutrition* 11: 23-31.

Katch, F.I., and McArdle, W.D. 1973. Prediction of body density from simple anthropometric measurements in college-age men and women. *Human Biology* 45: 445-454.

Katch, F.I., Solomon, R.T., Shayevitz, M., and Shayevitz, B. 1986. Validity of bioelectrical impedance to estimate body composition in cardiac and pulmonary patients. *American Journal of Clinical Nutrition* 43: 972-973.

Kerruish, K.P., O'Connor, J., Humphries, I.R.J., Kohn, M.R., Clarke, S.D., Briody, J.N., Thomson, E.J., Wright, K.A., Gaskin, K.J., and Baur, L.A. 2002. Body composition in adolescents with anorexia nervosa. *American Journal of Clinical Nutrition* 75: 31-37.

Keteyian, S.J., Marks, C.R.C., Fedel, F.J., Ehrman, J.K., Goslin, B.R., Connolly, A.M., Fachnie, J.D., Levine, T.B., and O'Neil, M.J. 1992. Assessment of body composition in heart transplant patients. *Medicine & Science in Sports & Exercise* 24: 247-252.

Keys, A., and Brozek, J. 1953. Body fat in adult man. *Physiological Reviews* 33: 245-325.

Keys, A., Fidanza, F., Karvonen, M., Kimura, N., and Taylor, H.L. 1972. Indices of relative weight and obesity. *Journal of Chronic Diseases* 25: 329-343.

Khaled, M.A., McCutcheon, M.J., Reddy, S., Pearman, P.L., Hunter, G.R., and Weinsier, R.L. 1988. Electrical impedance in assessing human body composition: The BIA method. *American Journal of Clinical Nutrition* 47: 789-792.

Kim, H.K., Tanaka, K., Nakadomo, F., and Watanabe, K. 1994. Fat-free mass in Japanese boys predicted from bioelectrical impedance and anthropometric variables. *European Journal of Clinical Nutrition* 48: 482-489.

Kistorp, C.N., and Svendsen, O.L. 1997. Body composition analysis by dual energy X-ray absorptiometry in female diabetics differ between manufacturers. *European Journal of Clinical Nutrition* 51: 449-454.

Kistorp, C.N., and Svendsen, O.L. 1998. Body composition results by DXA differ with manufacturer, instrument generation and software version. *Applied Radiation and Isotopes* 49: 515-516.

Klimis-Tavantis, D., Oulare, M., Lehnhard, H., and Cook, R.A. 1992. Near-infrared interactance: Validity and use in estimating body composition of adolescents. *Nutrition Research* 12: 427-439.

Koch, J. 1998. The role of body composition measurements in wasting syndromes. *Seminars in Oncology* 25: 12-19.

Kocina, P. 1997. Body composition of spinal cord injured adults. *Sports Medicine* 23: 48-60.

Kocina, P.S., and Heyward, V.H. 1997. Validation of a bio-impedance equation for estimating fat-free mass of spinal cord injured adults. *Medicine & Science in Sports & Exercise* 29: S55 [abstract].

Koda, M., Tsuzuku, S., Ando, F., Niino, N., and Shimokata, H. 2000. Body composition by air displacement plethysmography in middle-aged and elderly Japanese. *Annals of the New York Academy of Sciences* 904: 484-488.

Kohrt, W.M. 1995. Body composition by DEXA: Tried and true? *Medicine & Science in Sports & Exercise* 27: 1349-1353.

Kohrt, W.M. 1998. Preliminary evidence that DEXA provides an accurate assessment of body composition. *Journal of Applied Physiology* 84: 372-377.

Kooh, S.W., Noriega, E., Leslie, K., Muller, C., and Harrison, J.E. 1996. Bone mass and soft tissue composition in adolescents with anorexia nervosa. *Bone* 19: 181-188.

Kormas, N., Diamond, T., O'Sullivan, A., and Smerdely, P. 1998. Body mass and body composition after total thyroidectomy for benign goiters. *Thyroid* 8: 773-776.

Kotler, D.P., Burastero, S., Wang, J., and Pierson, R.N. 1996. Prediction of body cell mass, and total body water with bioelectrical impedance analysis: Effects of race, sex, and disease. *American Journal of Clinical Nutrition* 64: 489S-497S.

Kotler, D.P., Rosenbaum, K., Allison, D.B., Wang, J., and Pierson, R.N. 1999. Validation of bioimpedance analysis as a measure of change in body cell mass as estimated by whole-body counting of potassium in adults. *Journal of Parenteral and Enteral Nutrition* 23: 345-349.

Kotler, D.P., Thea, D.M., Heo, M., Allison, D.B., Engelson, E.S., Wang, J., Pierson, R.N., St Louis, M., and Keusch, G.T. 1999. Relative influences of sex, race, environment, and HIV infection on composition in adults. *American Journal of Clinical Nutrition* 69: 432-439.

Kotler, D.P., Tierney, A.R., Wang, J., and Pierson, R.N. 1989. Magnitude of body-cell-mass depletion and the timing of death from wasting in AIDS. *American Journal of Clinical Nutrition* 50: 444-447.

Koulmann, N., Jimenez, C., Regal, D., Bolliet, P., Launay, J., Savourey, G., and Melin, B. 2000. Use of bioelectrical impedance analysis to estimate body fluid compartments after acute variations of the body hydration level. *Medicine & Science in Sports & Exercise* 32: 857-864.

Kriska, A.M., Blair, S.N., and Pereira, M.A. 1994. The potential role of physical activity in the prevention of non-insulin dependent diabetes mellitus: The epidemiological evidence. In *Exercise and Sport Sciences Reviews,* ed. J.O. Holloszy, 22: 121-143.

Krosnick, A. 2000. The diabetes and obesity epidemic among the Pima Indians. *New Jersey Medicine* 97(8): 31-37.

Kuczmarski, R.J. 1989. Need for body composition information in elderly subjects. *American Journal of Clinical Nutrition* 50: 1150-1157.

Kuczmarski, R.J., Flegal, K.M., Campbell, S.M., and Johnson, C.L. 1994. Increasing prevalence of overweight among U.S. adults: The National Health and Nutrition Examination Surveys, 1960-1991. *Journal of the American Medical Association* 272: 205-211.

Kushner, R.F. 1992. Bioelectrical impedance analysis: A review of principles and applications. *Journal of the American College of Nutrition* 11: 199-209.

Kushner, R.F., Gudivaka, R., and Schoeller, D.A. 1996. Clinical characteristics influencing bioelectrical impedance analysis measurements. *American Journal of Clinical Nutrition* 64: 423S-427S.

Kushner, R.F., Kunigk, A., Alspaugh, M., Andronis, P.T., Leitch, C.A., and Schoeller, D.A. 1990. Validation of bioelectrical-impedance analysis as a measurement of change in body composition in obesity. *American Journal of Clinical Nutrition* 52: 219-223.

Kushner, R.F., and Schoeller, D.A. 1986. Estimation of total body water in bioelectrical impedance analysis. *American Journal of Clinical Nutrition* 44: 417-424.

Kushner, R.F., Schoeller, D.A., Fjeld, C.R., and Danford, L. 1992. Is the impedance index (ht^2/R) significant in predicting total body water? *American Journal of Clinical Nutrition* 56: 835-839.

Kwok, T., Woo, J., and Lau, E. 2001. Prediction of body fat by anthropometry in older Chinese people. *Obesity Research* 9: 97-101.

Kyle, U.G., Genton, L., Karsegard, L., Slosman, D.O., and Pichard, C. 2001. Single prediction equation for bioelectrical impedance analysis in adults aged 20-94 years. *Nutrition* 17: 248-253.

Kyle, U.G., Genton, L., Mentha, G., Nicod, L., Slosman, D.O., and Pichard, C. 2001. Reliable bioelectrical impedance analysis estimate of fat-free mass in liver, lung, and heart transplant patients. *Journal of Parenteral and Enteral Nutrition* 25: 45-51.

Kyle, U.G., and Pichard, C. 2000. Dynamic assessment of fat-free mass during catabolism and recovery. *Current Opinion in Clinical Nutrition and Metabolic Care* 3: 317-322.

Kyle, U.G., Pichard, C., Rochat, T., Slosman, D.O., Fitting, J-W., and Thiebaud, D. 1998. New bioelectrical impedance formula for patients with respiratory insufficiency: Comparison to dual-energy X-ray absorptiometry. *European Respiratory Journal* 12: 960-966.

Lambert, C.P., Archer, R.L., and Evans, W.J. 2002. Body composition in ambulatory women with multiple sclerosis. *Archives of Physical Medicine and Rehabilitation* 83: 1559-1561.

Lands, L.C., Gordon, C., Bar-Or, O., Blimkie, C.J., Hanning, R.M., Jones, N.L., Moss, L.A., Webber, C.E., Wilson, W.M., and Heigenhauser, G.J.F. 1993. Comparison of three techniques for body composition analysis in cystic fibrosis. *Journal of Applied Physiology* 75: 162-166.

Lang, P., Steiger, P., Faulkner, K., Gluer, C., and Genant, H.K. 1991. Osteoporosis: Current techniques and recent developments in quantitative bone densitometry. *Radiologic Clinics of North America* 29: 49-76.

Lean, M.E.J., Han, T.S., and Deurenberg, P. 1996. Predicting body composition by densitometry from simple anthropometric measurements. *American Journal of Clinical Nutrition* 63: 4-14.

Lee, J., and Hinds, M.W. 1981. Relative merits of the weight-corrected-for-height indices. *American Journal of Clinical Nutrition* 34: 2521-2529.

Leger, L.A., Lambert, J., and Martin, P. 1982. Validity of plastic skinfold caliper measurements. *Human Biology* 54: 667-675.

Lehnert, M.E., Clarke, D.D., Gibbons, J.G., Ward, L.C., Golding, S.M., Shepherd, R.W., Cornish, B.H., and Crawford, D.H.G. 2001. Estimation of body water compartments in cirrhosis by multiple-frequency bioelectrical impedance analysis. *Nutrition* 17: 31-34.

Leiter, L.A., and the Diabetes Control and Complications Trial Research Group. 1996. Use of bioelectrical impedance analysis measurements in patients with diabetes. *American Journal of Clinical Nutrition* 64 (Suppl.): 515S-518S.

Leiter, L.A., Lukaski, H.C., Kenny, D.J., Barnie, A., Camelon, K., Ferguson, R.S., MacLean, S., Simkins, S., Zinman, B., and Cleary, P.A. 1994. The use of bioelectrical impedance analysis (BIA) to estimate body composition in the Diabetes Control and Complications Trial (DCCT). *International Journal of Obesity and Related Metabolic Disorders* 18: 829-835.

Lemieux, S., Prud'homme, D., Bouchard, C., Tremblay, A., and Despres, J.P. 1996. A single threshold value of waist girth identifies normal-weight and overweight subjects with excess visceral adipose tissue. *American Journal of Clinical Nutrition* 64: 685-693.

Leroy-Willig, A., Willig, T.N., Henry-Feugeas, M.C., Frouin, V., Marinier, E., Boulier, A., Barzic, F., Schouman-Claeys, E., and Syrota, A. 1997. Body composition determined with MR in patients with Duchenne muscular dystrophy, spinal muscular atrophy, and normal subjects. *Magnetic Resonance Imaging* 15: 737-744.

Levenhagen, D.K., Borel, M.J., Welch, D.C., Piasecki, J.H., Piasecki, D.P., Chen, K.Y., and Flakoll, P.J. 1999. A comparison of air displacement plethysmography with three other techniques to determine body fat in healthy adults. *Journal of Parenteral and Enteral Nutrition* 23: 293-299.

Levin, M.E., Vincenza, C., Boisseau, V.C., and Avioli, L.V. 1976. Effects of diabetes mellitus on bone mass in juvenile and adult-onset diabetes. *New England Journal of Medicine* 294: 241-245.

Lewy, V.D., Danadian, K., and Arslanian, S. 1999. Determination of body composition in African-American children: Validation of bioelectrical impedance from dual energy X-ray absorptiometry. *Journal of Pediatric Endocrinology & Metabolism* 12: 443-448.

Liang, M.Y., and Norris, S. 1993. Effects of skin blood flow and temperature on bioelectrical impedance after exercise. *Medicine & Science in Sports & Exercise* 25: 1231-1239.

Liang, M.T.C., Su, H., and Lee, N. 2000. Skin temperature and skin blood flow affect bioelectrical impedance study of female fat-free mass. *Medicine & Science in Sports & Exercise* 32: 221-227.

Livingstone, B. 2000. Epidemiology of childhood obesity in Europe. *European Journal of Pediatrics* 159 (Suppl. 1): S14-34.

Lo, W.K., Prowant, B.F., Moore, H.L., Gamboa, S.B., Nolph, K.D., Flynn, M.A., Londeree, B., Keshaviah, P., and Emerson, P. 1994. Comparison of different measurements of lean body mass in normal individuals and in chronic peritoneal dialysis patients. *American Journal of Kidney Diseases* 23: 74-85.

Lockner, D.W., Heyward, V.H., Baumgartner, R.N., and Jenkins, K.A. 2000. Comparison of air-displacement plethysmography, hydrodensitometry, and dual X-ray absorptiometry for assessing body composition of children 10 to 18 years of age. *Annals of the New York Academy of Sciences* 904: 72-78.

Lockner, D.W., Heyward, V.H., Griffin, S.E., Marques, M.B., Stolarczyk, L.M., and Wagner, D.R. 1999. Cross-validation of modified fatness-specific bioelectrical impedance equations. *International Journal of Sport Nutrition* 9: 48-59.

Lockner, D.W., Heyward, V.H., and Kocina, P.S. 1998. Comparison of the body composition of spinal cord injured and able-bodied adults. *Medicine & Science in Sports & Exercise* 30: S237 [abstract].

Lohman, T.G. 1981. Skinfolds and body density and their relation to body fatness: A review. *Human Biology* 53: 181-225.

Lohman, T.G. 1986. Applicability of body composition techniques and constants for children and youth. In *Exercise and Sports Sciences Reviews,* ed. K.B. Pandolf, 325-357. New York: Macmillan.

Lohman, T.G. 1987. *Measuring body fat using skinfolds.* Champaign, IL: Human Kinetics [videotape].

Lohman, T.G. 1989a. Assessment of body composition in children. *Pediatric Exercise Science* 1: 19-30.

Lohman, T.G. 1989b. Bioelectrical impedance. In *Applying new technology to nutrition: Report of the ninth roundtable on medical issues,* 22-25. Columbus, OH: Ross Laboratories.

Lohman, T.G. 1992. *Advances in body composition assessment.* Current Issues in Exercise Science Series. Monograph No. 3. Champaign, IL: Human Kinetics.

Lohman, T.G. 1996. Dual energy X-ray absorptiometry. In *Human body composition,* ed. A.F. Roche, S.B. Heymsfield, and T.G. Lohman, 63-78. Champaign, IL: Human Kinetics.

Lohman, T.G., Boileau, R.A., and Slaughter, M.H. 1984. Body composition in children and youth. In *Advances in pediatric sport sciences,* ed. R.A. Boileau, 29-57. Champaign, IL: Human Kinetics.

Lohman, T.G., Caballero, B., Himes, J.H., Davis, C.E., Stewart, D., Houtkooper, L., Going, S.B., Hunsberger, S., Weber, J.L., Reid, R., and Stephenson, L. 2000. Estimation of body fat from anthropometry and bioelectrical impedance in Native American children. *International Journal of Obesity and Related Metabolic Disorders* 24: 982-988.

Lohman, T.G., Caballero, B., Himes, J.H., Hunsberger, S., Reid, R., Stewart, D., and Skipper, B. 1999. Body composition assessment in American Indian children. *American Journal of Clinical Nutrition* 69: 764S-766S.

Lohman, T.G., Harris, M., Teixeira, P.J., and Weiss, L. 2000. Assessing body composition and changes in body composition: Another look at dual-energy X-ray absorptiometry. *Annals of the New York Academy of Sciences* 904: 45-54.

Lohman, T.G., Houtkooper, L., and Going, S.B. 1997. Body fat measurement goes high-tech: Not all are created equal. *ACSM's Health & Fitness Journal* 7: 30-35.

Lohman, T.G., Pollock, M.L., Slaughter, M.H., Brandon, L.J., and Boileau, R.A. 1984. Methodological factors and the prediction of body fat in female athletes. *Medicine & Science in Sports & Exercise* 16: 92-96.

Lohman, T.G., Roche, A.F., and Martorell, R., eds. 1988. *Anthropometric standardization reference manual.* Champaign, IL: Human Kinetics.

Lohman, T.G., Slaughter, M.H., Boileau, R.A., Bunt, J., and Lussier, L. 1984. Bone mineral measurements and their relation to body density in children, youth, and adults. *Human Biology* 56: 667-679.

Lonn, L., Stenlof, K., Ottosson, M., Lindroos, A.K., Nystrom, E., and Sjostrom, L. 1998. Body weight and body composition changes after treatment of hyperthyroidism. *Journal of Clinical Endocrinology and Metabolism* 83: 4269-4273.

Loy, S.F., Likes, E.A., Andrews, P.M., Vincent, W.J., Holland, G.J., Kawai, H., Cen, S., Swenberger, J., Van Loan, M., Tanaka, K., Heyward, V., Stolarczyk, L., Lohman, T.G., and Going, S.B. 1998. Easy grip on body composition measurements. *ACSM's Health & Fitness Journal* 2(5): 16-19.

Lozano, A., Rosell, J., and Pallas-Areny, R. 1995. Errors in prolonged electrical impedance measurements due to electrode repositioning and postural changes. *Physiological Measurement* 16: 121-130.

Lukaski, H.C. 1986. Use of the tetrapolar bioelectrical impedance method to assess human body composition. In *Human body composition and fat patterning,* ed. N.G. Norgan, 143-158. Waginegen, Netherlands: Euronut.

Lukaski, H.C. 1987. Methods for the assessment of human body composition: Traditional and new. *American Journal of Clinical Nutrition* 46: 537-556.

Lukaski, H.C. 1993. Soft tissue composition and bone mineral status: Evaluation by dual-energy X-ray absorptiometry. *Journal of Nutrition* 123: 438-443.

Lukaski, H.C. 1996. Biological indexes considered in the derivation of the bioelectrical impedance analysis. *American Journal of Clinical Nutrition* 64: 397S-404S.

Lukaski, H.C. 1997. Validation of body composition assessment techniques in the dialysis population. *ASAIO Journal* 43: 251-255.

Lukaski, H.C., and Bolonchuk, W.W. 1987. Theory and validation of the tetrapolar bioelectrical impedance method to assess human body composition. In *In vivo body composition studies,* ed. K.J. Ellis, S. Yasamura, and W.D. Morgan, 410-414. London: Institute of Physical Sciences in Medicine.

Lukaski, H.C., and Bolonchuk, W.W. 1988. Estimation of body fluid volumes using tetrapolar impedance measurements. *Aviation, Space, and Environmental Medicine* 59: 1163-1169.

Lukaski, H.C., Bolonchuk, W.W., Hall, C.B., and Siders, W.A. 1986. Validation of tetrapolar bioelectrical impedance method to assess human body composition. *Journal of Applied Physiology* 60: 1327-1332.

Lukaski, H.C., Bolonchuk, W.W., Siders, W.A., and Hall, C.B. 1990. Body composition assessment of athletes using bioelectrical impedance measurements. *Journal of Sports Medicine and Physical Fitness* 30: 434-440.

Lukaski, H.C., Johnson, P.E., Bolonchuk, W.W., and Lykken, G.I. 1985. Assessment of fat-free mass using bioelectric impedance measurements of the human body. *American Journal of Clinical Nutrition* 41: 810-817.

Luke, A., Durazo-Arvizu, R., Rotimi, C., Prewitt, T.E., Forrester, T., Wilks, R., Ogunbiyi, O.J., Schoeller, D.A., McGee, D., and Cooper, R.S. 1997. Relation between body mass index and body fat in black population samples from Nigeria, Jamaica, and the United States. *American Journal of Epidemiology* 145: 620-628.

Maimoun, L., Couret, I., Micallef, J.P., Peruchon, E., Mariano-Goulart, D., Rossi, M., Leroux, J.L., and Ohanna, F. 2002. Use of bone biochemical markers with dual-energy X-ray absorptiometry for early determination of bone loss in persons with spinal cord injury. *Metabolism* 51: 958-963.

Malina, R.M., Huang, Y., and Brown, K.H. 1995. Subcutaneous adipose tissue distribution in adolescent girls of four ethnic groups. *International Journal of Obesity and Related Metabolic Disorders* 19: 793-797.

Marks, C., and Katch, V. 1986. Biological and technological variability of residual lung volume and the effect on body fat calculations. *Medicine & Science in Sports & Exercise* 18: 485-488.

Martin, A.D., Drinkwater, D.T., and Clarys, J.P. 1992. Effects of skin thickness and skinfold compressibility on skinfold thickness measurements. *American Journal of Human Biology* 4: 453-460.

Martin, A.D., Ross, W.D., Drinkwater, D.T., and Clarys, J.P. 1985. Prediction of body fat by skinfold caliper: Assumptions and cadaver evidence. *International Journal of Obesity and Related Metabolic Disorders* 9 (Suppl. 1): 31-39.

Massari, F., Mastropasqua, F., Guida, P., De Tommasi, E., Rizzon, B., Pontraldolfo, G., Pitzalis, M.V., and Rizzon, P. 2001. Whole-body bioelectrical impedance analysis in patients with chronic heart failure: Reproducibility of the method and effects of body size. *Italian Heart Journal* 2: 594-598.

Mayhew, J.L., Piper, F.C., Koss, M.A., and Montaldi, D.H. 1983. Prediction of body composition in female athletes. *Journal of Sports Medicine and Physical Fitness* 23: 333-340.

Mazariegos, M., Wang, Z.M., Gallagher, D., Baumgartner, R.N., Allison, D.B., Wang, J., Pierson, R.N. Jr., and Heymsfield, S.B. 1994. Differences between young and old females in the five levels of body composition and their relevance to the two-compartment chemical model. *Journal of Gerontology* 49: M201-M208.

Mazess, R.B., Barden, H.S., Bisek, J.P., and Hanson, J. 1990. Dual-energy X-ray absorptiometry for total-body and regional bone-mineral and soft-tissue composition. *American Journal of Clinical Nutrition* 51: 1106-1112.

Mazess, R.B., Barden, H.S., and Ohlrich, E.S. 1990. Skeletal and body-composition effects of anorexia nervosa. *American Journal of Clinical Nutrition* 52: 438-441.

Mazess, R.B., Peppler, W.W., and Gibbons, M. 1984. Total body composition by dual-photon ^{153}Gd absorptiometry. *American Journal of Clinical Nutrition* 40: 834-839.

McCrory, M.A., Gomez, T.D., Bernauer, E.M., and Mole, P.A. 1995. Evaluation of a new air displacement plethysmograph for measuring human body composition. *Medicine & Science in Sports & Exercise* 27: 1686-1691.

McCrory, M.A., Mole, P.A., Gomez, T.D., Dewey, K.G., and Bernauer, E.M. 1998. Body composition by air displacement plethysmography by using predicted and measured thoracic gas volumes. *Journal of Applied Physiology* 84: 1475-1479.

McCullough, A.J., Mullen, K.D., and Kalhan, S.C. 1991. Measurements of total body and extracellular water in cirrhotic patients with and without ascites. *Hepatology* 14: 1102-1111.

McHugh, D., Baumgartner, R.N., Stauber, P.M., Wayne, S., Hicks, V.L., and Heyward, V.H. 1993. Bone mineral in Native American women from New Mexico. In *Human body composition: In vivo methods, models and assessment,* ed. K.J. Ellis and J.D. Eastman, 87-88. New York: Plenum Press.

McLean, K.P., and Skinner, J.S. 1992. Validity of Futrex-5000 for body composition determination. *Medicine & Science in Sports & Exercise* 24: 253-258.

McNair, P., Christiansen, C., Christenssen, M.S., Madsbad, S., Faber, O.K., Binder, C., and Transbol, I. 1981. Development of bone mineral loss in insulin-treated diabetes: A 1 1/2 years follow-up study in sixty patients. *European Journal of Clinical Investigation* 11: 55-59.

McNaughton, S.A., Shepherd, R.W., Greer, R.G., Clegborn, G.J., and Thomas, B.J. 2000. Nutritional status of children with cystic fibrosis measured by total body potassium as a marker of body cell mass: Lack of sensitivity of anthropometric measures. *Journal of Pediatrics* 136: 188-194.

McNeill, G., Fowler, P.A., Maughan, R.J., McGaw, B.A., Fuller, M.F., Gvozdanovic, D., and Gvozdanovic, S. 1991. Body fat in lean and overweight women estimated by six methods. *British Journal of Nutrition* 65: 95-103.

McNulty, K.Y., Adams, C.H., Anderson, J.M., and Affenito, S.G. 2001. Development and validation of a screening tool to identify eating disorders in female athletes. *Journal of the American Dietetic Association* 101: 886-892.

Micozzi, M.S., Albanes, D., Jones, Y., and Chumlea, W.C. 1986. Correlations of body mass indices with weight, stature, and body composition in men and women in NHANES I and II. *American Journal of Clinical Nutrition* 44: 725-731.

Millard-Stafford, M.L., Collins, M.A., Evans, E.M., Snow, T.K., Cureton, K.J., and Rosskopf, L.B. 2001. Use of air displacement plethysmography for estimating body fat in a four-component model. *Medicine & Science in Sports & Exercise* 33: 1311-1317.

Miller, W.C., Swensen, T., and Wallace, J.P. 1998. Derivation of prediction equations for RV in overweight men and women. *Medicine & Science in Sports & Exercise* 30: 322-327.

Milliken, L.A., Going, S.B., and Lohman, T.G. 1996. Effects of variations in regional composition on soft tissue measurements by dual-energy X-ray absorptiometry. *International Journal of Obesity and Related Metabolic Disorders* 20: 677-682.

Mitchell, C.O., Rose, J., Familoni, B., Winters, S., and Ling, F. 1993. The use of multifrequency bioelectrical impedance analysis to estimate fluid volume changes as a function of the menstrual cycle. In *Human body composition: In vivo methods, models and assessment,* ed. K.J. Ellis and J.D. Eastman, 189-191. New York: Plenum Press.

Miyakawa, M., Tsushima, T., Murakami, H., Isozaki, O., and Takano, K. 1999. Serum leptin levels and bioelectrical impedance assessment of body composition in patients with Graves' disease and hypothyroidism. *Endocrine Journal* 46: 665-673.

Miyatake, N., Nonaka, K., and Fujii, M. 1999. A new air displacement plethysmograph for the determination of Japanese body composition. *Diabetes, Obesity and Metabolism* 1: 347-351.

Modlesky, C.M., Cureton, K.J., Lewis, R.D., Prior, B.M., Sloniger, M.A., and Rowe, D.A. 1996. Density of the fat-free mass and estimates of body composition in male weight trainers. *Journal of Applied Physiology* 80: 2085-2096.

Modlesky, C.M., Lewis, R.D., Yetman, K.A., Rose, B., Rosskopf, L.B., Snow, T.K., and Sparling, P.B. 1996. Comparison of body composition and bone mineral measurements from two DXA instruments in young men. *American Journal of Clinical Nutrition* 64: 669-676.

Moley, J.F., Aamodt, R., Rumble, W., Kaye, W., and Norton, J.A. 1987. Body cell mass in cancer-bearing and anorexic patients. *Journal of Parenteral and Enteral Nutrition* 11: 219-222.

Morgan, M.Y., and Madden, A.M. 1996. The assessment of body composition in patients with cirrhosis. *European Journal of Nuclear Medicine* 23: 213-225.

Morrison, J.A., Guo, S.S., Specker, B., Chumlea, W.C., Yanovski, S.Z., and Yanovski, J.A. 2001. Assessing the body composition of 6-17-year-old black and white girls in field studies. *American Journal of Human Biology* 13: 249-254.

Morrow, J.R., Fridye, T., and Monaghen, S.D. 1986. Generalizability of the AAHPERD health-related skinfold test. *Research Quarterly for Exercise and Sport* 57: 187-195.

Morrow, J.R., Jackson, A.S., Bradley, P.W., and Hartung, G.H. 1986. Accuracy of measured and predicted residual lung volume on body density measurement. *Medicine & Science in Sports & Exercise* 18: 647-652.

Motley, H.L. 1957. Comparison of a simple helium closed method with oxygen circuit method for measuring residual air. *American Review of Tuberculosis and Pulmonary Diseases* 76: 601-615.

Mueller, W.H., Shoup, R.F., and Malina, R.M. 1982. Fat patterning in athletes in relation to ethnic origin and sport. *Annals of Human Biology* 9: 371-376.

Munoz-Torres, M., Jodar, E., Escobar-Jimenez, F., Lopez-Ibarra, P.J., and Luna, J.D. 1996. Bone mineral density measured by dual X-ray absorptiometry in Spanish patients with insulin-dependent diabetes mellitus. *Calcified Tissue International* 58: 316-319.

Nagamine, S., and Suzuki, S. 1964. Anthropometry and body composition of Japanese young men and women. *Human Biology* 36: 8-15.

Nakadomo, F., Tanaka, K., Hazama, T., and Maeda, K. 1990. Validation of body composition assessed by bioelectrical impedance analysis. *Japanese Journal of Applied Physiology* 20: 321-330.

National Center for Health Statistics. 2001. *Prevalence of overweight among children and adolescents: United States, 1999.* Washington, DC: Author.

National Center for Health Statistics. 2002. *Health, United States, 2002 with chartbook on trends in the health of Americans.* Hyattsville, MD: U.S. Government Printing Office.

National Cholesterol Education Program (NCEP). 2001. Executive summary of the third report of the National Cholesterol Education Program (NCEP) Expert Panel on Detection, Evaluation, and Treatment of High Cholesterol in Adults (adult treatment panel III). *Journal of the American Medical Association* 285: 2486-2497.

National Institutes of Health. 1994. Bioelectrical impedance analysis in body composition measurement. *NIH technology assessment statement.* Bethesda, MD: Author.

National Research Council. 1989. *Diet and health. Implications for reducing chronic disease risk.* Washington, DC: National Academy Press.

Nattiv, A., Agostina, R., Drinkwater, B., and Yeager, K. 1994. The female athlete triad. *Clinics in Sports Medicine* 13: 405-418.

Nelson, M.E., Fiatarone, M.A., Layne, J.E., Trice, I., Economos, C.D., Fielding, R.A., Ma, R., Pierson, R.N., and Evans, W.J. 1996. Analysis of body-composition techniques and models for detecting change in soft tissue with strength training. *American Journal of Clinical Nutrition* 63: 678-686.

Newby, M.J., Keim, N.L., and Brown, D.L. 1990. Body composition of adult cystic fibrosis patients and control subjects as determined by densitometry, bioelectrical impedance, total-body electrical conductivity, skinfold measurements, and deuterium oxide dilution. *American Journal of Clinical Nutrition* 52: 209-213.

Nichols, D.L., Sanborn, C.F., Bonnick, S.L., Ben-Ezra, V., Gench, B., and DiMarco, N.M. 1994. The effects of gymnastics training on bone mineral density. *Medicine & Science in Sports & Exercise* 26: 1220-1225.

Nicholson, J.C., McDuffie, J.R., Bonat, S.H., Russell, D.L., Boyce, K.A., McCann, S., Michael, M., Sebring, N.G., Reynolds, J.C., and Yanovski, J.A. 2001. Estimation of body fatness by air displacement plethysmography in African American and white children. *Pediatric Research* 50: 467-473.

Nielsen, D.H., Cassady, S.L., Janz, K.F., Cook, J.S., Hansen, J.R., and Wu, Y. 1993. Criterion methods of body composition analysis for children and adolescents. *American Journal of Human Biology* 5: 211-223.

Nielsen, D.H., Cassady, S.L., Wacker, L.M., Wessels, A.K., Wheelock, B.J., and Oppliger, R.A. 1992. Validation of the Futrex-5000 near-infrared spectrophotometer analyzer for assessment of body composition. *Journal of Orthopaedic & Sports Physical Therapy* 16: 281-287.

Norris, K.H. 1983. Instrumental techniques for measuring quality of agricultural crops. In *Post-harvest physiology and crop preservation,* ed. M. Lieberman, 471-484. New York: Plenum Press.

Norton, K., Marfell-Jones, M., Whittingham, N., Kerr, D., Carter, J.E.L., Saddington, K., and Gore, C. 2000. Anthropometric assessment protocols. In *Physiological tests for elite athletes,* ed. C. Gore, 66-85. Champaign, IL: Human Kinetics.

Nuhlicek, D.N., Spurr, G.B., Barboriak, J.J., Rooney, C.B., el Ghatit, A.Z., and Bongard, R.D. 1988. Body composition of patients with spinal cord injury. *European Journal of Clinical Nutrition* 42: 765-773.

Nuñez, C., Gallagher, D., Visser, M., Pi-Sunyer, F.X., Wang, Z., and Heymsfield, S.B. 1997. Bioimpedance analysis: Evaluation of leg-to-leg system based on pressure contact foot pad electrodes. *Medicine & Science in Sports & Exercise* 29: 524-531.

Nuñez, C., Kovera, A.J., Pietrobelli, A., Heshka, S., Horlick, M., Kehayias, J.J., Wang, Z., and Heymsfield, S.B. 1999. Body composition in children and adults by air displacement plethysmography. *European Journal of Clinical Nutrition* 53: 382-387.

O'Brien, C., Baker-Fulco, C.J., Young, A.J., and Sawka, M.N. 1999. Bioimpedance assessment of hypohydration. *Medicine & Science in Sports & Exercise* 31: 1466-1471.

Ogden, C.L., Flegal, K.M., Carroll, M.D., and Johnson, C.L. 2002. Prevalence and trends in overweight among US children and adolescents. *Journal of the American Medical Association* 288(14): 1728-1732.

Ohrvall, M., Berglund, L., and Vessby, B. 2000. Sagittal abdominal diameter compared with other anthropometric measurements in relation to cardiovascular risk *International Journal of Obesity and Related Metabolic Disorders* 24: 497-501.

Okasora, K., Takaya, R., Tokuda, M., Fukunaga, Y., Oguni, T., Tanaka, H., Konishi, K., and Tamai, H. 1999. Comparison of bioelectrical impedance analysis and dual energy X-ray absorptiometry for assessment of body composition in children. *Pediatrics International* 41: 121-125.

Oldham, N.M. 1996. Overview of bioelectrical impedance analyzers. *American Journal of Clinical Nutrition* 64: 405S-412S.

Olds, T., and Norton, K. 1999. *LifeSize.* Champaign, IL: Human Kinetics [software].

Olle, M.M., Pivarnik, J.M., Klish, W.J., and Morrow, J.R. 1993. Body composition of sedentary and physically active spinal cord injured individuals estimated from total body electrical conductivity. *Archives of Physical Medicine and Rehabilitation* 74: 706-710.

Omron Institute of Life Science. 2002. Adolescent data analysis [personal communication].

Oppliger, R.A., and Cassady, S.L. 1994. Body composition assessment in women: Special considerations for athletes. *Sports Medicine* 17: 353-357.

Oppliger, R.A., Clark, R.R., and Nielsen, D.H. 2000. New equations improve NIR prediction of body fat among high school wrestlers. *Journal of Orthopaedic & Sports Physical Therapy* 30: 536-543.

Oppliger, R.A., Harms, R.D., Herrmann, D.E., Streich, C.M., and Clark, R.R. 1995. The Wisconsin wrestling minimum weight project: A model for weight control among high school wrestlers. *Medicine & Science in Sports & Exercise* 27: 1220-1224.

Oppliger, R.A., Nielsen, D.H., Hoegh, J.E., and Vance, C.G. 1991. Bioelectrical impedance prediction of fat-free mass for high school wrestlers validated. *Medicine & Science in Sports & Exercise* 23: S73 [abstract].

Oppliger, R.A., Nielsen, D.H., Shetler, A.C., Crowley, E.T., and Albright, J.P. 1992. Body composition of collegiate football players: Bioelectrical impedance and skinfolds compared to hydrostatic weighing. *Journal of Orthopaedic & Sports Physical Therapy* 15: 187-192.

Oppliger, R.A., Nielsen, D.H., and Vance, C.G. 1991. Wrestlers' minimal weight: Anthropometry, bioimpedance, and hydrostatic weighing compared. *Medicine & Science in Sports & Exercise* 23: 247-253.

Organ, L.W., Bradham, G.B., Gore, D.T., and Lozier, S.L. 1994. Segmental bioelectrical impedance analysis: Theory and application of a new technique. *Journal of Applied Physiology* 77: 98-112.

Organ, L.W., Eklund, A.D., and Ledbetter, J.D. 1994. An automated real time underwater weighing system. *Medicine & Science in Sports & Exercise* 26: 383-391.

Orphanidou, C., McCargar, L., Birmingham, C.L., Mathieson, J., and Goldner, E. 1994. Accuracy of subcutaneous fat measurement: Comparison of skinfold calipers, ultrasound, and computed tomography. *Journal of the American Dietetic Association* 94: 855-858.

Ortiz, O., Russell, M., Daley, T.L., Baumgartner, R.N., Waki, M., Lichtman, S., Wang, J., Pierson, R.N., and Heymsfield, S.B. 1992. Differences in skeletal muscle and bone mineral mass between black and white females and their relevance to estimates of body composition. *American Journal of Clinical Nutrition* 55: 8-13.

Pace, N., and Rathbun, E.N. 1945. Studies on body composition. III. The body water and chemically combined nitrogen content in relation to fat content. *Journal of Biological Chemistry* 158: 685-691.

Pacy, P.J., Quevedo, M., Gibson, N.R., Cox, M., Koutedakis, Y., and Millward, J. 1995. Body composition measurement in elite heavyweight oarswomen: A comparison of five methods. *Journal of Sports Medicine and Physical Fitness* 35: 67-74.

Paijmans, I.J.M., Wilmore, K.M., and Wilmore, J.H. 1992. Use of skinfolds and bioelectrical impedance for body composition assessment after weight reduction. *Journal of the American College of Nutrition* 11: 145-151.

Panotopoulos, G., Ruiz, J.C., Guy-Grand, B., and Basdevant, A. 2001. Dual X-ray absorptiometry, bioelectrical impedance, and near-infrared interactance in obese women. *Medicine & Science in Sports & Exercise* 33: 665-670.

Paton, N.I., Elia, M., Jennings, G., Ward, L.C., and Griffin, G.E. 1998. Bioelectrical impedance analysis in human immunodeficiency virus-infected patients: Comparison of single frequency with multifrequency, spectroscopy, and other novel approaches. *Nutrition* 14: 658-666.

Paton, N.I.J., Macallan, D.C., Jebb, S.A., Noble, C., Baldwin, C., Pazianas, M., and Griffin, G.E. 1997. Longitudinal changes in body composition measured with a variety of methods in patients with AIDS. *Journal of Acquired Immune Deficiency Syndromes and Human Retrovirology* 14: 119-127.

Pawson, I.G., Martorell, R., and Mendoza, F.E. 1991. Prevalence of overweight and obesity in US Hispanic populations. *American Journal of Clinical Nutrition* 53: 1522S-1528S.

Paxton, A., Lederman, S.A., Heymsfield, S.B., Wang, J., Thornton, J.C., and Pierson, R.N. 1998. Anthropometric equations for studying body fat in pregnant women. *American Journal of Clinical Nutrition* 67: 104-110.

Pedhazuer, E.J. 1982. *Multiple regression in behavioral research.* New York: CBS College.

Penn, I., Wang, Z., Buhl, K.M., Allison, D.B., Burastero, S.E., and Heymsfield, S.B. 1994. Body composition and two-compartment model assumptions in male long distance runners. *Medicine & Science in Sports & Exercise* 26: 392-397.

Peppler, W.W., and Mazess, R.B. 1981. Total body bone mineral and lean body mass by dual-photon absorptiometry. I. Theory and measurement procedure. *Calcified Tissue International* 33: 353-359.

Peters, D., Fox, K., Armstrong, N., Sharpe, P., and Bell, M. 1992. Assessment of children's abdominal fat distribution by magnetic resonance imaging and anthropometry. *International Journal of Obesity and Related Metabolic Disorders* 16 (Suppl. 2): S35 [abstract].

Pichard, C., Kyle, U.G., Gremion, G., Gerbase, M., and Slosman, D.O. 1997. Body composition by X-ray

absorptiometry and bioelectrical impedance in female runners. *Medicine & Science in Sports & Exercise* 29: 1527-1534.

Pichard, C., Kyle, U.G., Janssens, J-P., Burdet, L., Rochat, T., Slosman, D.O., Fitting, J-W., Thiebaud, D., Roulet, M., Tschopp, J-M., Landry, M., and Schutz, Y. 1997. Body composition by X-ray absorptiometry and bioelectrical impedance in chronic respiratory insufficiency patients. *Nutrition* 13: 952-958.

Pichard, C., Kyle, U.G., Slosman, D.O. 1999. Fat-free mass in chronic illness: Comparison of bioelectrical impedance and dual-energy X-ray absorptiometry in 480 chronically ill and healthy subjects. *Nutrition* 15: 668-676.

Pichiecchio, A., Uggetti, C., Egitto, M.G., Berardinelli, A., Orcesi, S., Gorni, K.O.T., Zanardi, C., and Tagliabue, A. 2002. Quantitative MR evaluation of body composition in patients with Duchenne muscular dystrophy. *European Radiology* 12: 2704-2709.

Pierson, R.N., Wang, J., Heymsfield, S.B., Russell-Aulet, M., Mazariegos, M., Tierney, M., Smith, R., Thornton, J.C., Kehayias, J., Weber, D.A., and Dilmanian, F.A. 1991. Measuring body fat: Calibrating the rulers. Intermethod comparisons in 389 normal Caucasian subjects. *American Journal of Physiology* 261: E103-E108.

Pietrobelli, A., Formica, C., Wang, Z., and Heymsfield, S.B. 1996. Dual-energy X-ray absorptiometry body composition model: Review of physical concepts. *American Journal of Physiology* 271: E941-E951.

Pietrobelli, A., Wang, Z., Formica, C., and Heymsfield, S.B. 1998. Dual-energy X-ray absorptiometry: Fat estimation errors due to variation in soft tissue hydration. *American Journal of Physiology* 274: E808-E816.

Pi-Sunyer, F.X. 1999. Comorbidities of overweight and obesity: Current evidence and research issues. *Medicine & Science in Sports & Exercise* 31 (Suppl.): S602-S608.

Pi-Sunyer, F.X. 2000. Obesity: Criteria and classification. *Proceedings of the Nutrition Society* 59: 505-509.

Polito, A., Cuzzolaro, M., Raguzzini, A., Censi, L., and Ferro-Luzzi, A. 1998. Body composition changes in anorexia nervosa. *European Journal of Clinical Nutrition* 52: 655-662.

Pollock, M.L., and Jackson, A.S. 1984. Research progress in validation of clinical methods of assessing body composition. *Medicine & Science in Sports & Exercise* 16: 606-613.

Pollock, M.L., Schmidt, D.H., and Jackson, A.S. 1980. Measurement of cardiorespiratory fitness and body composition in the clinical setting. *Comprehensive Therapy* 6: 12-27.

Preece, M.A., Law, C.M., and Davies, P.S. 1986. The growth of children with chronic paediatric disease. *Clinical Endocrinology and Metabolism* 15: 453-477.

Prijatmoko, D., Strauss, B.J.G., Lambert, J.R., Sievert, W., Stroud, D.B., Wahlqvist, M.L., Katz, B., Colman, J., Jones, P., and Korman, M.G. 1993. Early detection of protein depletion in alcoholic cirrhosis: Role of body composition analysis. *Gastroenterology* 105: 1839-1845.

Prior, B.M., Cureton, K.J., Modlesky, C.M., Evans, E.M., Sloniger, M.A., Saunders, M., and Lewis, R.D. 1997. In vivo validation of whole body composition estimates from dual-energy X-ray absorptiometry. *Journal of Applied Physiology* 83: 623-630.

Prior, B.M., Modlesky, C.M., Evans, E.M., Sloniger, M.A., Saunders, M.J., Lewis, R.D., and Cureton, K.J. 2001. Muscularity and the density of the fat-free mass in athletes. *Journal of Applied Physiology* 90: 1523-1531.

Pugh, R.N.H., Murray-Lyon, I.M., Dawson, J.L., Pietroni, M.C., and Williams, R. 1973. Transection of the esophagus for bleeding esophageal varices. *British Journal of Surgery* 60: 646-649.

Quatrochi, J.A., Hicks, V.L., Heyward, V.H., Colville, B.C., Cook, K.L., Jenkins, K.A., and Wilson, W. 1992. Relationship of optical density and skinfold measurements: Effects of age and level of body fatness. *Research Quarterly for Exercise and Sport* 63: 402-409.

Ragsdale, D. 1987. Nutritional program for heart transplantation. *Journal of Heart Transplantation* 6: 228-233.

Ravaglia, G., Forti, P., Maioli, F., Boschi, F., Cicognani, A., and Gasbarrini, G. 1999. Measurement of body fat in healthy elderly men: A comparison of methods. *Journal of Gerontology* 54A(2): M70-M76.

Reilly, J.J., Murray, L.A., Wilson, J., and Durnin, J.V.G.A. 1994. Measuring the body composition of elderly subjects: A comparison of methods. *British Journal of Nutrition* 72: 33-44.

Reilly, J.J., Wilson, J., and Durnin, J.V.G.A. 1995. Determination of body composition from skinfold thickness: A validation study. *Archives of Disease in Childhood* 73: 305-310.

Reilly, J.J., Wilson, J., McColl, J.H., Carmichael, M., and Durnin, J.V.G. 1996. Ability of bioelectric impedance to predict fat-free mass in prepubertal children. *Pediatric Research* 39: 176-179.

Resch, H., Newrkla, S., Grampp, S., Resch, A., Zapf, S., Piringer, S., Hockl, A., and Weiss, P. 2000. Ultrasound and X-ray-based bone densitometry in patients with anorexia nervosa. *Calcified Tissue International* 66: 338-341.

Riggio, O., Andreoli, A., Diana, F., Fiore, P., Meddi, P., Lionetti, R., Montagnese, F., Merli, M., Capocaccia, L., and DeLorenzo, A. 1997. Whole body and regional body composition analysis by dual-energy X-ray absorptiometry in cirrhotic patients. *European Journal of Clinical Nutrition* 51: 810-814.

Rishaug, U., Birkeland, K.I., Falch, J.A., and Vaaler, S. 1995. Bone mass in non-insulin-dependent diabetes mellitus. *Scandinavian Journal of Clinical Laboratory Investigation* 55: 257-262.

Rising, R., Swinburn, B., Larson, K., and Ravussin, E. 1991. Body composition in Pima Indians: Validation of bioelectrical resistance. *American Journal of Clinical Nutrition* 53: 594-598.

Risser, J.M.H., Rabeneck, L., Foote, L.W., and Klish, W.J. 1995. A comparison of fat-free mass estimates in men infected with the human immunodeficiency virus. *Journal of Parenteral and Enteral Nutrition* 19: 28-32.

Roemmich, J.N., Clark, P.A., Weltman, A., and Rogol, A.D. 1997. Alterations in growth and body composition during puberty. I. Comparing multicompartment body composition models. *Journal of Applied Physiology* 83: 927-935.

Rosenfalck, A.M., Almdal, T., Gotfredsen, A., Hojgaard, L.L., and Hilsted, J. 1995. Validity of dual X-ray absorptiometry scanning for determination of body composition in IDDM patients. *Scandinavian Journal of Clinical Laboratory Investigation* 55: 691-699.

Rosenfalck, A.M., Almdal, T., Gotfredsen, A., Viggers, L., and Hilsted, J. 1997. Increased lean body mass in patients with type 1 insulin-dependent diabetes mellitus: The relationship with metabolic control. *Endocrinology and Metabolism* 4: 241-246.

Rosenfalck, A.M., Almdal, T., Hilsted, J., and Madsbad, S. 2002. Body composition in adults with type 1 diabetes at onset and during the first year of insulin therapy. *Diabetic Medicine* 19: 417-423.

Ross, R., Leger, L., Marliss, E.B., Morris, D.V., and Gougeon, R. 1991. Adipose tissue distribution changes during rapid weight loss in obese adults. *International Journal of Obesity and Related Metabolic Disorders* 15: 733-739.

Ross, R., Leger, L., Martin, P., and Roy, R. 1989. Sensitivity of bioelectrical impedance to detect changes in human body composition. *Journal of Applied Physiology* 67: 1643-1648.

Ross, W.D., and Marfell-Jones, M.J. 1991. Kinanthropometry. In *Physiological testing of the high-performance athlete,* ed. J.D. MacDougall, H.A. Wenger, and H.J. Green, 75-115. Champaign, IL: Human Kinetics.

Roubenoff, R. 2000. Sarcopenia and its implications for the elderly. *European Journal of Clinical Nutrition* 54: S40-S47.

Roubenoff, R., Baumgartner, R.N., Harris, T.B., Dallal, G.E., Hannan, M.T., Economos, C.D., Stauber, P.M., Wilson, P.W.F., and Kiel, D.P. 1997. Application of bioelectrical impedance analysis to elderly populations. *Journal of Gerontology* 52A(3): M129-M136.

Roubenoff, R., Kehayias, J.J., Dawson-Hughes, B., and Heymsfield, S.B. 1993. Use of dual-energy X-ray absorptiometry in body-composition studies: Not yet a "gold standard." *American Journal of Clinical Nutrition* 58: 589-591.

Rush, E.C., Plank, L.D., Laulu, M.S., and Robinson, S.M. 1997. Prediction of percentage body fat from anthropometric measurements: Comparison of New Zealand European and Polynesian young women. *American Journal of Clinical Nutrition* 66: 2-7.

Sampei, M.A., Novo, N.F., Juliano, Y., and Sigulem, D.M. 2001. Comparison of the body mass index to other methods of body fat evaluation in ethnic Japanese and Caucasian adolescent girls. *International Journal of Obesity and Related Metabolic Disorders* 25: 400-408.

Sanborn, C.F. 1986. Etiology of athletic amenorrhea. In *The menstrual cycle and physical activity,* ed. J.L. Puhl and C.H. Brown, 45-58. Champaign, IL: Human Kinetics.

Santolaria, F., Gonzalez-Reimers, E., Perez-Manzano, J.L., Milena, A., Gomez-Rodriguez, M.A., Gonzalez-Diaz, A., de la Vega, M.J., and Martinez-Riera, A. 2000. Osteopenia assessed by body composition analysis is related to malnutrition in alcoholic patients. *Alcohol* 22: 147-157.

Sardinha, L.B., Lohman, T.G., Teixeira, P.J., Guedes, D.P., and Going, S.B. 1998. Comparison of air displacement plethysmography with dual-energy X-ray absorptiometry and 3 field methods for estimating body composition in middle-aged men. *American Journal of Clinical Nutrition* 68: 786-793.

Saunders, M.J., Blevins, J.E., and Broeder, C.E. 1998. Effects of hydration changes on bioelectrical impedance in endurance trained individuals. *Medicine & Science in Sports & Exercise* 30: 885-892.

Sawai, Mutoh, and Miyashita. 1990. Study on effectiveness, when applied to Japanese Natives, of an instrument for measurement of body fat with near-infrared interactance technology. Translated and reprinted from *The Annals of Tokyo University* by Futrex, Inc.

Scalfi, L., Bedogni, G., Marra, M., DiBiase, G., Caldara, A., Severi, S., Contaldo, F., and Battistini, N. 1997. The prediction of total body water from bioelectrical impedance in patients with anorexia nervosa. *British Journal of Nutrition* 78: 357-365.

Scalfi, L., Marra, M., Caldara, A., Silvestri, E., and Contaldo, F. 1999. Changes in bioimpedance analysis after stable refeeding of undernourished anorexic patients. *International Journal of Obesity and Related Metabolic Disorders* 23: 133-137.

Scalfi, L., Polito, A., Bianchi, L., Marra, M., Caldara, A., Nicolai, E., and Contaldo, F. 2002. Body composition changes in patients with anorexia nervosa after complete weight recovery. *European Journal of Clinical Nutrition* 56: 15-20.

Schaefer, F., Wuhl, E., Feneberg, R., Mehls, O., and Scharer, K. 2000. Assessment of body composition in children with chronic renal failure. *Pediatric Nephrology* 14: 673-678.

Scherf, J., Franklin, B.A., Lucas, C.P., Stevenson, D., and Rubenfire, M. 1986. Validity of skinfold thickness measures of formerly obese adults. *American Journal of Clinical Nutrition* 43: 128-135.

Schmidt, P.K., and Carter, J.E.L. 1990. Static and dynamic differences among five types of skinfold calipers. *Human Biology* 62: 369-388.

Schoeller, D.A. 1996. Hydrometry. In *Human body composition,* ed. A.F. Roche, S.B. Heymsfield, and T.G. Lohman, 25-43. Champaign, IL: Human Kinetics.

Schoeller, D.A. 2000. Bioelectrical impedance analysis: What does it measure? *Annals of the New York Academy of Sciences* 904: 159-162.

Schoeller, D.A., Kushner, R.F., Taylor, P., Dietz, W.H., and Bandini, L. 1985. Measurement of total body water: Isotope dilution techniques. *Report of the sixth Ross conference on medical research,* 24-29. Columbus, OH: Ross Laboratories.

Schoeller, D.A., and Luke, A. 2000. Bioelectrical impedance analysis prediction equations differ between African Americans and Caucasians, but it is not clear why. *Annals of the New York Academy of Sciences* 904: 225-226.

Schoenborn, C.A., Adams, P.F., and Barnes, P.M. 2002. *Body weight status of adults: United States, 1997-98.* Advanced data from vital and health statistics; No. 330. Hyattsville, MD: National Center for Health Statistics.

Schols, A.M., Wouters, E.F., Soeters, P.B., and Westerterp, K.R. 1991. Body composition by bioelectrical-impedance analysis compared with deuterium dilution and skinfold anthropometry in patients with chronic obstructive pulmonary disease. *American Journal of Clinical Nutrition* 53: 421-424.

Schreiner, P.J., Pitkaniemi, J., Pekkanen, J., and Salomaa, V.V. 1995. Reliability of near-infrared interactance body fat assessment relative to standard anthropometric techniques. *Journal of Clinical Epidemiology* 48: 1361-1367.

Schutte, J.E., Townsend, E.J., Hugg, J., Shoup, R.F., Malina, R.M., and Blomqvist, C.G. 1984. Density of lean body mass is greater in blacks than in whites. *Journal of Applied Physiology* 56: 1647-1649.

Schwenk, A. 2002. Methods of assessing body shape and composition in HIV-associated lipodystrophy. *Current Opinion in Infectious Diseases* 15: 9-16.

Schwenk, A., Breuer, P., Kremer, G., and Ward, L. 2001. Clinical assessment of HIV-associated lipodystrophy syndrome: Bioelectrical impedance analysis, anthropometry and clinical scores. *Clinical Nutrition* 20: 243-249.

Segal, K.R. 1996. Use of bioelectrical impedance analysis measurements as an evaluation for participating in sports. *American Journal of Clinical Nutrition* 64: 469S-471S.

Segal, K.R., Borastero, S., Chun, A., Coronel, P., Pierson, R.L., and Wang, J. 1991. Estimation of extra-cellular and total body water by multiple frequency bioelectrical-impedance measurement. *American Journal of Clinical Nutrition* 54: 26-29.

Segal, K.R., Gutin, B., Presta, E., Wang, J., and Van Itallie, T.B. 1985. Estimation of human body composition by electrical impedance methods: A comparative study. *Journal of Applied Physiology* 58: 1565-1571.

Segal, K.R., Van Loan, M., Fitzgerald, P.I., Hodgdon, J.A., and Van Itallie, T.B. 1988. Lean body mass estimation by bioelectrical impedance analysis: A four-site cross-validation study. *American Journal of Clinical Nutrition* 47: 7-14.

Segal, K.R., Wang, J., Gutin, B., Pierson, R.N., and Van Itallie, T.B. 1987. Hydration and potassium content of lean body mass: Effects of body fat, sex, and age. *American Journal of Clinical Nutrition* 45: 865 [abstract].

Seip, R., and Weltman, A. 1991. Validity of skinfold and girth based regression equations for the prediction of body composition in obese adults. *American Journal of Human Biology* 3: 91-95.

Selinger, A. 1977. *The body as a three component system.* Unpublished doctoral dissertation. University of Illinois, Urbana.

Seppel, T., Kosel, A., and Schlaghecke, R. 1997. Bioelectrical impedance assessment of body composition in thyroid disease. *European Journal of Endocrinology* 136: 493-498.

Sheng, H-P., and Huggins, R.A. 1979. A review of body composition studies with emphasis on total body water and fat. *American Journal of Clinical Nutrition* 32: 630-647.

Shypailo, R.J., Posada, J.K.J., and Ellis, K.J. 1998. Whole-body phantoms with anthropomorphic-shaped skeletons for evaluation of dual-energy X-ray absorptiometry measurements. *Applied Radiation and Isotopes* 49: 503-505.

Simons, J.P., Schols, A., Westerterp, K.R., ten Velde, G., and Wouters, E. 1995. The use of bioelectrical impedance analysis to predict total body water in patients with cancer cachexia. *American Journal of Clinical Nutrition* 61: 741-745.

Simpson, J.A.D., Lobo, D.N., Anderson, J.A., MacDonald, I.A., Perkins, A.C., Neal, K.R., Allison, S.P., and Rowlands, B.J. 2001. Body water compartment measurements: A comparison of bioelectrical impedance analysis with tritium and sodium bromide dilution techniques. *Clinical Nutrition* 20: 339-343.

Sinning, W.E., Dolny, D.G., Little, K.D., Cunningham, L.N., Racaniello, A., Siconolfi, S.F., and Sholes, J.L. 1985. Validity of "generalized" equations for body composition analysis in male athletes. *Medicine & Science in Sports & Exercise* 17: 124-130.

Sinning, W.E., and Wilson, J.R. 1984. Validity of "generalized" equations for body composition analysis in women athletes. *Research Quarterly for Exercise and Sport* 55: 153-160.

Siri, W.E. 1956. The gross composition of the body. In *Advances in biological and medical physics,* ed. C.A. Tobias and J.H. Lawrence, 239-280. New York: Academic Press.

Siri, W.E. 1961. Body composition from fluid spaces and density: Analysis of methods. In *Techniques for measuring body composition,* ed. J. Brozek and A. Henschel, 223-244. Washington, DC: National Academy of Sciences.

Slaughter, M.H., Lohman, T.G., Boileau, R.A., Horswill, C.A., Stillman, R.J., Van Loan, M.D., and Bemben, D.A. 1988. Skinfold equations for estimation of body fatness in children and youth. *Human Biology* 60: 709-723.

Sloan, A.W. 1967. Estimation of body fat in young men. *Journal of Applied Physiology* 23: 311-315.

Smalley, K.J., Knerr, A.N., Kendrick, Z.V., Colliver, J.A., and Owens, O.E. 1990. Reassessment of body mass indices. *American Journal of Clinical Nutrition* 52: 405-408.

Smith, D.B., Johnson, G.O., Stout, J.R., Housh, T.J., Housh, D.J., and Evetovich, T.K. 1997. Validity of near-infrared interactance for estimating relative body fat in female high school gymnasts. *International Journal of Sports Medicine* 18: 531-537.

Smith, M.R., Fuchs, V., Anderson, E.J., Fallon, M.A., and Manola, J. 2002. Measurement of body fat by dual-energy X-ray absorptiometry and bioimpedance analysis in men with prostate cancer. *Nutrition* 18: 574-577.

Smye, S.W., Sutcliffe, J., and Pitt, E. 1993. A comparison of four commercial systems used to measure whole-body electrical impedance. *Physiological Measurement* 14: 473-478.

Snead, D.B., Birge, S.J., and Kohrt, W.M. 1993. Age-related differences in body composition by hydrodensitometry and dual-energy X-ray absorptiometry. *Journal of Applied Physiology* 74: 770-775.

Snijder, M.B., Kuyf, B.E., and Deurenberg, P. 1999. Effect of body build on the validity of predicted body fat from body mass index and bioelectrical impedance. *Annals of Nutrition & Metabolism* 43: 277-285.

Song, J.H., Lee, S.W., Kim, G.A., and Kim, M.J. 1999. Measurement of fluid shift in CAPD patients using segmental bioelectrical impedance analysis. *Peritoneal Dialysis International* 19: 386-390.

Sparling, P.B., Millard-Stafford, M.L., Rosskopf, L.B., DiCarlo, L.J., and Hinson, B.T. 1993. Body composition by bioelectric impedance and densitometry in black women. *American Journal of Human Biology* 5: 111-117.

Sparling, P.B., Snow, T.K., Rosskopf, L.B., O'Donnell, E.M., Freedson, P.S., and Byrnes, W.C. 1998. Bone mineral density and body composition of the United States Olympic women's field hockey team. *British Journal of Sports Medicine* 32: 315-318.

Spungen, A.M., Wang, J., Pierson, R.N., and Bauman, W.A. 2000. Soft tissue body composition differences in monozygotic twins discordant for spinal cord injury. *Journal of Applied Physiology* 88: 1310-1315.

Stall, S., DeVita, M.V., Ginsberg, N.S., Frumkin, D., Lynn, R.I., and Michelis, M.F. 2000. Body composition assessed by neutron activation analysis in dialysis patients. *Annals of the New York Academy of Sciences* 904: 558-563.

Steel, S.A., Baker, A.J., and Saunderson, J.R. 1998. An assessment of the radiation dose to patients and staff from a Lunar Expert-XL fan beam densitometer. *Physiological Measurement* 19: 17-26.

Steele, I.C., Young, I.S., Stevenson, H.P., Maguire, S., Livingstone, M.B., Rollo, M., Scrimgeour, C., Rennie, M.J., and Nicholls, D.P. 1998. Body composition and energy expenditure of patients with chronic cardiac failure. *European Journal of Clinical Investigation* 28: 33-40.

Stellato, D., Cirillo, M., De Santo, L.S., Anastasio, P., Frangiosa, A., Cotrufo, M., De Santo, N.G., and Dilorio, B. 2001. Bioelectrical impedance analysis in heart transplantation: Early and late changes. *Seminars in Nephrology* 21: 282-285.

Stevens, J., Keil, J.E., Rust, P.F., Tyroler, H.A., Davis, C.E., and Gazes, P.C. 1992. Body mass index and body girth as predictors of mortality in black and white women. *Archives of Internal Medicine* 152: 1257-1262.

Stewart, A.D., and Hannan, W.J. 2000. Prediction of fat and fat-free mass in male athletes using dual X-ray absorptiometry as the reference method. *Journal of Sports Sciences* 18: 263-274.

Stokes, T.M., Sinning, W.E., Morgan, A.L., and Ellison, J.D. 1993. Bioimpedance (BI) estimation of fat free mass in black and white women. *Medicine & Science in Sports & Exercise* 25: S162 [abstract].

Stolarczyk, L.M., and Heyward, V.H. 1999. Assessing body composition of adults with diabetes. *Diabetes Technology & Therapeutics* 1: 289-296.

Stolarczyk, L.M., Heyward, V.H., Goodman, J.A., Grant, D.J., Kessler, K.L., Kocina, P.S., and Wilmerding, V. 1995. Predictive accuracy of bioimpedance equations in estimating fat-free mass of Hispanic women. *Medicine & Science in Sports & Exercise* 27: 1450-1456.

Stolarcyzk, L.M., Heyward, V.H., Hicks, V.L., and Baumgartner, R.N. 1994. Predictive accuracy of bioelectrical impedance in estimating body composition of Native American women. *American Journal of Clinical Nutrition* 59: 964-970.

Stolarczyk, L.M., Heyward, V.H., Van Loan, M.D., Hicks, V.L., Wilson, W.L., and Reano, L.M. 1997. The fatness-specific bioelectrical impedance analysis equations of Segal et al: Are they generalizable and practical? *American Journal of Clinical Nutrition* 66: 8-17.

Stout, J.R., Eckerson, J.M., Housh, T.J., and Johnson, G.O. 1994. Validity of methods for estimating percent body fat in black males. *Journal of Strength and Conditioning Research* 8: 243-246.

Stout, J.R., Eckerson, J.M., Housh, T.J., Johnson, G.O., and Betts, N.M. 1994. Validity of percent body fat estimations in males. *Medicine & Science in Sports & Exercise* 26: 632-636.

Stout, J.R., Housh, T.J., Eckerson, J.M., Johnson, G.O., and Betts, N.M. 1996. Validity of methods for estimating percent body fat in young women. *Journal of Strength and Conditioning Research* 10: 25-29.

Stout, J.R., Housh, T.J., Johnson, G.O., Housh, D.J., Evans, S.A., and Eckerson, J.M. 1995. Validity of skinfold equations for estimating body density in youth wrestlers. *Medicine & Science in Sports & Exercise* 27: 1321-1325.

Strain, G.W., and Zumoff, B. 1992. The relationship of weight-height indices of obesity to body fat content. *Journal of the American College of Nutrition* 11: 715-718.

Strauss, B.J.G., Gibson, P.R., Stroud, D.B., Borovnicar, D.J., Xiong, D.W., and Keogh, J. 2000. Total body dual X-ray absorptiometry is a good measure of both fat mass and fat-free mass in liver cirrhosis compared to "gold-standard" techniques. *Annals of the New York Academy of Sciences* 904: 55-62.

Strauss, R.S., and Pollack, H.A. 2001. Epidemic increase in childhood overweight, 1986-1998. *Journal of the American Medical Association* 286(22): 2845-2848.

Styne, D.M. 2001. Childhood and adolescent obesity. Prevalence and significance. *Pediatric Clinics of North America* 48: 823-854.

Sundgot-Borgen, J. 1994. Risk and trigger factors for the development of eating disorders in female elite athletes. *Medicine & Science in Sports & Exercise* 26: 414-419.

Sung, R.Y.T., Lau, P., Yu, C.W., Lam, P.K.W., and Nelson, E.A.S. 2001. Measurement of body fat using leg to leg bioimpedance. *Archives of Disease in Childhood* 85: 263-267.

Sutcliffe, J.F. 1996. A review of in vivo experimental methods to determine the composition of the human body. *Physics in Medicine and Biology* 41: 791-833.

Svendsen, O.L., and Hassager, C. 1998. Body composition and fat distribution measured by dual-energy X-ray absorptiometry in premenopausal and postmenopausal insulin-dependent and non-insulin-dependent diabetes mellitus patients. *Metabolism* 47: 212-216.

Svendsen, O.L., Hassager, C., Bergmann, I., and Christiansen, C. 1992. Measurement of abdominal and intra-abdominal fat in postmenopausal women by dual energy X-ray absorptiometry and anthropometry: Comparison with computerized tomography. *International Journal of Obesity and Related Metabolic Disorders* 17: 45-51.

Taaffe, D.R., Robinson, T.L., Snow, C.M., and Marcus, R. 1997. High-impact exercise promotes bone gain in well-trained female athletes. *Journal of Bone and Mineral Research* 12: 255-260.

Taaffe, D.R., Villa, M.L., Holloway, L., and Marcus, R. 2000. Bone mineral density in older non-Hispanic Caucasian and Mexican-American women: Relationship to lean and fat mass. *Annals of Human Biology* 27: 331-344.

Tabachnick, B.G., and Fidell, L.S. 1983. *Using multivariate statistics.* New York: Harper & Row.

Talluri, T., Lietdke, R.J., Evangelisti, A., Talluri, J., and Maggia, G. 1999. Fat-free mass qualitative assessment with bioelectric impedance analysis (BIA). *Annals of the New York Academy of Sciences* 873: 94-98.

Tan, Y.X., Nuñez, C., Sun, Y., Zhang, K., Wang, Z., and Heymsfield, S.B. 1997. New electrode system for rapid whole-body and segmental bioimpedance assessment. *Medicine & Science in Sports & Exercise* 29: 1269-1273.

Tanaka, K., Nakadomo, F., Watanabe, K., Inagaki, A., Kim, H.K., and Matsuura, Y. 1992. Body composition prediction equations based on bioelectrical impedance and anthropometric variables for Japanese obese women. *American Journal of Human Biology* 4: 739-745.

Tataranni, P.A., Pettit, D.J., and Ravussin, E. 1996. Dual energy X-ray absorptiometry: Inter-machine variability. *International Journal of Obesity and Related Metabolic Disorders* 20: 1048-1050.

Taylor, A., Scopes, J.W., du Mont, G., and Taylor, B.A. 1985. Development of an air displacement method for whole body volume measurement of infants. *Journal of Biomedical Engineering* 7: 9-17.

Taylor, R.W., Keil, D., Gold, E.J., Williams, S.M., and Goulding, A. 1998. Body mass index, waist girth, and waist-to-hip ratio as indexes of total and regional adiposity in women: Evaluation using receiver operating characteristic curves. *American Journal of Clinical Nutrition* 67: 44-49.

Teran, J.C., Sparks, K.E., Quinn, L.M., Fernandez, B.S., Krey, S.H., and Steffee, W.P. 1991. Percent body fat in obese white females predicted by anthropometric measurements. *American Journal of Clinical Nutrition* 53: 7-13.

Thomas, B.J., Ward, L.C., and Cornish, B.H. 1998. Bioimpedance spectrometry in the determination of body water compartments: Accuracy and clinical significance. *Applied Radiation and Isotopes* 49: 447-455.

Thomas, D.Q., and Whitehead, J.R. 1993. Body composition assessment: Some practical answers to teachers' questions. *Journal of Physical Education, Recreation and Dance* 63: 16-19.

Thomas, K.T., Keller, C.S., and Holbert, K.E. 1997a. Ethnic and age trends for body composition in women residing in the U.S. Southwest: I. Regional fat. *Medicine & Science in Sports & Exercise* 29: 82-89.

Thomas, K.T., Keller, C.S., and Holbert, K.E. 1997b. Ethnic and age trends for body composition in women residing in the U.S. Southwest: II. Total fat. *Medicine & Science in Sports & Exercise* 29: 90-98.

Thomas, T.R., and Etheridge, G.L. 1980. Hydrostatic weighing at residual volume and functional residual capacity. *Journal of Applied Physiology* 49: 157-159.

Thomasett, A. 1962. Bio-electrical properties of tissue impedance measurements. *Lyon Medical* 207: 107-118.

Thompson, D.L., and Moreau, K.L. 2000. Brozek two-compartment model under-estimates body fat in black female athletes. *Clinical Physiology* 20: 311-314.

Thorland, W.G., Johnson, G.O., Cisar, C.J., and Housh, T.J. 1987. Estimation of minimal wrestling weight using measures of body build and body composition. *International Journal of Sports Medicine* 8: 365-370.

Thorland, W.G., Johnson, G.O., Tharp, G.D., Fagot, T.G., and Hammer, R.W. 1984. Validity of anthropometric equations for the estimation of body density in adolescent athletes. *Medicine & Science in Sports & Exercise* 16: 77-81.

Thorland, W.G., Tipton, C.M., Lohman, T.G., Bowers, R.W., Housh, T.J., Johnson, G.O., Kelly, J.M., Oppliger, R.A., and Tcheng, T. 1991. Midwest wrestling study: Prediction of minimal weight for high school wrestlers. *Medicine & Science in Sports & Exercise* 23: 1102-1110.

Timson, B.F., and Coffman, J.L. 1984. Body composition by hydrostatic weighing at total lung capacity and residual volume. *Medicine & Science in Sports & Exercise* 16: 411-414.

Tipton, C.M., and Oppliger, R.A. 1984. The Iowa wrestling study: Lessons for physicians. *Iowa Medicine* 74: 381-385.

Tisdale, M.J. 1999. Wasting in cancer. *Journal of Nutrition* 129: 243S-246S.

Tothill, P., Avenell, A., Love, J., and Reid, D.M. 1994. Comparison between Hologic, Lunar and Norland dual-energy X-ray absorptiometers and other techniques used for whole-body soft tissue measurements. *European Journal of Clinical Nutrition* 48: 781-794.

Tothill, P., and Hannan, W.J. 2000. Comparisons between Hologic QDR 1000W, QDR 4500A, and Lunar Expert dual-energy X-ray absorptiometry scanners used for measuring total body bone and soft tissue. *Annals of the New York Academy of Sciences* 904: 63-71.

Tran, Z.V., and Weltman, A. 1988. Predicting body composition of men from girth measurements. *Human Biology* 60: 167-175.

Tran, Z.V., and Weltman, A. 1989. Generalized equation for predicting body density of women from girth measurements. *Medicine & Science in Sports & Exercise* 21: 101-104.

Tremblay, M.S., and Willms, J.D. 2000. Secular trends in the body mass index of Canadian children. *Canadian Medical Association Journal* 163(11): 1429-1433.

Tsui, E.Y.L., Gao, X.J., and Zinman, B. 1998. Bioelectrical impedance analysis (BIA) using bipolar foot electrodes in the assessment of body composition in type 2 diabetes mellitus. *Diabetic Medicine* 15: 125-128.

Turcato, E., Bosello, O., Francesco, V.D., Harris, T.B., Zoico, E., Bissoli, L., Fracassi, E., and Zamboni, M. 2000. Waist circumference and abdominal sagittal diameter as surrogates of body fat distribution in the elderly: Their relation with cardiovascular risk factors. *International Journal of Obesity and Related Metabolic Disorders* 24: 1005-1010.

Tyrrell, V.J., Richards, G., Hofman, P., Gillies, G.F., Robinson, E., and Cutfield, W.S. 2001. Foot-to-foot bioelectrical impedance analysis: A valuable tool for the measurement of body composition in children. *International Journal of Obesity and Related Metabolic Disorders* 25: 273-278.

U.S. Department of Health and Human Services. 1988. *The Surgeon General's report on nutrition and health.* Washington, DC: U.S. Government Printing Office. DHHS (PHS) Pub. No. 88-50210.

U.S. Department of Health and Human Services. 2000a. *Dietary guidelines for Americans.* Washington, DC: U.S. Government Printing Office.

U.S. Department of Health and Human Services. 2000b. *Healthy people 2010—conference edition: Physical activity and fitness* (22). Atlanta: Author.

U.S. Department of Health and Human Services. 2000c. *Healthy people 2010: Understanding and improving health—overweight and obesity.* Washington, DC: U.S. Government Printing Office.

Utter, A.C., Nieman, D.C., Ward, A.N., and Butterworth, D.E. 1999. Use of the leg-to-leg bioelectrical impedance method in assessing body-composition change in obese women. *American Journal of Clinical Nutrition* 69: 603-607.

Utter, A.C., Scott, J.R., Oppliger, R.A., Visich, P.S., Goss, F.L., Marks, B.L., Nieman, D.C., and Smith, B.W. 2001. A comparison of leg-to-leg bioelectrical impedance and skinfolds in assessing body fat in collegiate wrestlers. *Journal of Strength and Conditioning Research* 15: 157-160.

Vache, C., Rousset, P., Gachon, P., Gachon, A.M., Morio, B., Boulier, A., Coudert, J., Beaufrere, B., and Ritz, P. 1998. Bioelectrical impedance analysis measurements of total body water and extracellular water in healthy elderly subjects. *International Journal of Obesity and Related Metabolic Disorders* 22: 537-543.

Vague, J. 1947. La differenciation sexuelle facteur determinant des formes de l'obesite. *La Presse Medicale* 30: 339-340.

Vaisman, N., Corey, M., Rossi, M.F., Goldberg, E., and Pencharz, P. 1988. Changes in body composition during refeeding of patients with anorexia nervosa. *Journal of Pediatrics* 113: 925-929.

van der Kooy, K., Leenen, R., Deurenberg, P., Seidell, J.C., Westerterp, K.R., and Hautvast, J.G. 1992. Changes in fat-free mass in obese subjects after weight loss: A comparison of body composition measures. *International*

Journal of Obesity and Related Metabolic Disorders 16: 675-683.

van der Kooy, K., Leenen, R., Seidell, J.C., Deurenberg, P., Droop, A., and Bakker, C.J.G. 1993. Waist-hip ratio is a poor predictor of changes in visceral fat. *American Journal of Clinical Nutrition* 57: 327-333.

van der Ploeg, G.E., Brooks, A.G., Withers, R.T., Dollman, J., Leaney, F., and Chatterton, B.E. 2001. Body composition changes in female bodybuilders during preparation for competition. *European Journal of Clinical Nutrition* 55: 268-277.

Van Loan, M.D. 1990. Bioelectrical impedance analysis to determine fat-free mass, total body water and body fat. *Sports Medicine* 10: 205-217.

Van Loan, M.D. 1998. Estimates of fat-free mass (FFM) by densitometry, dual energy X-ray absorptiometry (DXA), and bioimpedance spectroscopy (BIS) in Caucasian and Chinese-American women. *Applied Radiation and Isotopes* 49: 751-752.

Van Loan, M.D., Boileau, R.A., Slaughter, M.H., Stillman, R.J., Lohman, T.G., Going, S.B., and Carswell, C. 1990. Association of bioelectrical resistance with estimates of fat-free mass determined by densitometry and hydrometry. *American Journal of Human Biology* 2: 219-226.

Van Loan, M.D., and Mayclin, P.L. 1987. Bioelectrical impedance analysis: Is it a reliable estimator of lean body mass and total body water? *Human Biology* 59: 299-309.

Van Loan, M.D., and Mayclin, P.L. 1992. Body composition assessment: Dual-energy X-ray absorptiometry (DEXA) compared to reference methods. *European Journal of Clinical Nutrition* 46: 125-130.

Vazquez, J.A., and Janosky, J.E. 1991. Validity of bioelectrical-impedance analysis in measuring changes in lean body mass during weight reduction. *American Journal of Clinical Nutrition* 54: 970-975.

Vehrs, P., Morrow, J.R., and Butte, N. 1998. Reliability and concurrent validity of Futrex and bioelectrical impedance. *International Journal of Sports Medicine* 19: 560-566.

Vescovi, J.D., Zimmerman, S.L., Miller, W.C., Hildebrandt, L., Hammer, R.L., and Fernhall, B. 2001. Evaluation of the Bod Pod for estimating percentage body fat in a heterogeneous group of adult humans. *European Journal of Applied Physiology* 85: 326-332.

Vickery, M.C., Cureton, K.J., and Collins, M.A. 1988. Prediction of body density from skinfolds in black and white young men. *Human Biology* 60: 135-149.

Visser, M., Gallagher, D., Deurenberg, P., Wang, J., Pierson, R.N., and Heymsfield, S.B. 1997. Density of fat-free body mass: Relationship with race, age, and level of body fatness. *American Journal of Physiology* 272: E781-E787.

Visser, M., Van Den Heuvel, E., and Deurenberg, P. 1994. Prediction equations for the estimation of body composition in elderly using anthropometric data. *British Journal of Nutrition* 71: 823-833.

Wagner, D.R. 1996. Body composition assessment and minimal weight recommendations for high school wrestlers. *Journal of Athletic Training* 31: 262-265.

Wagner, D.R., and Heyward, V.H. 2000. Measures of body composition in blacks and whites: A comparative review. *American Journal of Clinical Nutrition* 71: 1392-1402.

Wagner, D.R., and Heyward, V.H. 2001. Validity of two-component models of estimating body fat of black men. *Journal of Applied Physiology* 90: 649-656.

Wagner, D.R., Heyward, V.H., and Gibson, A.L. 2000. Validation of air displacement plethysmography for assessing body composition. *Medicine & Science in Sports & Exercise* 32: 1339-1344.

Wagner, D.R., Heyward, V.H., Kocina, P.S., Stolarczyk, L.M., and Wilson, W.L. 1997. Predictive accuracy of BIA equations for estimating fat-free mass of black men. *Medicine & Science in Sports & Exercise* 29: 969-974.

Wang, J., and Deurenberg, P. 1996. The validity of predicted body composition in Chinese adults from anthropometry and bioelectrical impedance in comparison with densitometry. *British Journal of Nutrition* 76: 175-182.

Wang, J., Kotler, D.P., Russell, M., Burastero, S., Mazariegos, M., Thornton, J., Dilmanian, F.A., and Pierson, R.N. 1992. Body-fat measurement in patients with acquired immunodeficiency syndrome: Which method should be used? *American Journal of Clinical Nutrition* 56: 963-967.

Wang, J., Thornton, J.C., Russell, M., Burastero, S., Heymsfield, S., and Pierson, R.N. 1994. Asians have lower body mass index (BMI) but higher percent body fat than do Whites: Comparison of anthropometric measurements. *American Journal of Clinical Nutrition* 60: 23-28.

Wang, Y., Ge, K., and Popkin, B.M. 2000. Tracking of body mass index from childhood to adolescence: A 6-y follow-up study in China. *American Journal of Clinical Nutrition* 72: 1018-1024.

Wang, Z.M., Deurenberg, P., Guo, S.S., Pietrobelli, A., Wang, J., Pierson, R.N., and Heymsfield, S.B. 1998. Six-compartment body composition model: Inter-method comparisons of total body fat measurement. *International Journal of Obesity and Related Metabolic Disorders* 22: 329-337.

Wang, Z.M., Deurenberg, P., Wang, W., Pietrobelli, A., Baumgartner, R.N., and Heymsfield, S.B. 1999. Hydration of fat-free mass: Review and critique of a classic body composition constant. *American Journal of Clinical Nutrition* 69: 833-841.

Wang, Z.M., Pierson, R.N. Jr., and Heymsfield, S.B. 1992. The five level model: A new approach to organizing body composition research. *American Journal of Clinical Nutrition* 56: 19-28.

Wanke, C., Polsky, B., and Kotler, D. 2002. Guidelines for using body composition measurement in patients with human immunodeficiency virus infection. *AIDS Patient Care and STDs* 16: 375-388.

Ward, L.C., Byrne, N.M., Rutter, K., Hennoste, L., Hills, A.P., Cornish, B.H., and Thomas, B.J. 1997. Reliability of multiple frequency bioelectrical impedance analysis: An intermachine comparison. *American Journal of Human Biology* 9: 63-72.

Ward, L.C., Heitmann, B.L., Craig, P., Stroud, D., Azinge, E.C., Jebb, S., Cornish, B.H., Swinburn, B., O'Dea, K., Rowley,

K., McDermott, R., Thomas, B.J., and Leonard, D. 2000. Association between ethnicity, body mass index, and bioelectrical impedance: Implications for the population specificity of prediction equations. *Annals of the New York Academy of Sciences* 904: 199-202.

Ward, R., and Anderson, G.S. 1998. Resilience of anthropometric data assembly strategies to imposed error. *Journal of Sports Sciences* 16: 755-759.

Ward, R., Rempel, R., and Anderson, G.S. 1999. Modeling dynamic skinfold compression. *American Journal of Human Biology* 11: 521-537.

Watanabe, K., Nakadomo, F., Tanaka, K., Kim, K., and Maeda, K. 1993. Estimation of fat-free mass from bioelectrical impedance and anthropometric variables in Japanese girls. *Medicine & Science in Sports & Exercise* 25: S163 [abstract].

Watson, P.E., Watson, I.D., and Batt, R.D. 1980. Total body volumes for adult males and females estimated from simple anthropometric measurements. *American Journal of Clinical Nutrition* 33: 27-39.

Weimann, E. 2002. Gender-related differences in elite gymnasts: The female athlete triad. *Journal of Applied Physiology* 92: 2146-2152.

Weits, T., Van der Beek, E.J., Wedel, M., and Ter Haar Romeny, B.M. 1988. Computed tomography measurement of abdominal fat deposition in relation to anthropometry. *International Journal of Obesity and Related Metabolic Disorders* 12: 217-225.

Wells, J.C.K., Douros, I., Fuller, N.J., Elia, M., and Dekker, L. 2000. Assessment of body volume using three-dimensional photonic scanning. *Annals of the New York Academy of Sciences* 904: 247-254.

Wells, J.C.K., and Fuller, N.J. 2001. Precision of measurement and body size in whole-body air-displacement plethysmography. *International Journal of Obesity and Related Metabolic Disorders* 25: 1161-1167.

Wells, J.C.K., Fuller, N.J., Dewit, O., Fewtrell, M.S., Elia, M., and Cole, T.J. 1999. Four-component model of body composition in children: Density and hydration of fat-free mass and comparison with simpler models. *American Journal of Clinical Nutrition* 69: 904-912.

Weltman, A., Levine, S., Seip, R.L., and Tran, Z.V. 1988. Accurate assessment of body composition in obese females. *American Journal of Clinical Nutrition* 48: 1179-1183.

Weltman, A., Seip, R.L., and Tran, Z.V. 1987. Practical assessment of body composition in adult obese males. *Human Biology* 59: 523-535.

Werkman, A., Deurenberg-Yap, M., Schmidt, G., and Deurenberg, P. 2000. A comparison between composition and density of the fat-free mass of young adult Singaporean Chinese and Dutch Caucasians. *Annals of Nutrition & Metabolism* 44: 235-242.

Weststrate, J.A., and Deurenberg, P. 1989. Body composition in children: Proposal for a method for calculating body fat percentage from total body density or skinfold-thickness measurements. *American Journal of Clinical Nutrition* 50: 1104-1115.

Widdowson, E.M., McCance, R.A., and Spray, C.M. 1951. The chemical composition of the human body. *Clinical Science* 10: 113-125.

Williams, C.A., and Bale, P. 1998. Bias and limits of agreement between hydrodensitometry, bioelectrical impedance and skinfold calipers measures of percentage body fat. *European Journal of Applied Physiology* 77: 271-277.

Williams, D.P., Going, S.B., Lohman, T.G., Harsha, D.W., Srinivasan, S.R., Webber, L.S., and Berenson, G.S. 1992. Body fatness and risk of elevated blood pressure, total cholesterol, and serum lipoprotein ratios in children and adolescents. *American Journal of Public Health* 82: 358-363.

Williams, D.P., Going, S.B., Lohman, T.G., Hewitt, M.J., and Haber, A.E. 1992. Estimation of body fat from skinfold thicknesses in middle-aged and older men and women: A multiple component approach. *American Journal of Human Biology* 4: 595-605.

Williams, D.P., Going, S.B., Milliken, L.A., Hall, M.C., and Lohman, T.G. 1995. Practical techniques for assessing body composition in middle-aged and older adults. *Medicine & Science in Sports & Exercise* 27: 776-783.

Wilmore, J.H. 1969. A simplified method for determination of residual lung volumes. *Journal of Applied Physiology* 27: 96-100.

Wilmore, J.H. 1983. Body composition in sport and exercise: Directions for future research. *Medicine & Science in Sports & Exercise* 15: 21-31.

Wilmore, J.H., and Behnke, A.R. 1969. An anthropometric estimation of body density and lean body weight in young men. *Journal of Applied Physiology* 27: 25-31.

Wilmore, J.H., and Behnke, A.R. 1970. An anthropometric estimation of body density and lean body weight in young women. *American Journal of Clinical Nutrition* 23: 267-274.

Wilmore, J.H., Frisancho, R.A., Gordon, C.C., Himes, J.H., Martin, A.D., Martorell, R., and Seefeldt, R.D. 1988. Body breadth equipment and measurement techniques. In *Anthropometric standardization reference manual*, ed. T.G. Lohman, A.F. Roche, and R. Martorell, 27-38. Champaign, IL: Human Kinetics.

Wilmore, J.H., Vodak, P.A., Parr, R.B., Girandola, R.N., and Billing, J.E. 1980. Further simplification of a method for determination of residual volume. *Medicine & Science in Sports & Exercise* 12: 216-218.

Wilmore, K.M., McBride, P.J., and Wilmore, J.H. 1994. Comparison of bioelectric impedance and near-infrared interactance for human body composition assessment in a population of self-perceived overweight adults. *International Journal of Obesity and Related Metabolic Disorders* 18: 375-381.

Wilson, W.L., and Heyward, V.H. 1993. Effects of skintone, skinfold, and mid-arm muscle area on optical density measurements at the biceps site. *Medicine & Science in Sports & Exercise* 25: S60 [abstract].

Withers, R.T., Craig, N.P., Bourdon, P.C., and Norton, K.I. 1987. Relative body fat and anthropometric prediction of body density of male athletes. *European Journal of Applied Physiology* 56: 191-200.

Withers, R.T., LaForgia, J., Pillans, R.K., Shipp, N.J., Chatterton, B.E., Schultz, C.G., and Leaney, F. 1998. Comparisons of two-, three-, and four-compartment models of body composition analysis in men and women. *Journal of Applied Physiology* 85: 238-245.

Withers, R.T., Whittingham, N.O., Norton, K.I., La Forgia, J., Ellis, M.W., and Crockett, A. 1987. Relative body fat and anthropometric prediction of body density of female athletes. *European Journal of Applied Physiology* 56: 169-180.

Wong, J.C.H., Lewindon, P., Mortimer, R., and Shepherd, R. 2001. Bone mineral density in adolescent females with recently diagnosed anorexia nervosa. *International Journal of Eating Disorders* 29: 11-16.

Wong, W.W., Cochran, W.J., Klish, W.J., Smith, E.O., Lee, L.S., and Klein, P.D. 1988. In vivo isotope-fractionation factors and the measurement of deuterium- and oxygen-18-dilution spaces from plasma, urine, saliva, respiratory water vapor, and carbon dioxide. *American Journal of Clinical Nutrition* 47: 1-6.

Wong, W.W., Hergenroeder, A.C., Stuff, J.E., Butte, N.F., Smith, E.O., and Ellis, K.J. 2002. Evaluating body fat in girls and female adolescents: Advantages and disadvantages of dual-energy X-ray absorptiometry. *American Journal of Clinical Nutrition* 76: 384-389.

Wong, W.W., Stuff, J.E., Butte, N.F., O'Brian Smith, E., and Ellis, K.J. 2000. Estimating body fat in African American and white adolescent girls: A comparison of skinfold-thickness equations with a 4-compartment criterion method. *American Journal of Clinical Nutrition* 72: 348-354.

Woodrow, G., Oldroyd, B., Turney, J.H., Davies, P.S., Day, J.M., and Smith, M.A. 1996. Four-component model of body composition in chronic renal failure comprising dual-energy X-ray absorptiometry and measurement of total body water by deuterium oxide dilution. *Clinical Science* 91: 763-769.

World Health Organization. 1988. Measuring obesity—classification and description of anthropometric data. *Report of a WHO Regional Office Consultation on the Epidemiology of Obesity.* Copenhagen, Denmark: WHO Regional Office for Europe, Nutrition Unit. (Document EUR/ICP/NUT 125).

World Health Organization (WHO). 1998. Obesity: Preventing and managing a global epidemic. *Report of a WHO Consultation on Obesity.* Geneva: Author.

World Health Organization. 2001. Global database on obesity and body mass index (BMI) in adults. http://www.who.int/nut/db_bmi.

Wotton, M.J., Thomas, B.J., Cornish, B.H., and Ward, L.C. 2000. Comparison of whole body and segmental bioimpedance methodologies for estimating total body water. *Annals of the New York Academy of Sciences* 904: 181-186.

Wotton, M.J., Trocki, O., Thomas, B.J., Hammond, P., Shepherd, R.W., Lewindon, P.J., Wilcox, J., Murphy, A.J., and Cleghorn,

G.J. 2000. Changes in body composition in adolescents with anorexia nervosa. Comparison of bioelectrical impedance analysis and total body potassium. *Annals of the New York Academy of Sciences* 904: 418-419.

Wuhl, E., Fusch, C., Scharer, K., Mehls, O., and Schaefer, F. 1996. Assessment of total body water in paediatric patients on dialysis. *Nephrology Dialysis Transplantation* 11: 75-80.

Yamanoto, K. 2002. Omron Institute of Life Science [personal communication].

Yannakoulia, M., Keramopoulos, A., Tsakalakos, N., and Matalas, A. 2000. Body composition in dancers: The bioelectrical impedance method. *Medicine & Science in Sports & Exercise* 32: 228-234.

Yeager, K.K., Agostini, R., Nattiv, A., and Drinkwater, B. 1993. The female athlete triad: Disordered eating, amenorrhea, osteoporosis. *Medicine & Science in Sports & Exercise* 25: 775-777.

Yee, A.J., Fuerst, T., Salamone, L., Visser, M., Dockrell, M., Van Loan, M., and Kern, M. 2001. Calibration and validation of an air-displacement plethysmography method for estimating percentage body fat in an elderly population: A comparison among compartmental models. *American Journal of Clinical Nutrition* 74: 637-642.

Young, H., Porcari, J., Terry, L., and Brice, G. 1998. Validity of body composition assessment methods for older men with cardiac disease. *Journal of Cardiopulmonary Rehabilitation* 18: 221-227.

Zamboni, M., Turcato, E., Armellini, F., Kahn, H.S., Zivelonghi, A., Santana, H., Bergamo-Andreis, I.A., and Bosello, O. 1998. Sagittal abdominal diameter as a practical predictor of visceral fat. *International Journal of Obesity and Related Metabolic Disorders* 22: 655-660.

Zando, K.A., and Robertson, R.J. 1987. The validity and reliability of the Cramer Skyndex caliper in the estimation of percent body fat. *Athletic Training* 22: 23-25, 79.

Zhu, F., Schneditz, D., Kaufman, A.M., and Levin, N.W. 2000. Estimation of body fluid changes during peritoneal dialysis by segmental bioimpedance analysis. *Kidney International* 57: 299-306.

Zhu, F., Schneditz, D., and Levin, N.W. 1999. Sum of segmental bioimpedance analysis during ultrafiltration and hemodialysis reduces sensitivity to changes in body position. *Kidney International* 56: 692-699.

Zhu, F., Schneditz, D., Wang, E., Martin, K., Morris, A.T., and Levin, N.W. 1998. Validation of changes in extracellular volume measured during hemodialysis using a segmental bioimpedance technique. *ASAIO Journal* 44: M541-M545.

Zhu, S., Wang, Z., Heshka, S., Heo, M., Faith, M.S., and Heymsfield, S.B. 2002. Waist circumference and obesity-associated risk factors among whites in the third National Health and Nutrition Examination Survey: Clinical action thresholds. *American Journal of Clinical Nutrition* 76: 743-749.

Zillikens, M.C., and Conway, J.M. 1990. Anthropometry in blacks: Applicability of generalized skinfold equations and differences in fat patterning between blacks and whites. *American Journal of Clinical Nutrition* 52: 45-51.

Zillikens, M.C., and Conway, J.M. 1991. Estimation of total body water by bioelectrical impedance analysis in blacks. *American Journal of Human Biology* 3: 25-32.

Zillikens, M.C., van den Berg, J.W.O., Wilson, J.H.P., Rietveld, T., and Swart, G.R. 1992. The validity of bioelectrical imped-ance analysis in estimating total body water in patients with cirrhosis. *Journal of Hepatology* 16: 59-65.

Zillikens, M.C., van den Berg, J.W.O., Wilson, J.H.P., and Swart, G.R. 1992. Whole-body and segmental bioelec-trical-impedance analysis in patients with cirrhosis of the liver: Changes after treatment of ascites. *American Journal of Clinical Nutrition* 55: 621-625.

Index

Note: The italicized *f* and *t* following page numbers refer to figures and tables, respectively.

About the Authors

Vivian H. Heyward, PhD, is a regents professor emeritus at the University of New Mexico, where she taught exercise science, body composition, and physical fitness assessment courses for 26 years. Extensively published, Dr. Heyward is the author of four editions of *Advanced Fitness Assessment and Exercise Prescription* as well as the first and second editions of *Applied Body Composition Assessment.* She also has written numerous articles for research and professional journals dealing with physical fitness and body composition assessment. Dr. Heyward has given many presentations at international, national, and regional meetings of professional organizations in the field. The American College of Sports Medicine (ACSM) named her a fellow in 1997. She also was named a fellow by the American Society of Clinical Nutrition.

Dale R. Wagner, PhD, studied body composition under Dr. Heyward's direction and was an assistant professor of exercise science at Vanguard University in Costa Mesa, California, for six years. Dr. Wagner was invited to present at the Sixth International In Vivo Body Composition Symposium in Rome in 2002. He is the author and coauthor of a variety of body composition articles that have appeared in highly respected journals and has also been a reviewer for several internationally recognized journals, including the *American Journal of Clinical Nutrition, Obesity Research,* and the *International Journal of Sports Medicine.* He earned his PhD in exercise physiology from the University of New Mexico. Dr. Wagner is a certified health and fitness instructor, exercise physiologist, and strength and conditioning specialist.